Mammals

of the

World

A Checklist

Mammals
of the
World
A Checklist

Andrew Duff and Ann Lawson

A & C BLACK • LONDON

Published by A & C Black Publishers Ltd., 37 Soho Square, London W1D 3QZ

Copyright © 2004 Andrew Duff and Ann Lawson

ISBN 0-7136-6021-X

A CIP catalogue record for this book is available from the British Library

A & C Black uses paper produced with elemental chlorine-free pulp, harvested from managed sustainable forests.

Printed in Italy through G. Canale and C. S.p.a

Produced and designed by Fluke Art, Cornwall

10 9 8 7 6 5 4 3 2 1

www.acblack.com

CONTENTS

INTRODUCTION

This checklist gives the English and scientific names of every extant mammal species in the world. Mammal species which are known only from fossil or subfossil remains, or bones in ancient middens, as well as all mammals which certainly became extinct before 1800, are not included in the list.

Against each species entry there is a checkbox so that you can keep a handy visual record of which mammals you have personally seen on your travels. A space is provided where you can record the details of your first sighting, or other notes. There is also a brief distribution summary, with habitat information where this is known, so that you can corroborate your sightings of wild mammals with the currently accepted distributional and ecological data.

The checklist has been made as up-to-date as possible by including every new species that has been described or elevated from synonymy in the last decade. Since many of these names will be unfamiliar, we provide an appendix in which they are all listed. Where a species has been 'split', we give the name of the species that it has been split from, so that you can be aware of the new restricted status of the original species. Also, a bibliography of literature references should enable you to find out more detailed information about the new species listed.

At the end of the checklist are comprehensive indexes of English and scientific names. These are intended to help you find a species which you may have confidently named from your field guide, but which appears in the checklist under another English or scientific name. This is unavoidable, as field guides covering different regions frequently use different names for the same species, and lack of space did not permit us to include many alternative English names.

The checklist will be of use to anyone who needs to keep comprehensive records of mammal species, including mammalogists, birdwatchers with a side interest in mammals, general naturalists and professional zoologists. Well-travelled safari and whale-watching tourists will find it helpful to be able to keep a personal record of their mammal sightings. Museum curators and zoo-keepers may also find the checklist useful for helping to manage their collections.

HIGHER CLASSIFICATION

The higher classification—that is, the arrangement of subgroups of the class Mammalia down to the level of genus— is taken from McKenna & Bell's *Classification of Mammals Above the Species Level* (1997) including the updated version dated 31 October 2002 (McKenna & Bell, 2002), but with some differences mentioned below.

As far as possible the higher classification is a phylogenetic arrangement. One advantage of a phylogenetic approach is that groups of mammals which are thought to be related to one another usually are listed close to each other. A common alternative is to adopt an alphabetical listing of genera, within a phylogenetic arrangement of order and families, but inevitably this often conspires to keep related genera at some distance from each other. A phylogenetic classification is generally regarded as being more 'natural' and informative, because related mammals are kept together in the sequence, but its main disadvantage is that information retrieval is slightly slower.

Although most previous classifications of mammals have been at least partly phylogenetic, the McKenna-Bell higher classification differs in recognising evolutionary branching as occurring at many different levels, which in turn has required them to use a much larger number of categories than is usual. Whereas previous classifications have usually managed with just the categories *order*, *suborder*, *family* and *subfamily*, the new arrangement of McKenna & Bell uses many more categories in order to represent more fully the complexity of their proposed evolutionary tree. In addition to the familiar order, suborder, family and subfamily, they employ a whole range of less familiar categories such as *legion*, *cohort*, and an eight-fold series of categories based on the order, from *magnorder* all the way down to *parvorder*. It so happens that some of these categories are only used to discriminate groups of fossil species, so in our checklist of purely modern forms we have chosen to show only those categories that are strictly necessary, with the other categories being discretely omitted. If you are interested in understanding the way McKenna & Bell show the relationships of all mammals, including the many fossil forms, then we recommend that you do not depend on the categories we have listed but refer to their original work.

The McKenna-Bell classification proposes a rather radical rearrangement of the main mammalian orders. Their new arrangement has the considerable merit of including all of the fossil mammals, and we believe that it will become widely accepted. At least one recent field guide (Kays & Wilson, 2002) has already begun to move in this direction, and we expect others will do so in due course.

We have departed from McKenna & Bell's higher classification in three respects, two of them fairly major and one very minor. Firstly, for the Primates families our genus-level classification is taken from Groves (2001). Secondly, our

7

genera of the marine mammal families are taken from Reeves *et al.* (2002). Finally, for reasons mentioned in Jones *et al.* (1997) we use the spelling 'Aplodontiidae' (and not Aplodontidae).

At the time this checklist was completed ready for publication, five new genera had not yet been incorporated into the McKenna-Bell classification. The genus *Juliomys* González, 2000, includes the species hitherto named *Wilfredomys pictipes* and the new species *J. rimofrons* de Oliveira & Bonvicino, 2002, and has been placed by us immediately after *Wilfredomys*. The genus *Sommeromys* Musser & Durden, 2002, has been placed by us after *Crunomys*, to which it appears to be closely related. The genus *Voalavo* Carleton & Goodman, 1998, is a sister to *Eliurus* (Carleton & Goodman, 1998) and so has been placed before that genus in the list. The remaining two genera, *Callistomys* Emmons & Vucetich, 1998, and *Handleyomys* Voss, Gómez-Laverde & Pacheco, 2002, were created for species which are already listed and have not been included in the checklist, pending their formal placement in the McKenna-Bell classification.

SPECIES CLASSIFICATION

The checklist's species-level classification is based upon Wilson & Reeder's *Mammal Species of the World: a taxonomic and geographic reference* (2nd edition 1993), but again we have made some changes that are explained below.

Our first task was to assign all of Wilson & Reeder's species to the genera used in McKenna & Bell's higher classification. This proved to be comparatively straightforward because the two works were published only four years apart and use similar genus-level classifications.

To help keep related species close to each other in the checklist, and to simplify the situation where at some future date a subgenus is upgraded to a genus, we have followed Wilson & Reeder's subgeneric assignments, at least for those genera where every species is unambiguously placed in a subgenus (this is not true of some large genera such as *Sorex*, for example). Subgenera are arranged alphabetically, except in the cats (Felidae) where the subgenera of *Felis* and *Panthera* follow the arrangement of McKenna & Bell (2002).

Our checklist does not include species which are known only from subfossil remains or which certainly became extinct before 1800, whereas Wilson & Reeder include every species thought to have been alive since 1500. A list of the Wilson & Reeder species we have excluded on these grounds is given in the Appendix section 'Extinct species not included' on page 210. In this context it should be noted that some mammals are very poorly known, often from just a single holotype specimen, or a short series of specimens, collected in the 19th Century; however, we continue to list these species as extant unless there is good evidence that the species is certainly extinct. Species which are thought to have probably, or have certainly, become extinct since 1800 are included but marked with a cross (†) before the English name in the checklist. Table 2 on page 11 provides a list of these recently extinct species.

Our taxonomic treatment of domesticated mammals follows that of Clutton-Brock (1999), except that *Mustela putorius* European Polecat is regarded as the wild ancestor of *M. furo* Ferret (Davison *et al.*, 1999). This departs from Wilson & Reeder (1993) in giving separate species status for distinctive domesticated forms which do not, in fact, interbreed to any significant extent with their wild ancestors. Although the taxonomic merits of this position may be debated, there are good practical grounds for making this distinction, not least the fact that wild and domestic forms are often given very different legal statuses.

We have tried to include every species which has been newly described or elevated from synonymy since Wilson & Reeder's catalogue was published in 1993. Our main sources were the *Zoological Record* for 1993-2001, recent issues of mammalogy journals not yet abstracted in *Zoological Record* and recently published field guides of mammals. Field guides covering Europe (Macdonald & Barrett, 1993; Mitchell-Jones *et al.*, 1999), Africa (Kingdon, 1997; Stuart & Stuart, 2001), Madagascar (Garbutt, 1999), India (Gurung & Singh, 1996), Borneo (Payne & Francis, 1998), New Guinea (Flannery, 1995a), Moluccan and Southwest Pacific Ocean islands (Flannery, 1995b), Australia (Menkhorst & Knight, 2001), North America (Kays & Wilson, 2002), Southern Mexico with Central America (Reid, 1997) and South America (Eisenberg, 1989; Eisenberg & Redford, 1999; Emmons & Feer, 1997; Redford & Eisenberg, 1992) were all consulted, plus recent monographs on cetaceans (Carwardine & Camm, 1995) and marine mammals (Reeves *et al.*, 2002).

In the case of the rodents, the discovery of many new species and several new genera reflects the greatly increased interest in this order in recent years, especially among American mammalogists. Moreover, growing adoption of the Phylogenetic Species Concept (for example Groves, 2001), and the use of cytogenetics in classification, have together resulted in the recognition of many more sibling species than hitherto, particularly in the marsupials, insectivores, bats, primates and cetaceans. Even such familiar animals as the House Mouse *Mus musculus*, Gorilla *Gorilla gorilla*, Common Dolphin *Delphinus delphis* and African Elephant *Loxodonta africana* are now considered to comprise more than one species.

Species which are recognised by taxonomists but have not yet been formally described are not included. We are aware of the existence of two species of *Phalanger* (Phalangeridae), one from Ternate and one from Mt Karimui in New Guinea (Flannery & Schouten, 1994), a *Scotorepens* (Vespertitionidae) and at least seven species of *Mormopterus* (Molossidae) from Australia (Menkhorst & Knight, 2001), three species of *Galago* (Loridae) from East Africa (Groves, 2000) and a *Tarsius* species (Tarsiidae) from Pulau Salayar, SW Sulawesi (Groves, 2000).

Much to our surprise, we calculate that there has been a net increase of 481 in the number of recognised extant mammal species, compared to Wilson & Reeder's list, or an average of 48 species per year for the decade 1993–2002. Table 1 compares the number of extant mammal species in Wilson & Reeder's (1993) catalogue with the present checklist.

Table 1: Number of extant mammal species, 1993 vs 2002

Total species per Wilson & Reeder (1993)	4629
Pre-1800 extinct species excluded	-41
Total extant species per Wilson & Reeder (1993)	4588
New species described 1993–2002	+522
Species demoted into synonymy 1993–2002	-41
Total extant species in this checklist	**5069**

For the benefit of those who are already familiar with Wilson & Reeder's catalogue, a complete list of the new species is given in the Appendix section 'New species' on page 211. We have also taken account of the much smaller number of species which have been demoted into synonymy since 1993. These are tabulated in the Appendix section 'Species demoted into synonymy' on page 230.

ENGLISH NAMES

English names have been compiled from a variety of sources. Our starting point was Wilson & Cole's *Common Names of Mammals of the World* (2000), which rather conveniently uses the Wilson & Reeder (1993) species-level taxonomy. We first corrected a number of errors, and changed other names which we felt to be either clearly inappropriate or were unnecessary departures from well-established names. In a number of cases we introduced modifying adjectives where a species had the same name as that of its group common name. In some cases we also added alternative common names in brackets, recognising that such alternatives are widely used and are a potential source of confusion.

We then used a variety of recently published mammal field guides, adopting some (but by no means all) of their English names where it is clear that the field guide usage will become the *de facto* standard for the species concerned.

Finally, for new species described since 1993 which are not yet included in any field guide, we sought English names both from the primary literature and from the Internet. Only as a last resort have we coined new English names ourselves; where this was necessary, we have tried to choose names which are both appropriate and euphonious.

HABITAT AND DISTRIBUTIONAL DATA

The habitat and distribution summaries are based on information taken from Wilson & Reeder (1993), supplemented and corrected by referring to regional field guides and monographs which are listed in the bibliography.

Habitat information is much more comprehensively described for species from Europe, southern Africa, Australasia and the Americas than for other regions, simply because their mammal faunas are much better known. We hope to redress this imbalance in later editions of the checklist, as new field guides and monographs are published.

Where a species has died out in the wild but has since become re-established through reintroduction to its original range, this is indicated by "reintrod." in the summary. Established deliberately released (or feral) populations, are listed separately, after the main range.

The distributional summary is generally described from a northwesterly to a southeasterly direction. For widespread species we start with Europe and Africa and work eastwards. Where the overall range is divided into several widely separated populations, the distributions are separated by semicolons.

STATUS INFORMATION

After the habitat and distributional data we include the status grade for every species which is listed by the International Union for Conservation of Nature and Natural Resources (IUCN) as Extinct, Extinct in the Wild, Critically Endangered or Endangered. This information was obtained from the IUCN website '2003 IUCN Red List of Threatened Species' (http://www.redlist.org).

Table 2 lists 46 mammal species which are now presumed to be extinct but which are considered to have been definitely still alive during the 19th or 20th Centuries. These recently extinct species are included in the main Systematic List starting on page 13, whereas species which we believe certainly became extinct before 1800 are listed in the Appendix section 'Extinct species not included' on page 210.

Table 2: Extinct mammal species included in this checklist

Species	Date last recorded
Peromyscus pembertoni Pemberton's Deer Mouse	?
Oryzomys nelsoni Nelson's Rice Rat	?
Paulamys naso Flores Long-nosed Rat	?
Gazella saudiya Saudi Gazelle	?
Notomys mordax Darling Downs Hopping Mouse	1840s
Notomys macrotis Big-eared Hopping Mouse	before 1844
Pseudomys goudii Gould's Mouse	1857
Pteropus brunneus Dusky Flying Fox	1859
Pteropus subniger Lesser Mascarene Flying Fox	1860s
Conilurus albipes White-footed Rabbit Rat	1862
Potorous platyops Broad-faced Potoroo	1865
Equus quagga Quagga	1872
Pteropus pilosus Large Palau Flying Fox	before 1874
Dusicyon australis Falkland Island Wolf	1876
Lagorchestes leporides Eastern Hare-Wallaby	1891
Acerodon lucifer Panay Golden-capped Acerodon	1892
Nyctimene sanctacrucis Nendö Tube-nosed Fruit Bat	1892
Gazella rufina Red Gazelle	before 1894
Notomys amplus Short-tailed Hopping Mouse	1895
Megalomys luciae St Lucia Giant Rice Rat	before 1900
Chaeropus ecaudatus Pig-footed Bandicoot	1901
Notomys longicaudatus Long-tailed Hopping Mouse	1901
Megalomys desmarestii Antillean Giant Rice Rat	1902
Rattus macleari MacLear's Rat	1908
Rattus nativitatis Bulldog Rat	1908
Pipistrellus sturdeei Sturdee's Pipistrelle	1915
Macropus greyi Toolache Wallaby	1927
Nesoryzomys darwini Darwin's Galápagos Mouse	1930
Macrotis leucura Lesser Bilby	1931
Lagorchestes asomatus Central Hare-Wallaby	1932

Cervus schomburgki Schomburgk's Deer	1932
Leporillus apicalis Lesser Stick-nest Rat	1933
Caloprymnus campestris Desert Rat-Kangaroo	c.1935
Thylacinus cynocephalus Thylacine	1936
Onychogalea lunata Crescent Nail-tailed Wallaby	1940s
Perameles eremiana Desert Bandicoot	1943
Nesoryzomys indeffesus Indefatigable Galápagos Mouse	c.1945
Geocapromys thoracatus Swan Island Hutia	c.1950
Gazella arabica Arabian Gazelle (incl. *G. arabica bilkis* Queen of Sheba's Gazelle)	1951
Monachus tropicalis West Indian Monk Seal	1952
Uromys imperator Emperor Rat	c.1960
Uromys porculus Guadalcanal Rat	c.1960
Procyon gloveralleni Barbados Raccoon	1964
Mystacina robusta Greater New Zealand Short-tailed Bat	1965
Pteropus tokudae Guam Flying Fox	1967
Dobsonia chapmani Negros Bare-backed Fruit Bat	c.1970

A much greater number of mammal species are now facing extinction: 525 species, or over 10% of the surviving world fauna, are now listed by the IUCN as Extinct in the Wild, Critically Endangered or Endangered. Many other, less threatened species, which are now graded as Vulnerable, Near Threatened or Data Deficient by the IUCN, may in their turn soon be endangered if current rates of habitat destruction and human population growth continue unabated.

Lest it be thought that these are nearly all obscure species of bats and rodents, which in truth nobody apart from mammalogists would miss, it should be pointed out that for eighteen families of mammals—over 14% of the families that remain—*every* species in that family is graded by the IUCN as Near Threatened, Vulnerable, Endangered or Critically Endangered. These families are: Microbiotheriidae (Monito del Monte), Notoryctidae (marsupial-moles), Pedetidae (Spring-Hare), Dinomyidae (Pacarana), Solenodontidae (solenodons), Craseonycteridae (Bumblebee Bat), Mystacinidae (New Zealand short-tailed bats), Myzopodidae (Sucker-footed Bat), Daubentoniidae (Aye-aye), Physeteridae (Great Sperm Whale), Platanistidae (Indian river dolphins), Iniidae (Amazon River Dolphin), Lipotidae (Yangtze River Dolphin), Rhinocerotidae (rhinoceroses), Tapiridae (tapirs), Trichechidae (manatees), Dugongidae (Dugong) and Elephantidae (elephants). This roll call of threatened families represents an astonishingly wide range of animal body form and way of life: terrestrial, freshwater and marine, carnivorous as well as vegetarian, from the smallest known mammal (the Bumblebee Bat *Craseonycteris thonglongyai*) to one of our largest (the Great Sperm Whale *Physeter catodon*), and including some of our most cherished and emblematic species (manatees, rhinoceroses and elephants). A clearer example of biodiversity in a state of crisis would be hard to imagine.

It is bad enough to realise that overhunting by humans was the probable cause of the extinction, during prehistoric times, of many unique forms of large mammal which made up the so-called mammalian megafaunas of North America, Europe and Australia. Now it seems that, by both direct and indirect means—but always as the result of many small acts of self-interest or indifference—we are intent on ensuring the eradication of the remaining wild mammalian faunas around the globe.

We hope that these disturbing statistics and this checklist may help to draw attention to the dire plight of many mammal species, at the start of the 21st Century, and that everyone who cares about wild animals will be encouraged to try and conserve for posterity what remains of our already greatly impoverished mammalian wildlife heritage.

SOFTWARE VERSION

We started this project by approaching Robert Eisberg, Founder of Santa Barbara Software Products, Inc., whose popular BirdBase™ and BirdArea™ listing software products are used by many of the world's top birdwatchers. We suggested that he might consider adding to BirdBase/BirdArea the ability to record mammals, and it soon fell to us to provide the checklist of mammal species, as well as the distributional data. Prof. Eisberg willingly granted us the right to produce the mammals checklist in book form, which you are now holding, and in turn we undertook to keep the Mammal Data Add-on™ software in close agreement with the book. Users of the printed checklist may therefore be interested to know that a software version is available to facilitate keeping computerised records. In addition to the checklist itself, the software provides detailed country by country distributional information so that mammal checklists for particular regions or countries (also US states and Canadian provinces) can quickly and easily be produced.

The Mammal Data Add-on™ can be obtained from Santa Barbara Software Products, Inc., 1400 Dover Road, Santa Barbara, CA 93103, USA; website: http://members.aol.com/sbsp; e-mail: sbsp@aol.com.

ACKNOWLEDGEMENTS

Producing the Mammal Data Add-on™ checklist of world mammals in book form was suggested to us by Robert Eisberg, Founder and President of Santa Barbara Software Products, Inc. We are very grateful for his original suggestion and for his constant encouragement.

We are greatly indebted to Dr Andrew Kitchener (National Museums of Scotland, Edinburgh, UK) for painstakingly reviewing the manuscript and pointing out several important errors and omissions, as well as advising us on the treatment of domesticated mammals and suggesting other improvements to the checklist's scope and layout.

Susan Bell (American Museum of Natural History, New York, USA) provided invaluable taxonomic advice and helpful comments on a draft version of the checklist. Sarah Brewer (Biology Library, Bristol University, Bristol, UK) and Barbara Craven (General Library, Natural History Museum, London, UK) are thanked for arranging access to library facilities. Dr Pedro Durant (Universidad de Los Andes, Mérida, Venezuela) advised us on the preferred English name for a species described by him, *Sylvilagus varynaensis*. Michael Grigson kindly allowed us to borrow reference works from his extensive natural history library. Jan Hammond provided useful comments on a draft of the introductory sections. Dr Don McFarlane (W.M. Keck Science Center, The Claremont Colleges, Claremont, California, USA) provided detailed information on extinct mammals of the West Indies, including advice on appropriate English names. Thanks also to Dr Anthony Rylands (Center for Applied Biodiversity Science, Conservation International, Washington DC, USA) for sending us a copy of the beautifully illustrated revision of the titi monkeys by Van Roosmalen *et al.* (2002).

Finally, we are very grateful to Nigel Redman at A&C Black for enthusiastically taking the project on, to Mike Unwin for his delightful cover illustrations, and to Julie Dando of Fluke Art for the design, layout and indexing.

SYSTEMATIC LIST

Class: MAMMALIA (Mammals)
Subclass: PROTOTHERIA (Monotremes)
Order: PLATYPODA
Family: ORNITHORHYNCHIDAE (Platypus—1)

❏ *Ornithorhynchus anatinus* Platypus

E Australia incl. Tasmania

Order: TACHYGLOSSA
Family: TACHYGLOSSIDAE (Echidnas—4)

❏ *Zaglossus attenboroughi*
Attenborough's Echidna

Jayapura region (Cyclops Mtns, C New Guinea); possibly extinct, only known from holotype specimen

❏ *Zaglossus bartoni* Barton's Echidna

New Guinea

❏ *Zaglossus bruijni* Long-beaked Echidna

Salawati and New Guinea; Endangered

❏ *Tachyglossus aculeatus* Short-beaked Echidna

E New Guinea and Australia incl. Tasmania

Subclass: THERIIFORMES
Cohort: MARSUPIALIA (Marsupials)
Magnorder: AUSTRALIDELPHIA
Superorder: MICROBIOTHERIA
Family: MICROBIOTHERIIDAE (Monito del Monte—1)

❏ *Dromiciops gliroides* Monito del Monte

Cool moist forests with bamboo thickets of Concepción region to Chiloé I. (C Chile) and WC Argentina

Superorder: EOMETATHERIA
Order: NOTORYCTEMORPHIA
Family: NOTORYCTIDAE (Marsupial-Moles—2)

❏ *Notoryctes caurinus* Northern Marsupial-Mole

Arid sandy deserts of NW Western Australia; Endangered

❏ *Notoryctes typhlops* Southern Marsupial-Mole

Arid sandy deserts of W Australia; Endangered

Grandorder: DASYUROMORPHIA
Family: THYLACINIDAE (Thylacine—1)

❏ †*Thylacinus cynocephalus* Thylacine

Formerly Tasmania (Australia); Extinct (last captive animal died in 1936)

Family: DASYURIDAE (Dasyurids—70)

❏ *Dasyurus albopunctatus* New Guinea Quoll

New Guinea

❏ *Dasyurus geoffroii* Western Quoll

Formerly widespread in W & C Australia, now confined to forests of SW Western Australia

❏ *Dasyurus hallucatus* Northern Quoll

N Australia

❏ *Dasyurus maculatus* Spot-tailed Quoll

E Queensland to SE South Australia and Tasmania (Australia)

❏ *Dasyurus spartacus* Bronze Quoll

Morehead region (Fly R. plains, SC New Guinea)

❏ *Dasyurus viverrinus* Eastern Quoll

Formerly widespread in SE Australia, now confined to Tasmania (Australia)

❏ *Planigale gilesi* Paucident Planigale

Deserts of EC Australia

❏ *Planigale ingrami* Long-tailed Planigale

NE Western Australia to E Queensland (Australia)

❏ *Planigale maculata* Pygmy Planigale

N Northern Territory to NE New South Wales (Australia)

❏ *Planigale novaeguineae* Papuan Planigale

Lowlands of S & SE New Guinea

❏ *Planigale tenuirostris* Narrow-nosed Planigale

SC Queensland to NW New South Wales (Australia)

❏ *Phascolosorex doriae*
Red-bellied Marsupial-Shrew

Vogelkop Peninsula, Weyland Range and NW Central Cordillera (W New Guinea)

❏ *Phascolosorex dorsalis*
Narrow-striped Marsupial-Shrew

Montane rainforests of W & E New Guinea

❏ *Neophascogale lorentzi* **Speckled Dasyure**

High mountain forests of W & C New Guinea

Antechinus (Antechinus)

❏ *Antechinus adustus* **Rusty Antechinus**

Vine forests in coastal NE Queensland, Australia

❏ *Antechinus agilis* **Agile Antechinus**

S Victoria and SE New South Wales (Australia)

❏ *Antechinus bellus* **Fawn Antechinus**

Tropical forests of N Northern Territory (Australia)

❏ *Antechinus flavipes* **Yellow-footed Antechinus**

Queensland to SE South Australia; SW Western Australia

❏ *Antechinus godmani* **Atherton Antechinus**

Rainforests of NE Queensland, Australia

❏ *Antechinus habbema* **Habbema Antechinus**

Highlands of New Guinea

❏ *Antechinus leo* **Cinnamon Antechinus**

Rainforests of N Cape York Peninsula, N Queensland, Australia

❏ *Antechinus melanurus* **Black-tailed Antechinus**

New Guinea

❏ *Antechinus minimus* **Swamp Antechinus**

Wet heaths and grasslands of extreme SE Australia, and Tasmania

❏ *Antechinus naso* **Long-nosed Antechinus**

Primary rainforests in highlands of New Guinea

❏ *Antechinus stuartii* **Brown Antechinus**

Coastal SE Queensland and New South Wales (Australia)

❏ *Antechinus subtropicus* **Subtropical Antechinus**

SE Queensland and NE New South Wales (EC Australia)

❏ *Antechinus swainsonii* **Dusky Antechinus**

SE Queensland to SE Victoria and Tasmania (Australia)

❏ *Antechinus wilhelmina* **Lesser Antechinus**

Lake Habbema region (W New Guinea)

Antechinus (Parantechinus)

❏ *Antechinus apicalis* **Southern Dibbler**

SW Western Australia; Endangered

❏ *Antechinus bilarni* **Sandstone Dibbler**

Northern Territory (Australia)

Antechinus (Pseudantechinus)

❏ *Antechinus macdonnellensis*
Fat-tailed Pseudantechinus

Deserts of N Western Australia and Northern Territory

❏ *Antechinus mimulus* **Carpentarian Pseudantechinus**

Locally in NE Northern Territory and NW Queensland (Australia)

❏ *Antechinus ningbing* **Ningbing Pseudantechinus**

Kimberley Plateau (NE Western Australia)

❏ *Antechinus roryi* **Tan Pseudantechinus**

NC Western Australia

❏ *Antechinus woolleyae* **Woolley's Pseudantechinus**

WC Western Australia

Antechinus (Dasykaluta)

❏ *Antechinus rosamondae* **Little Red Kaluta**

NW Western Australia

❏ *Phascogale calura* **Red-tailed Phascogale**

Now confined to S Western Australia; Endangered

❏ *Phascogale tapoatafa* **Brush-tailed Phascogale**

Coastal NW, SW & E Australia

❏ *Sarcophilus laniarius* **Tasmanian Devil**

Tasmania (Australia)

❏ *Dasyuroides byrnei* **Kowari**

Lake Eyre basin (SW Queensland and NE South Australia)

❏ *Dasycercus cristicauda* **Mulgara**

Arid deserts of interior W & C Australia

❏ *Dasycercus hillieri* **Ampurta**

Simpson Desert region (C Australia); Endangered

Sminthopsis (Antechinomys)

❏ *Sminthopsis laniger* Kultarr

Western Australia E in interior deserts to SW Queensland (Australia)

Sminthopsis (Sminthopsis)

❏ *Sminthopsis aitkeni* Kangaroo Island Dunnart

Kangaroo I. (South Australia); Endangered

❏ *Sminthopsis archeri* Chestnut Dunnart

Lowlands of S New Guinea and Cape York Peninsula (N Queensland, Australia)

❏ *Sminthopsis bindi* Kakadu Dunnart

NC Northern Territory (Australia)

❏ *Sminthopsis butleri* Carpentarian Dunnart

New Guinea and near Kalumburu (Western Australia)

❏ *Sminthopsis crassicaudata* Fat-tailed Dunnart

SW & C Australia

❏ *Sminthopsis dolichura* Little Long-tailed Dunnart

SW & SC Australia

❏ *Sminthopsis douglasi* Julia Creek Dunnart

Cloncurry R. watershed (WC Queensland, Australia); Endangered

❏ *Sminthopsis fuliginosus* Sooty Dunnart

SW Western Australia

❏ *Sminthopsis gilberti* Gilbert's Dunnart

SW Western Australia

❏ *Sminthopsis granulipes* White-tailed Dunnart

SW Western Australia

❏ *Sminthopsis griseoventer* Grey-bellied Dunnart

SW Western Australia

❏ *Sminthopsis hirtipes* Hairy-footed Dunnart

Deserts of W & C Australia

❏ *Sminthopsis leucopus* White-footed Dunnart

Queensland to S Victoria and Tasmania (Australia)

❏ *Sminthopsis longicaudata* Long-tailed Dunnart

Western Australia

❏ *Sminthopsis macroura* Stripe-faced Dunnart

NW & C Australia

❏ *Sminthopsis murina* Slender-tailed Dunnart

SW & E Australia

❏ *Sminthopsis ooldea* Ooldea Dunnart

C Australia

❏ *Sminthopsis psammophila* Sandhill Dunnart

SW Northern Territory and Eyre Peninsula (South Australia); Endangered

❏ *Sminthopsis virginiae* Red-cheeked Dunnart

Aru Is. (SE Moluccas), S New Guinea and coastal NC & NE Australia

❏ *Sminthopsis youngsoni* Lesser Hairy-footed Dunnart

Western Australia and Northern Territory (Australia)

❏ *Murexia longicaudata* Short-furred Dasyure

Primary rainforests of Aru Is. (SE Moluccas), Japen and New Guinea

❏ *Murexia rothschildi* Broad-striped Dasyure

Primary forests of SE New Guinea

❏ *Myoictis melas* Three-striped Dasyure

Lowland rainforests of Aru Is. (SE Moluccas), Salawati, Waigeo, Japen and New Guinea

❏ *Ningaui ridei* Wongai Ningaui

Deserts of W & C Australia

❏ *Ningaui timealeyi* Pilbara Ningaui

NW Western Australia

❏ *Ningaui yvonnae* Southern Ningaui

S Australia

❏ *Myrmecobius fasciatus* Numbat

Formerly widespread in SW & SC Australia, now very local in SW Western Australia

Grandorder: SYNDACTYLI
Order: PERAMELIA
Family: PERAMELIDAE
(Australian Bandicoots and Bilbies—10)

❏ *Isoodon auratus* Golden Bandicoot

NW Western Australia and Barrow I. (Australia)

❏ *Isoodon macrourus* Northern Brown Bandicoot

S & E New Guinea and N & E Australia

❏ *Isoodon obesulus* Southern Brown Bandicoot

SW & E Australia incl. Tasmania

❏ *Perameles bougainville*
Western Barred Bandicoot

Bernier and Dorre Is. (Western Australia); Endangered

❏ †*Perameles eremiana* Desert Bandicoot

Formerly deserts of C Australia; Extinct (last collected in 1943)

❏ *Perameles gunnii* Eastern Barred Bandicoot

Near Hamilton (S Victoria) and Tasmania (Australia)

❏ *Perameles nasuta* Long-nosed Bandicoot

E Queensland to E Victoria (Australia)

❏ *Macrotis lagotis* Bilby

NW & EC Australia

❏ †*Macrotis leucura* Lesser Bilby

Formerly C Australia; Extinct (last collected in 1931)

❏ †*Chaeropus ecaudatus* Pig-footed Bandicoot

Formerly Australia; Extinct (last collected in 1901)

Family: PERORYCTIDAE
(New Guinea Bandicoots—11)

❏ *Peroryctes broadbenti* Giant Bandicoot

Lowlands of SE New Guinea

❏ *Peroryctes raffrayana* Raffray's Bandicoot

Forests of Japen and New Guinea

❏ *Echymipera clara* Clara's Echymipera

Foothill forests of Japen and N New Guinea

❏ *Echymipera davidi* David's Echymipera

Kiriwana (D'Entrecasteaux Is., New Guinea)

❏ *Echymipera echinista* Menzies' Echymipera

Strickland and Fly Rs. region (S New Guinea)

❏ *Echymipera kalubu* Common Echymipera

Misoöl, Salawati, Waigeo, Geelvinck Is., New Guinea and Bismarck Archipelago

❏ *Echymipera rufescens*
Long-nosed Echymipera

Kai and Aru Is. (SE Moluccas), Misoöl, Japen, New Guinea, D'Entrecasteaux Is. and NE Australia

❏ *Rhynchomeles prattorum* Seram Bandicoot

Precipitous montane forests of Seram (S Moluccas); status uncertain, only known from type series collected in 1920

❏ *Microperoryctes longicauda*
Striped Bandicoot

New Guinea

❏ *Microperoryctes murina* Mouse Bandicoot

W New Guinea

❏ *Microperoryctes papuensis* Papua Bandicoot

SE New Guinea

Order: DIPROTODONTIA
Family: TARSIPEDIDAE
(Honey Possum—1)

❏ *Tarsipes rostratus* Honey Possum

SW Western Australia

Family: VOMBATIDAE (Wombats—3)

❏ *Vombatus ursinus* Coarse-haired Wombat

SE Australia incl. Tasmania

❏ *Lasiorhinus krefftii*
Northern Hairy-nosed Wombat

E Australia; Critically Endangered

❏ *Lasiorhinus latifrons*
Southern Hairy-nosed Wombat

S Australia

Family: PHALANGERIDAE
(Cuscuses and Brushtail Possums—23)

❏ *Ailurops ursinus* **Bear Cuscus**

Sangihe and Talaud Is. (N Moluccas), and Sulawesi

❏ *Strigocuscus celebensis* **Small Sulawesi Cuscus**

Sulawesi, Sangihe Is. (N Moluccas) and Taliabu (Sula Is.)

❏ *Strigocuscus pelengensis* **Peleng Cuscus**

Peleng and Sula Is. (W Moluccas)

❏ *Wyulda squamicaudata* **Scaly-tailed Possum**

Kimberley Plateau (NE Western Australia)

❏ *Trichosurus caninus*
Short-eared Brushtail Possum

SE Australia

❏ *Trichosurus cunninghamii*
Mountain Brushtail Possum

Subtropical forests of C Queensland to Victoria (E Australia)

❏ *Trichosurus vulpecula* **Common Brushtail Possum**

SW, NC & E Australia incl. Tasmania; feral New Zealand

❏ *Spilocuscus kraemeri* **Admiralty Cuscus**

Admiralty Is. (Bismarck Archipelago)

❏ *Spilocuscus maculatus* **Common Spotted Cuscus**

S Moluccas incl. Kai and Aru Is., New Guinea, Bismarck Arch. and N Cape York Pen. (N Queensland, Australia)

❏ *Spilocuscus papuensis* **Waigeo Cuscus**

Waigeo (New Guinea)

❏ *Spilocuscus rufoniger* **Black-spotted Cuscus**

Undisturbed forests of N New Guinea; Endangered

❏ *Phalanger alexandrae* **Gebe Cuscus**

Gebe (N Moluccas)

❏ *Phalanger carmelitae* **Mountain Cuscus**

Montane forests of New Guinea

❏ *Phalanger gymnotis* **Ground Cuscus**

Aru Is. (SE Moluccas) and New Guinea

❏ *Phalanger intercastellanus*
Southern Common Cuscus

Aru Is. (SE Moluccas), New Guinea, D'Entrecasteaux Is., Louisiade Archipelago and NE Australia

❏ *Phalanger lullulae* **Woodlark Cuscus**

Alcester and Woodlark Is. (D'Entrecasteaux Is., New Guinea)

❏ *Phalanger matanim* **Telefomin Cuscus**

Telefomin region (C New Guinea); Endangered

❏ *Phalanger mimicus* **Cryptic Cuscus**

New Guinea and NE Australia

❏ *Phalanger orientalis* **Northern Common Cuscus**

Timor (Lesser Sundas), S & E Moluccas, New Guinea, Bismarck Archipelago and Solomon Is.

❏ *Phalanger ornatus* **Ornate Cuscus**

Morotai, Halmahera, Ternate, Tidore and Bacan (N Moluccas)

❏ *Phalanger rothschildi* **Obi Cuscus**

Obi and Bisa (S Moluccas)

❏ *Phalanger sericeus* **Silky Cuscus**

Primary montane forests of New Guinea

❏ *Phalanger vestitus* **Stein's Cuscus**

Primary oak forests of New Guinea

Family: BURRAMYIDAE
(Pygmy Possums—5)

❏ *Burramys parvus* **Mountain Pygmy Possum**

SE Australia; Endangered

❏ *Cercartetus caudatus* **Long-tailed Pygmy Possum**

Montane forests and grasslands of Central Cordillera (New Guinea) and rainforests of NE Queensland (Australia)

❏ *Cercartetus concinnus* **Western Pygmy Possum**

S Australia

❏ *Cercartetus lepidus* **Little Pygmy Possum**

SE Australia incl. Tasmania

❏ *Cercartetus nanus* **Eastern Pygmy Possum**

SE Australia and Tasmania (Australia)

Family: MACROPODIDAE
(Kangaroos and Wallabies—73)

❏ *Hypsiprymnodon moschatus*
Musky Rat-Kangaroo

NE Queensland (Australia)

❏ *Bettongia gaimardi* Tasmanian Bettong

Tasmania (Australia)

❏ *Bettongia lesueur* Burrowing Bettong

Islands of W Australia

❏ *Bettongia penicillata* Brush-tailed Bettong

S & E Australia

❏ *Bettongia tropica* Northern Bettong

Forests of NE Queensland (Australia); Endangered

❏ *Potorous gilbertii* Gilbert's Potoroo

Albany region (SW Western Australia); Critically
Endangered

❏ *Potorous longipes* Long-footed Potoroo

NE Victoria (Australia); Endangered

❏ *†Potorous platyops* Broad-faced Potoroo

Formerly SW & SC Australia; Extinct (probably by 1865)

❏ *Potorous tridactylus* Long-nosed Potoroo

SE Australia incl. Tasmania

❏ *Aepyprymnus rufescens* Rufous Bettong

E Australia

❏ *†Caloprymnus campestris*
Desert Rat-Kangaroo

Formerly deserts of EC Australia; Extinct (by c.1935)

❏ *Dorcopsulus macleayi* Papuan Forest Wallaby

Mid-montane forests of E New Guinea

❏ *Dorcopsulus vanheurni* Lesser Forest Wallaby

Montane forests of New Guinea

❏ *Thylogale billardierii* Tasmanian Pademelon

SE Australia incl. Tasmania

❏ *Thylogale browni* New Guinea Pademelon

NC & NE New Guinea and Bismarck Archipelago

❏ *Thylogale brunii* Dusky Pademelon

Kai and Aru Is. (SE Moluccas) and S New Guinea

❏ *Thylogale calabyi* Calaby's Pademelon

Mts. Edward and Giluwe (E New Guinea); Endangered

❏ *Thylogale stigmatica* Red-legged Pademelon

S Fly R. plains region (SC New Guinea) and E Australia

❏ *Thylogale thetis* Red-necked Pademelon

E Australia

Macropus (Macropus)

❏ *Macropus fuliginosus* Western Grey Kangaroo

S Australia incl. Tasmania

❏ *Macropus giganteus* Eastern Grey Kangaroo

E Australia incl. Tasmania

Macropus (Notamacropus)

❏ *Macropus agilis* Agile Wallaby

S New Guinea, D'Entrecasteaux Is. and N Australia

❏ *Macropus dorsalis* Black-striped Wallaby

E Australia; feral Kawau I. (New Zealand, probably extinct)

❏ *Macropus eugenii* Tammar Wallaby

SW & SC Australia; feral New Zealand

❏ *†Macropus greyi* Toolache Wallaby

Formerly SE Australia; Extinct (last seen in 1927)

❏ *Macropus irma* Western Brush Wallaby

SW Western Australia (Australia)

❏ *Macropus parma* Parma Wallaby

E New South Wales (Australia); feral New Zealand

❏ *Macropus parryi* Whiptail Wallaby

E Australia

❏ *Macropus rufogriseus* Red-necked Wallaby

SE Australia incl. Tasmania; feral Britain and New Zealand

Macropus (Osphranter)

❏ *Macropus antilopinus* Antelopine Wallaroo

N Australia

❏ *Macropus bernardus* Black Wallaroo

NC Northern Territory (Australia)

❏ *Macropus robustus* Common Wallaroo (Euro)

Australia

❏ *Dorcopsis atrata* Black Dorcopsis

Montane forests of Goodenough (D'Entrecasteaux Is., New Guinea); Endangered

❏ *Dorcopsis hageni* White-striped Dorcopsis

Mixed alluvial forests and river banks in lowlands of N New Guinea

❏ *Dorcopsis luctuosa* Grey Dorcopsis

Lowland rainforests of S & SE New Guinea

❏ *Dorcopsis muelleri* Brown Dorcopsis

Misoöl, Salawati, Japen and lowlands of W New Guinea

❏ *Megaleia rufa* Red Kangaroo

Australia

❏ *Wallabia bicolor* Swamp Wallaby

E Australia; feral Kawau I. (New Zealand)

❏ *Onychogalea fraenata* Bridled Nail-tailed Wallaby

Taunton region (S Queensland, Australia); Endangered

❏ †*Onychogalea lunata*
Crescent Nail-tailed Wallaby

Formerly C & SE Australia; Extinct (last seen in 1940s)

❏ *Onychogalea unguifera*
Northern Nail-tailed Wallaby

N Australia

❏ *Setonix brachyurus* Quokka

SW Western Australia

❏ *Petrogale assimilis* Allied Rock Wallaby

Rocky areas of Townsville region incl. Magnetic and Palm Is. (NE Queensland, Australia)

❏ *Petrogale brachyotis* Short-eared Rock Wallaby

Rocky hills and cliffs in savanna woods of coastal NW & NC Australia

❏ *Petrogale burbidgei* Monjon

Open eucalypt woods of coastal Kimberley Plateau and Bonaparte Archipelago (NE Western Australia)

❏ *Petrogale coenensis* Cape York Rock Wallaby

Rocky areas of EC Cape York Peninsula from Musgrave to Pascoe R. (N Queensland, Australia)

❏ *Petrogale godmani* Godman's Rock Wallaby

Rocky areas of E Cape York Peninsula from Bathurst Head to Mt Carbine (N Queensland, Australia)

❏ *Petrogale herberti* Herbert's Rock Wallaby

Rocky areas of SE Queensland (Australia)

❏ *Petrogale inornata* Unadorned Rock Wallaby

Rocky areas from Home Hill near Townsville to S of Rockhampton (N Queensland, Australia)

❏ *Petrogale lateralis* Black-flanked Rock Wallaby

Very locally in rocky areas of W, C & S Australia incl. near islands

❏ *Petrogale mareeba* Mareeba Rock Wallaby

Rocky areas of E Cape York Peninsula from Mt Carbine to Mt Garnet (N Queensland, Australia)

❏ *Petrogale penicillata* Brush-tailed Rock Wallaby

Rocky areas of SE Queensland and E New South Wales (Australia); feral New Zealand

❏ *Petrogale persephone* Proserpine Rock Wallaby

Rocky areas in foothill forests of Whitsunday Shire region (coastal E Queensland, Australia); Endangered

❏ *Petrogale purpureicollis*
Purple-necked Rock Wallaby

Rocky areas and grasslands of WC Queensland (Australia)

❏ *Petrogale rothschildi* Rothschild's Rock Wallaby

Rocky slopes of Pilbara region and Dampier Archipelago (NW Western Australia, Australia)

❏ *Petrogale sharmani* Sharman's Rock Wallaby

Rocky areas of Seaview and Coane Ranges, E of Ingham (E Queensland, Australia)

❏ *Petrogale xanthopus*
Yellow-footed Rock Wallaby

Rocky areas in semi-arid deserts of EC Australia

❏ †*Lagorchestes asomatus* **Central Hare-Wallaby**

Formerly Mt Farewell region (Northern Territory, Australia); Extinct (one specimen only, collected in 1932)

❏ *Lagorchestes conspicillatus*
Spectacled Hare-Wallaby

N Australia

❏ *Lagorchestes hirsutus* **Rufous Hare-Wallaby**

Formerly widespread in W Australia, now confirmed to Bernier and Dorre Is. (WA); Alice Springs (S Northern Territory)

❏ †*Lagorchestes leporides* **Eastern Hare-Wallaby**

Formerly grassy plains of SE South Australia, NW Victoria and New South Wales (Australia); Extinct (last recorded 1891)

❏ *Lagostrophus fasciatus* **Banded Hare-Wallaby**

Bernier and Dorre Is. (Western Australia)

❏ *Peradorcas concinna*
Nabarlek (Pygmy Rock Wallaby)

Rocky scrub of NE Western Australia and N Northern Territory (Australia)

❏ *Dendrolagus bennettianus*
Bennett's Tree Kangaroo

NE Queensland (Australia)

❏ *Dendrolagus dorianus* **Doria's Tree Kangaroo**

New Guinea

❏ *Dendrolagus goodfellowi*
Goodfellow's Tree Kangaroo

Oak forests of N & E New Guinea; Endangered

❏ *Dendrolagus inustus* **Grizzled Tree Kangaroo**

Salawati?, Japen and NW & N New Guinea

❏ *Dendrolagus lumholtzi*
Lumholtz's Tree Kangaroo

NE Queensland (Australia)

❏ *Dendrolagus matschiei* **Huon Tree Kangaroo**

Huon Peninsula (NE New Guinea); feral Umboi (NE New Guinea); Endangered

❏ *Dendrolagus mbaiso* **Dingiso Tree Kangaroo**

S & W slopes of Surdirman Range (Snow Mtns, W New Guinea)

❏ *Dendrolagus scottae* **Tenkile Tree Kangaroo**

Torricelli Mtns (NC New Guinea); Endangered

❏ *Dendrolagus spadix* **Lowlands Tree Kangaroo**

Papuan Plateau (S New Guinea)

❏ *Dendrolagus ursinus* **Vogelkop Tree Kangaroo**

Vogelkop and Fak Fak Peninsulas (NW New Guinea)

Family: PETAURIDAE
(Ringtail Possums and Gliders—28)

❏ *Pseudocheirus canescens*
Lowland Ringtail Possum

Secondary lowland forests of Salawati, Japen and New Guinea

❏ *Pseudocheirus caroli* **Weyland Ringtail Possum**

Locally in W New Guinea

❏ *Pseudocheirus cinereus*
Daintree River Ringtail Possum

Montane tropical rainforests N of Cairns (NE Queensland, Australia)

❏ *Pseudocheirus forbesi* **Painted Ringtail Possum**

Montane and foothill forests of C & E New Guinea

❏ *Pseudocheirus herbertensis*
Herbert River Ringtail Possum

Montane tropical rainforests from Mt Lee to Lamb Range (NE Queensland, Australia)

❏ *Pseudocheirus mayeri* **Pygmy Ringtail Possum**

Montane moss forests of W & C New Guinea

❏ *Pseudocheirus occidentalis*
Western Ringtail Possum

Locally in forests and gardens of SW Western Australia

❏ *Pseudocheirus peregrinus*
Common Ringtail Possum

Forests, scrub and gardens of E Australia incl. Tasmania

❏ *Pseudocheirus schlegeli* **Arfak Ringtail Possum**

Arfak Mtns (Vogelkop Peninsula, NW New Guinea)

❏ *Pseudochirops albertisii*
D'Albertis' Ringtail Possum

Montane forests of Japen and NW & NC New Guinea

❏ *Pseudochirops archeri* Green Ringtail Possum

Locally in montane tropical rainforests between Paluma and Mt Windsor Tableland (NE Queensland, Australia)

❏ *Pseudochirops corinnae*
Plush-coated Ringtail Possum

Undisturbed mid-elevation forests of New Guinea

❏ *Pseudochirops coronatus*
Reclusive Ringtail Possum

Undisturbed primary forests of Arfak Mtns (Vogelkop Peninsula, W New Guinea)

❏ *Pseudochirops cupreus* Coppery Ringtail Possum

High-altitude primary montane forests and scrub of New Guinea

❏ *Petropseudes dahli* Rock Ringtail Possum

NW Western Australia, N Northern Territory and extreme N Queensland (Australia)

❏ *Hemibelideus lemuroides*
Lemuroid Ringtail Possum

Mature rainforests between Ingham and Cairns, and on Carbine Tableland (NE Queensland, Australia)

❏ *Petauroides volans* Greater Glider

Wet sclerophyll forest on ranges and coastal plains of E Australia

❏ *Petaurus abidi* Northern Glider

Mt Somoro region (Torricelli Mtns, NC New Guinea)

❏ *Petaurus australis* Yellow-bellied Glider

Locally in wet sclerophyll forests of coastal E Australia

❏ *Petaurus biacensis* Biak Glider

Biak (Geelvinck Is., New Guinea)

❏ *Petaurus breviceps* Sugar Glider

N Moluccas, Misoöl, Salawati, Japen, New Guinea, Bismarck Archipelago, D'Entrecasteaux Is. and N & E Australia

❏ *Petaurus gracilis* Mahogany Glider

Swampy coastal woods between Bambaru and Tully (NE Queensland, Australia); Endangered

❏ *Petaurus norfolcensis* Squirrel Glider

Forests of coastal E Australia

❏ *Gymnobelideus leadbeateri* Leadbeater's Possum

Montane wet sclerophyll forests NE of Melbourne and lowland swamp forests at Yellingbo (SC Victoria, Australia); Endangered

❏ *Dactylopsila megalura* Great-tailed Triok

Very few records from montane rainforests of W & C New Guinea

❏ *Dactylopsila palpator* Long-fingered Triok

Montane mossy forests of New Guinea

❏ *Dactylopsila tatei* Tate's Triok

Fergusson (D'Entrecasteaux Is., New Guinea); Endangered

❏ *Dactylopsila trivirgata* Striped Possum

Aru Is. (SE Moluccas), Waigeo, Japen, lowlands of New Guinea and near islands, and NE Queensland (Australia)

Family: PHASCOLARCTIDAE (Koala—1)

❏ *Phascolarctos cinereus* Koala

E Australia

Family: ACROBATIDAE (Feathertail Glider and Feather-tailed Possum—2)

❏ *Acrobates pygmaeus* Feathertail Glider

Cool temperate and tropical eucalypt forests of E Australia, S to SE South Australia

❏ *Distoechurus pennatus* Feather-tailed Possum

Early regrowth and disturbed habitats of New Guinea

Magnorder: AMERIDELPHIA
Order: DIDELPHIMORPHIA
Family: DIDELPHIDAE (American Opossums—78)

Marmosa (Marmosa)

❏ *Marmosa andersoni* Anderson's Mouse-Opossum

Cosnipata region (near Cusco, Peru); Critically Endangered

❏ *Marmosa canescens* Greyish Mouse-Opossum

Desert scrub and dry forests of W & SE Mexico incl. Tres Marías Is. (Nayarit)

❏ *Marmosa lepida* Little Rufous Mouse-Opossum

Very locally in rainforests of Amazon basin of E Colombia to Suriname, S to Peru, Bolivia and S Brazil

❏ *Marmosa mexicana* Mexican Mouse-Opossum

Forests, regrowth areas and grasslands of Mexico to W Panama

❏ *Marmosa murina* Murine Mouse-Opossum

Rainforests, roadside brush and gardens of E Colombia to French Guiana, S to Bolivia and C & SE Brazil; Trinidad & Tobago

❏ *Marmosa robinsoni* Robinson's Mouse-Opossum

Lowland forests, regrowth areas and llanos of Belize to Venezuela, S to N Peru; Roatán (Honduras); Trinidad & Tobago; Grenada

❏ *Marmosa rubra* Red Mouse-Opossum

Forests of Amazon basin of E Colombia, E Ecuador and NE Peru

❏ *Marmosa tyleriana* Tyler's Mouse-Opossum

Tepuis of S Venezuela

❏ *Marmosa xerophila* Dryland Mouse-Opossum

Dry lowland deciduous forests of NE Colombia and NW Venezuela; Endangered

Marmosa (Marmosops)

❏ *Marmosa bishopi*
Bishop's Slender Mouse-Opossum

Forests of Amazon basin of NE Peru

❏ *Marmosa cracens*
Slim-faced Slender Mouse-Opossum

Foothill rainforests of NW Venezuela; Endangered

❏ *Marmosa dorothea*
Dorothy's Slender Mouse-Opossum

Dense thickets in forests of S Amazonian Brazil to C Bolivia

❏ *Marmosa fuscata*
Grey-bellied Slender Mouse-Opossum

Montane forests and clearings of C Colombia and N Venezuela; Trinidad

❏ *Marmosa handleyi*
Handley's Slender Mouse-Opossum

Antioquia region (Colombia); Critically Endangered

❏ *Marmosa impavida*
Andean Slender Mouse-Opossum

Dense montane forests of E Panama to Colombia, Venezuela, Ecuador, S Peru and Bolivia

❏ *Marmosa incana*
Grey Slender Mouse-Opossum

Atlantic coastal forests of Bahia to Paraná (SE Brazil)

❏ *Marmosa invicta*
Slaty Slender Mouse-Opossum

Mid-elevation forests and regrowth areas of Panama

❏ *Marmosa juninensis*
Junín Slender Mouse-Opossum

Forests of C Peru

❏ *Marmosa neblina*
Cerro Neblina Slender Mouse-Opossum

Montane and lowland forests of Venezuela, E Ecuador and C Amazonian Brazil

❏ *Marmosa noctivaga*
White-bellied Slender Mouse-Opossum

Swamps and riparian habitats in forests of Amazon basin of E Ecuador, E Peru, Bolivia and W Brazil

❏ *Marmosa parvidens*
Delicate Slender Mouse-Opossum

Primary terra firme forests of Amazon basin of E Colombia to French Guiana, S to Peru and C Brazil

❏ *Marmosa paulensis*
São Paulo Slender Mouse-Opossum

Atlantic montane cloud forests of S Minas Gerais to São Paulo (SE Brazil)

❏ *Marmosa pinheiroi*
Pinheiro's Slender Mouse-Opossum

Forests of French Guiana

❏ *Gracilinanus aceramarcae*
Aceramarca Gracile Mouse-Opossum

S Peru and R. Aceramarca region (Yungas, Bolivia); Critically Endangered

❏ *Gracilinanus agilis*
Agile Gracile Mouse-Opossum

Forests of E Peru, Bolivia, S & E Brazil, Paraguay, Uruguay and Argentina

❏ *Gracilinanus dryas*
Wood Sprite Gracile Mouse-Opossum

Cloud forests and regrowth areas of mountains of E Colombia and W Venezuela

❏ *Gracilinanus emiliae*
Emilia's Gracile Mouse-Opossum

Very few records from lowland forests of Colombia, Suriname, French Guiana and NE Brazil

❏ *Gracilinanus ignitus*
Red-bellied Gracile Mouse-Opossum

Yuto region (Jujuy Prov., NW Argentina); only known from holotype specimen collected in 1962

❏ *Gracilinanus kalinowskii*
Kalinowski's Gracile Mouse-Opossum

Very few records from lowland and mid-elevation forests of Guyana and French Guiana, and E Peru

❏ *Gracilinanus marica*
Northern Gracile Mouse-Opossum

Forests and savannas of coastal N Colombia and N Venezuela

❏ *Gracilinanus microtarsus*
Brazilian Gracile Mouse-Opossum

Atlantic coastal forests and regrowth areas of Minas Gerais to Rio Grande do Sul (SE Brazil) and Misiones Prov. (NW Argentina)

❏ *Gracilinanus perijae*
Sierra de Perijá Gracile Mouse-Opossum

E Colombia

❏ *Monodelphis adusta* Sepia Short-tailed Opossum

Locally in wet forests and grasslands of E Panama to Colombia, Ecuador and E slope of Andes of Peru

❏ *Monodelphis americana*
Three-striped Short-tailed Opossum

Atlantic coastal forests of Pará to Santa Catarina (E Brazil) and Misiones Prov. (NE Argentina)

❏ *Monodelphis brevicaudata*
Red-legged Short-tailed Opossum

Forests and forest edge of Colombia to French Guiana, S to Peru, C Brazil, N Bolivia and N Argentina

❏ *Monodelphis dimidiata*
Southern Short-tailed Opossum

Grassland and marshes of C & SE Brazil, Uruguay and NE Argentina

❏ *Monodelphis domestica*
Grey Short-tailed Opossum

Dry savannas, scrub, caatinga and gardens of Bolivia, Paraguay and S & E Brazil

❏ *Monodelphis emiliae*
Emilia's Short-tailed Opossum

Lowland rainforests of Amazon basin of E Peru, N Bolivia and C Brazil N to R. Amazonas

❏ *Monodelphis iheringi*
Ihering's Short-tailed Opossum

SE Brazil

❏ *Monodelphis kunsi*
Pygmy Short-tailed Opossum

Very few records from forests and regrowth areas of Bolivia and SE Brazil; Endangered

❏ *Monodelphis maraxina*
Marajó Short-tailed Opossum

Marajó I. and Caldeirão I. (Pará, Brazil)

❏ *Monodelphis orinoci*
Orinoco Short-tailed Opossum

Llanos of N Venezuela

❏ *Monodelphis osgoodi*
Osgood's Short-tailed Opossum

SE Peru and C Bolivia

❏ *Monodelphis rubida*
Chestnut-striped Short-tailed Opossum

Rainforests of Goias to São Paulo (E Brazil)

❏ *Monodelphis scalops*
Long-nosed Short-tailed Opossum

Lowland Atlantic coastal forests of Espírito Santo to Santa Catarina (SE Brazil) and Misiones Prov. (NE Argentina)

❏ *Monodelphis sorex*
Shrewish Short-tailed Opossum

Atlantic coastal rainforests of SE Brazil, S Paraguay and NE Argentina

❏ *Monodelphis theresa*
Theresa's Short-tailed Opossum

Mountains of Peru and E Brazil

❏ *Monodelphis unistriata*
One-striped Short-tailed Opossum

Itararé region (São Paulo, Brazil); possibly extinct, only known from holotype collected prior to 1842

❏ *Thylamys cinderella*
Cinderella Fat-tailed Opossum

NW Argentina

❏ *Thylamys elegans* **Elegant Fat-tailed Opossum**

Forests and scrub of Andes from C Peru to Chile

❏ *Thylamys macrurus*
Long-tailed Fat-tailed Opossum

S Brazil and Paraguay

❏ *Thylamys pallidior* **Pallid Fat-tailed Opossum**

SE Bolivia and Argentina

❏ *Thylamys pusillus* **Small Fat-tailed Opossum**

Arid rocky slopes and thorn scrub of SW Bolivia, Paraguay, C & E Brazil and N & C Argentina

❏ *Thylamys sponsorius*
Sponsorial Fat-tailed Opossum

NW Argentina

❏ *Thylamys velutinus* **Velvety Fat-tailed Opossum**

Cerrado and caatinga of interior EC Brazil

❏ *Thylamys venustus* **Pretty Fat-tailed Opossum**

Bolivia and NW Argentina

❏ *Lestodelphys halli* **Patagonian Opossum**

WC & S Argentina

❏ *Micoureus alstoni*
Alston's Woolly Mouse-Opossum

Forests and gardens of S Belize to Costa Rica

❏ *Micoureus constantiae*
Pale-bellied Woolly Mouse-Opossum

Mid-elevation forests of SW & E Bolivia and S Brazil to NW Argentina

❏ *Micoureus demerarae*
Long-furred Woolly Mouse-Opossum

Forests and gardens of Colombia to French Guiana, S to E Paraguay, Misiones Prov. (NE Argentina) and S Brazil

❏ *Micoureus phaea* **Little Woolly Mouse-Opossum**

Mid-elevation forests of Panama to N Colombia and W Ecuador

❏ *Micoureus regina*
Short-furred Woolly Mouse-Opossum

Forests and gardens of Colombia to S Peru, W Bolivia and C Amazonian Brazil

❏ *Metachirus nudicaudatus*
Brown Four-eyed Opossum

Forests and forest edge of Chiapas (S Mexico) and Nicaragua to French Guiana, S to Peru, E Paraguay and Misiones Prov. (NE Argentina)

❏ *Lutreolina crassicaudata* **Lutrine Opossum**

Streams and lagoons of E Colombia to W Guyana; E Bolivia, Paraguay, N Argentina and S Brazil

❏ *Philander andersoni*
Anderson's Four-eyed Opossum

Lowland rainforests of Venezuela, W Amazonian Brazil, E Ecuador and E Peru

❏ *Philander frenata* **Bridled Four-eyed Opossum**

Forests of coastal SE Brazil

❏ *Philander mcilhennyi*
McIlhenny's Four-eyed Opossum

Rainforests of W Amazonian Brazil and NE Peru

❏ *Philander opossum* **Grey Four-eyed Opossum**

Lowland forests and gardens of NE Mexico to French Guiana, S to Paraguay, NE Argentina and SE Brazil

❏ *Chironectes minimus* **Water Opossum**

Locally in hill forest streams and rivers of S Mexico to Guyana, S to Peru, Paraguay, Misiones Prov. (NE Argentina) and SE Brazil

❏ *Didelphis albiventris* **White-eared Opossum**

Forests of Colombia to French Guiana, S to Peru, N Argentina, Uruguay and S Brazil

❏ *Didelphis aurita* **Big-eared Opossum**

Forests of Paraguay, Misiones Prov. (NE Argentina) and E Brazil

❏ *Didelphis marsupialis* **Southern Opossum**

Lowland forests and regrowth areas of NE Mexico to Peru and Bolivia

❏ *Didelphis virginiana* Virginia Opossum

Woods and gardens of extreme S Canada, USA (except interior W), and Mexico to NW Costa Rica

❏ *Caluromys derbianus*
Central American Woolly Opossum

Forests and regrowth areas of Mexico to W Colombia and N Ecuador

❏ *Caluromys lanatus* Western Woolly Opossum

Rainforests of E Colombia and SW Venezuela, S to Paraguay, Misiones Prov. (NE Argentina) and S Brazil

❏ *Caluromys philander*
Bare-tailed Woolly Opossum

Rainforests and plantations of Venezuela to French Guiana and E Brazil; Trinidad & Tobago

❏ *Caluromysiops irrupta*
Black-shouldered Opossum

Rainforests of Amazon basin of E Peru and W Brazil

❏ *Glironia venusta* Bushy-tailed Opossum

Lowland rainforests of Amazon basin of Ecuador, Peru, Bolivia and W Amazonian Brazil

Order: PAUCITUBERCULATA
Family: CAENOLESTIDAE
(Shrew-Opossums—6)

❏ *Caenolestes caniventer*
Grey-bellied Shrew-Opossum

Tumbes region of SW Ecuador and NW Peru

❏ *Caenolestes condorensis*
Condor Shrew-Opossum

Cordillera del Cóndor (Ecuador)

❏ *Caenolestes convelatus*
Blackish Shrew-Opossum

W Colombia and NW Ecuador

❏ *Caenolestes fuliginosus* Silky Shrew-Opossum

Wet montane forest edge habitats of Colombia, NW Venezuela and Ecuador

❏ *Rhyncholestes raphanurus*
Chilean Shrew-Opossum

Dense moist forests of SC Chile incl Chiloé I.

❏ *Lestoros inca* Incan Shrew-Opossum

Moist habitats of Andes of S Peru

Cohort: PLACENTALIA (Placentals)
Magnorder: XENARTHRA (Edentates)
Order: CINGULATA
Family: DASYPODIDAE (Armadillos—21)

❏ *Dasypus hybridus*
Southern Long-nosed Armadillo

Grasslands of S Paraguay, N Argentina, Uruguay and S Brazil

❏ *Dasypus kappleri* Great Long-nosed Armadillo

Forests of E Colombia to Suriname, S to Peru and S Amazonian Brazil

❏ *Dasypus novemcinctus* Nine-banded Armadillo

Forests, savannas, llanos and caatinga of SC & SE USA to Suriname, S to Uruguay and N Argentina; Margarita; Trinidad & Tobago; Grenada

❏ *Dasypus pilosus* Hairy Long-nosed Armadillo

High-elevation areas of W slope of Andes of SC Peru

❏ *Dasypus sabanicola*
Llanos Long-nosed Armadillo

Llanos and savannas of C Colombia and C Venezuela

❏ *Dasypus septemcinctus*
Seven-banded Armadillo

Grassland and forests of SE Amazonian & E Brazil to Bolivia, Paraguay and N Argentina

❏ *Dasypus yepesi* Yepes' Long-nosed Armadillo

Jujuy and Salta Provs. (NW Argentina)

❏ *Chaetophractus nationi* Andean Hairy Armadillo

Altiplano brush of Andes of N Chile and puna of Bolivia

❏ *Chaetophractus vellerosus*
Screaming Hairy Armadillo

Dry sandy areas of Gran Chaco region of Bolivia and NW Paraguay to C Argentina

❏ *Chaetophractus villosus* Larger Hairy Armadillo

S Bolivia and Paraguay to S Chile and Argentina

❏ *Zaedyus pichiy* Pichi

Open areas of C & S Chile and C & S Argentina

❏ *Euphractus sexcinctus* Six-banded Armadillo

Rainforests and savannas of E Suriname and NC Brazil;
Bolivia to NE Brazil, S to Uruguay and N Argentina

❏ *Chlamyphorus retusus* Chacoan Fairy Armadillo

Gran Chaco region of SE Bolivia, W Paraguay and N
Argentina

❏ *Chlamyphorus truncatus* Pink Fairy Armadillo

Dry sandy areas of C Argentina; Endangered

❏ *Tolypeutes matacus*
Southern Three-banded Armadillo

Dry areas of SE Bolivia and SC Brazil to C Argentina

❏ *Tolypeutes tricinctus*
Brazilian Three-banded Armadillo

Cerrado and caatinga of Bahia, Ceará and Pernambuco (EC
Brazil)

❏ *Cabassous centralis*
Northern Naked-tailed Armadillo

Forests and savannas of E Chiapas (SE Mexico) to NW
Venezuela, and W of Andes S to NW Ecuador

❏ *Cabassous chacoensis*
Chacoan Naked-tailed Armadillo

Gran Chaco region of W Paraguay and NW Argentina

❏ *Cabassous tatouay*
Greater Naked-tailed Armadillo

Open areas of SE Paraguay, NE Argentina, Uruguay and E
Brazil

❏ *Cabassous unicinctus*
Southern Naked-tailed Armadillo

Forests and savannas of E Colombia to French Guiana, and E
of Andes S to N Peru and S Brazil

❏ *Priodontes maximus* Giant Armadillo

Forests and savannas of Venezuela to French Guiana, and E of
Andes S to Paraguay and N Argentina; Endangered

Order: PILOSA
Family: MYRMECOPHAGIDAE
(American Anteaters—3)

❏ *Myrmecophaga tridactyla* Giant Anteater

Grasslands, savannas and forests of Costa Rica to Paraguay
and N Argentina (formerly Uruguay and N to Guatemala)

❏ *Tamandua mexicana* Northern Tamandua

Forests and savannas of E Mexico to NW Venezuela, and W of
Andes S to NW Peru

❏ *Tamandua tetradactyla* Southern Tamandua

Forests, savannas and grasslands of NW Venezuela to French
Guiana, and E of Andes S to N Argentina and Uruguay

Family: CYCLOPEDIDAE
(Silky Anteater—1)

❏ *Cyclopes didactylus* Silky Anteater

Rainforests and regrowth areas of EC Mexico to French
Guiana, S to C Brazil and Bolivia; Trinidad

Family: MEGALONYCHIDAE
(Two-toed Sloths—2)

❏ *Choloepus didactylus*
Southern Two-toed Sloth

Forests of Amazon basin of S Colombia to French Guiana, and
E of Andes S to Peru and Amazonian Brazil

❏ *Choloepus hoffmanni*
Hoffmann's Two-toed Sloth

Forests and regrowth areas of E Honduras to W Venezuela, S
to Peru, E Bolivia and W Amazonian Brazil

Family: BRADYPODIDAE
(Three-toed Sloths—4)

❏ *Bradypus pygmaeus* Pygmy Three-toed Sloth

Isla Escudo de Veraguas (Bocas del Toro Is., W Panama)

❏ *Bradypus torquatus* Maned Three-toed Sloth

Remnant patches of Atlantic coastal rainforests of Bahia,
Espírito Santo and Rio de Janeiro (SE Brazil); Endangered

❏ *Bradypus tridactylus*
Pale-throated Three-toed Sloth

Forests of E Venezuela to French Guiana, S to R. Negro and
both banks of lower R. Amazonas (E Amazonian Brazil)

❏ *Bradypus variegatus*
Brown-throated Three-toed Sloth

Lowland forests and regrowth areas of C Honduras to
Venezuela, S to E Peru and N Argentina

Magnorder: EPITHERIA
Grandorder: ANAGALIDA
Mirorder: MACROSCELIDEA
Family: MACROSCELIDIDAE
(Elephant-Shrews—15)

❏ *Macroscelides proboscideus*
Round-eared Elephant-Shrew

Karoo scrub of SW Namibia to W Cape Prov. (South Africa)

❏ *Elephantulus brachyrhynchus*
Short-snouted Elephant-Shrew

Sandy habitats of Kenya to N South Africa

❏ *Elephantulus edwardii*
Cape Rock Elephant-Shrew

Rocky habitats of W Northern Cape to W Eastern Cape Provs.
(South Africa)

❏ *Elephantulus fuscipes*
Dusky-footed Elephant-Shrew

S Sudan, NE D.R. Congo and Uganda

❏ *Elephantulus fuscus* Dusky Elephant-Shrew

Sandy habitats of SE Zambia, S Malawi and C Mozambique

❏ *Elephantulus intufi* Bushveld Elephant-Shrew

Sandy habitats of SW Angola, Namibia, Botswana and extreme
N South Africa

❏ *Elephantulus myurus*
Eastern Rock Elephant-Shrew

Rocky habitats of Zimbabwe, E Botswana, W Mozambique
and E South Africa

❏ *Elephantulus revoili* Somali Elephant-Shrew

N Somalia; Endangered

❏ *Elephantulus rozeti*
North African Elephant-Shrew

Morocco to W Libya

❏ *Elephantulus rufescens* Rufous Elephant-Shrew

S Sudan and W Tanzania to E Ethiopia and S Somalia

❏ *Elephantulus rupestris*
Smith's Rock Elephant-Shrew

Rocky habitats of Namibia to Eastern Cape Prov. (South
Africa)

❏ *Petrodromus tetradactylus*
Four-toed Elephant-Shrew

Dense forests and thickets of Rep. Congo and NE Angola to
SE Kenya and extreme E South Africa

❏ *Rhynchocyon chrysopygus*
Golden-rumped Elephant-Shrew

Very locally in dry coastal thickets of E Kenya; Endangered

❏ *Rhynchocyon cirnei* Checkered Elephant-Shrew

Forests and savannas of E D.R. Congo and Uganda to Mozambique

❏ *Rhynchocyon petersi*
Black-and-rufous Elephant-Shrew

Forests and coastal thickets of SE Kenya to E Tanzania incl.
Zanzibar and Mafia Is.; Endangered

Mirorder: DUPLICIDENTATA
Order: LAGOMORPHA
Family: OCHOTONIDAE (Pikas—26)

❏ *Ochotona alpina* Alpine Pika

S Russia and Kazakhstan to NW Mongolia and N China

❏ *Ochotona cansus* Gansu Pika

Gansu, Qinghai and Sichuan (C China)

❏ *Ochotona collaris* Collared Pika

E Alaska (USA), Yukon and NW British Columbia (Canada)

❏ *Ochotona curzoniae* Black-lipped Pika

E Nepal and Sikkim (India) to Tibet and N China

❏ *Ochotona dauurica* Daurian Pika

E Siberia and Mongolia to NC China

❏ *Ochotona erythrotis* Chinese Red Pika

Tibet and C China

❏ *Ochotona forresti* Forrest's Pika

Sikkim (India) and SE Tibet to N Myanmar and NW Yunnan
(SW China)

❏ *Ochotona gaoligongensis* Gaoligong Pika

Mt Gaoligong (NW Yunnan, SW China)

❏ *Ochotona gloveri* Glover's Pika

NE Tibet, SW Qinghai to NW Yunnan (WC China)

❏ *Ochotona himalayana* **Himalayan Pika**

Mt Everest region (S Tibet)

❏ *Ochotona hyperborea* **Northern Pika**

Siberia, NC Mongolia, NE China, N Korea and Hokkaido (N Japan)

❏ *Ochotona iliensis* **Ili Pika**

Borokhoro Shan (Xinjiang, China)

❏ *Ochotona koslowi* **Kozlov's Pika**

Kunlun Mtns (Xinjiang, China); Endangered

❏ *Ochotona ladacensis* **Ladak Pika**

Pakistan, E Tibet, Xinjiang and Qinghai (NW China)

❏ *Ochotona macrotis* **Large-eared Pika**

Mountains of Hindu Kush (NE Afghanistan), Karakoram (N Pakistan) and Tien Shan to Bhutan and Yunnan (SW China)

❏ *Ochotona muliensis* **Muli Pika**

Muli region (Sichuan, C China)

❏ *Ochotona nigritia* **Pianma Black Pika**

Yunnan (SW China)

❏ *Ochotona nubrica* **Nubra Pika**

Kashmir and Nepal to E Tibet

❏ *Ochotona pallasi* **Pallas' Pika**

Mountains of Kazakhstan and Russia to Mongolia and NE China

❏ *Ochotona princeps* **American Pika**

Mountains of SW Canada and W USA

❏ *Ochotona pusilla* **Steppe Pika**

Steppes of SW Russia and Kazakhstan

❏ *Ochotona roylei* **Royle's Pika**

Mountains of NW Pakistan and India to Nepal and Tibet

❏ *Ochotona rufescens* **Afghan Pika**

Armenia and Turkmenistan to Iran, Afghanistan and Pakistan

❏ *Ochotona rutila* **Turkestan Red Pika**

Mountains of Pamirs (Tajikistan) and Tien Shan (Kyrgyzstan, Kazakhstan and NW China)

❏ *Ochotona thibetana* **Moupin Pika**

S Tibet, C China, Sikkim (NE India) and N Myanmar

❏ *Ochotona thomasi* **Thomas' Pika**

Gansu, NE Qinghai and Sichuan (C China)

Family: LEPORIDAE
(Rabbits and Hares—57)

❏ *Oryctolagus cuniculus* **European Rabbit**

Grasslands and farmland, originally of Iberian Peninsula; feral worldwide except Antarctica

❏ *Pentalagus furnessi* **Amami Rabbit**

Amami Is. (Japan); Endangered

❏ *Pronolagus crassicaudatus* **Natal Red Rockhare**

Thickets on steep rocky slopes of SE South Africa, Swaziland and extreme S Mozambique

❏ *Pronolagus randensis* **Jameson's Red Rockhare**

Grassy areas on rocky slopes of NW & C Namibia; SE Botswana, NE South Africa, S Zimbabwe and WC Mozambique

❏ *Pronolagus rupestris* **Smith's Red Rockhare**

Rocky thickets of SW Kenya to E Zambia; S Namibia to S & C South Africa

❏ *Lepus alleni* **Antelope Jackrabbit**

Desert grasslands of SC Arizona (USA) to N Nayarit (WC Mexico)

❏ *Lepus americanus* **Snowshoe Hare**

Boreal and montane forest thickets from Alaska and Canada to W, NC & NE USA

❏ *Lepus arcticus* **Arctic Hare**

Tundra zone of N & NE Canada and Greenland

❏ *Lepus brachyurus* **Japanese Hare**

Honshu to Kyushu, Oki and Sado Is. (Japan)

❏ *Lepus californicus* **Black-tailed Jackrabbit**

Grasslands and farmland of W & SC USA to C Mexico; feral E USA

❏ *Lepus callotis* **White-sided Jackrabbit**

Desert grasslands of extreme SW New Mexico (USA) to C Oaxaca (S Mexico)

❏ *Lepus capensis* Cape Hare

Open savannas and dry grasslands of Sardinia (Italy), North Africa, Middle East and subsaharan Africa

❏ *Lepus castroviejoi* Broom Hare

Montane scrub, heathlands and forest clearings of Cantabrian Mtns (NW Spain)

❏ *Lepus comus* Yunnan Hare

Yunnan and W Guizhou (China)

❏ *Lepus coreanus* Korean Hare

NE China and Korea

❏ *Lepus corsicanus* Corsican Hare

Open grasslands and farmland of S Italy and Sicily; formerly feral Corsica (France, now extinct)

❏ *Lepus europaeus* European Hare

Open grasslands and farmland of Europe to W Siberia and Middle East; feral N Ireland, S Sweden, SE Canada, NE USA, S South America, Australasia etc.

❏ *Lepus fagani* Ethiopian Hare

SE Sudan, Ethiopia and NW Kenya

❏ *Lepus flavigularis* Tehuantepec Jackrabbit

Locally in sand dune scrub bordering coastal lagoons of SE Oaxaca and SW Chiapas (Mexico); Endangered

❏ *Lepus granatensis* Iberian Hare

Forests, grasslands and farmland of S Portugal and Spain incl. Mallorca

❏ *Lepus hainanus* Hainan Hare

Hainan (S China)

❏ *Lepus insularis* Black Jackrabbit

Espiritu Santo I. (Gulf of California, Mexico)

❏ *Lepus mandshuricus* Manchurian Hare

Ussuriland (NE Siberia), NE China and NE Korea

❏ *Lepus nigricollis* Indian Hare

Forests, grasslands and farmland of Pakistan, Nepal, India, Bhutan, Bangladesh and Sri Lanka; feral Java and Indian Ocean is.

❏ *Lepus oiostolus* Woolly Hare

NW & NE India, Nepal and Tibet to C China

❏ *Lepus othus* Alaskan Hare

Tundra zone of extreme E Siberia and W Alaska (USA)

❏ *Lepus peguensis* Burmese Hare

C Myanmar to Vietnam and N Malay Peninsula

❏ *Lepus saxatilis* Scrub Hare

Woods, scrubby grassland and farmland of S Namibia and South Africa

❏ *Lepus sinensis* Chinese Hare

SE China, Tiawan and NE Vietnam

❏ *Lepus starcki* Ethiopian Highland Hare

Montane moorlands of C Ethiopia

❏ *Lepus timidus* Mountain Hare

Montane heaths and open forests of N Europe (locally feral) and Alps, Belarus and Russia to N Mongolia, N China and Japan

❏ *Lepus tolai* Tolai Hare

Steppes from Caspian Sea E to NE China

❏ *Lepus townsendii* White-tailed Jackrabbit

Sagebrush and grasslands of S Canada and interior W & C USA

❏ *Lepus victoriae* African Savanna Hare

Savannas of NW and subsaharan Africa S to NE Namibia, Botswana and Natal (E South Africa)

❏ *Lepus yarkandensis* Yarkand Hare

Steppes of Tarim Basin (S Xinjiang, China)

Sylvilagus (Microlagus)

❏ *Sylvilagus bachmani* Brush Rabbit

Dense brush habitats of coastal Oregon and California (USA) and Baja California (Mexico)

❏ *Sylvilagus mansuetus* San José Brush Rabbit

San José I. (Gulf of California, Mexico)

Sylvilagus (Sylvilagus)

❏ *Sylvilagus audubonii* Desert Cottontail

Desert scrub and woods of SW & C USA, and N Mexico

❏ *Sylvilagus cunicularius* Mexican Cottontail

S Mexico

❏ *Sylvilagus floridanus* Eastern Cottontail

Fields and scrub of S Canada and USA (except much of interior W) to Costa Rica; llanos of C & N Colombia and N Venezuela; feral N Italy

❏ *Sylvilagus graysoni* Tres Marías Cottontail

Tres Marías Is. (Nayarit, Mexico); Endangered

❏ *Sylvilagus nuttallii* Mountain Cottontail

Sagebrush and montane forests of interior SW Canada and W USA

❏ *Sylvilagus obscurus* Appalachian Cottontail

Dense high-elevation scrub habitats in Appalachian Mtns (E USA)

❏ *Sylvilagus transitionalis* New England Cottontail

Forests of New England states of NE USA

Sylvilagus (Tapeti)

❏ *Sylvilagus aquaticus* Swamp Rabbit

Swamps of SC USA

❏ *Sylvilagus brasiliensis* Forest Rabbit (Tapiti)

Forests, plantations, grasslands and gardens of E Mexico to Suriname, S to Peru, Bolivia, N Argentina, Paraguay and S Brazil

❏ *Sylvilagus dicei* Dice's Rabbit

Montane forests and páramo of Cordillera de Talamanca (C Costa Rica to NW Panama); Endangered

❏ *Sylvilagus insonus* Omilteme Cottontail

Sierra Madre del Sur (C Guerrero, Mexico); Critically Endangered

❏ *Sylvilagus palustris* Marsh Rabbit

Wetlands of SE USA

❏ *Sylvilagus varynaensis* Barinas Rabbit

W Venezuela

❏ *Caprolagus hispidus* Hispid Hare

Hill forest edges and grasslands of N & NE India, Nepal, Bhutan and NW Bangladesh; Endangered

❏ *Bunolagus monticularis* Riverine Rabbit

Very locally in arid riparian scrub of C Karoo region (S Northern Cape Prov., South Africa); Critically Endangered

❏ *Brachylagus idahoensis* Pygmy Rabbit

Sagebrush deserts of Great Basin region of W USA

❏ *Poelagus marjorita* Bunyoro Rabbit

Rocky areas and thickets in moist savannas of S Chad and S Sudan to NE D.R. Congo and Burundi; Angola

❏ *Nesolagus netscheri* Sumatran Rabbit

Sumatra; Critically Endangered

❏ *Nesolagus timminsi* Annamite Striped Rabbit

Annamite Mtns (Laos and Vietnam)

❏ *Romerolagus diazi* Volcano Rabbit (Zacatuche)

Volcanoes of C Mexico (Distrito Federal and W Puebla); Endangered

Mirorder: SIMPLICIDENTATA
Order: RODENTIA
Family: APLODONTIIDAE (Sewellel—1)

❏ *Aplodontia rufa* Sewellel (Mountain Beaver)

Moist forests of SW British Columbia (Canada) to C California (USA)

Family: SCIURIDAE (Squirrels—276)

Sciurotamias (Rupestes)

❏ *Sciurotamias forresti* Forrest's Rock Squirrel

Sichuan and Yunnan (C & SW China)

Sciurotamias (Sciurotamias)

❏ *Sciurotamias davidianus*
Père David's Rock Squirrel

Sichuan and Guizhou to Hebei (C China)

❏ *Rheithrosciurus macrotis* Tufted Ground Squirrel

Borneo

Sciurus (Guerlinguetus)

❏ *Sciurus aestuans* Guianan Squirrel

Forests, plantations and gardens of Venezuela to French Guiana, S to Misiones Prov. (NE Argentina) and SE Brazil

❏ *Sciurus argentinius* South Yungas Red Squirrel

Montane moss forests of Chuquisaca Dept. (SE Bolivia) to Jujuy Prov. (N Argentina)

❏ *Sciurus gilvigularis* Yellow-throated Squirrel

S Venezuela, Guyana and E Amazonian Brazil

❏ *Sciurus granatensis* Red-tailed Squirrel

Forests of Costa Rica to N Venezuela, and W of Andes to S Ecuador; Margarita; Trinidad & Tobago

❏ *Sciurus ignitus* Bolivian Squirrel

Rainforests of Amazon basin of E Peru, SW Brazil and N Bolivia

❏ *Sciurus pucheranii* Andean Squirrel

Andes of Colombia

❏ *Sciurus richmondi* Richmond's Squirrel

Forests and plantations of EC Nicaragua

❏ *Sciurus sanborni* Sanborn's Squirrel

Lowland rainforests between Rs. Manu and Tambopata (Madre de Dios Dept., SE Peru)

❏ *Sciurus stramineus* Guayaquil Squirrel

Lowland forests of Tumbes region of SW Ecuador and NW Peru

Sciurus (Hadrosciurus)

❏ *Sciurus flammifer* Fiery Squirrel

Few localities in Bolívar state, S of R. Orinoco (C Venezuela)

❏ *Sciurus pyrrhinus* Junín Red Squirrel

Montane foothill forests of E slope of Andes of C Peru

Sciurus (Hesperosciurus)

❏ *Sciurus griseus* Western Grey Squirrel

Oak-conifer forests of C Washington (USA) to Baja California Norte (Mexico)

Sciurus (Otosciurus)

❏ *Sciurus aberti* Abert's Squirrel

Pine forests of SW USA and NW Mexico

Sciurus (Sciurus)

❏ *Sciurus alleni* Allen's Squirrel

NE Mexico

❏ *Sciurus arizonensis* Arizona Grey Squirrel

Broadleaf forests of SE Arizona and WC New Mexico (USA), and NE Sonora (Mexico)

❏ *Sciurus aureogaster*
Mexican Grey Squirrel (Red-bellied Squirrel)

Forests, plantations and scrub of Mexico and W Guatemala; feral Elliot Key (S Florida, USA)

❏ *Sciurus carolinensis* Eastern Grey Squirrel

Forests, parks and gardens of SC & SE Canada and E USA; feral Ireland, Britain, NW Italy, SW South Africa, SW Canada and NW USA

❏ *Sciurus colliaei* Collie's Squirrel

Sonora to Colima (NW Mexico)

❏ *Sciurus deppei* Deppe's Squirrel

Forests and forest edge of EC Mexico to N Costa Rica

❏ *Sciurus lis* Japanese Squirrel

Honshu, Shikoku and Kyushu (Japan)

❏ *Sciurus nayaritensis* Mexican Fox Squirrel

Pine-oak forests of extreme SE Arizona (USA) to WC Mexico

❏ *Sciurus niger* Eastern Fox Squirrel

Parks and woods of SC Canada, E USA and N Coahuila (NE Mexico)

❏ *Sciurus oculatus* Peters' Squirrel

EC Mexico (San Luis Potosi to Veracruz)

❏ *Sciurus variegatoides* Variegated Squirrel

Forests, plantations and regrowth areas of extreme S Chiapas (S Mexico) to C Panama

❏ *Sciurus vulgaris* Eurasian Red Squirrel

Forests, parks and gardens of Europe to E Siberia, NE China and Korea

❏ *Sciurus yucatanensis* Yucatán Squirrel

Lowland forests, plantations and regrowth areas of Yucatán Peninsula (SE Mexico, Belize and N Guatemala)

Sciurus (Syntheosciurus)

❏ *Sciurus brochus* Montane Squirrel

Cool wet montane forests of Cordillera de Talamanca (Costa Rica and N Panama)

Sciurus (Tenes)

❏ *Sciurus anomalus* Persian Squirrel

Forests of Lesbos (Greece), Turkey and Transcaucasia S to
Syria, Israel, Jordan, Iraq and W Iran

Sciurus (Urosciurus)

❏ *Sciurus igniventris*
Northern Amazon Red Squirrel

Rainforests of Amazon basin of SE Colombia, S Venezuela, E
Ecuador, E Peru and NW Brazil

❏ *Sciurus spadiceus*
Southern Amazon Red Squirrel

Lowland rainforests of Amazon basin of SE Colombia,
Ecuador, Peru, Bolivia and WC Brazil

❏ *Microsciurus alfari*
Central American Dwarf Squirrel

Rain and cloud forests of S Nicaragua to W Colombia

❏ *Microsciurus flaviventer*
Amazon Dwarf Squirrel

Rainforests of Amazon basin of SE Colombia, E Ecuador, E
Peru and between Rs. Negro and Madeira (W Amazonian
Brazil)

❏ *Microsciurus mimulus* Western Dwarf Squirrel

Rainforests of Panama, N Colombia and NW Ecuador

❏ *Microsciurus santanderensis*
Santander Dwarf Squirrel

Santander state (C Colombia)

❏ *Sciurillus pusillus* Neotropical Pygmy Squirrel

Locally in lowland rainforest canopy of S Colombia and NE
Peru; Suriname, French Guiana and NC Brazil; between Rs.
Madeira and Tapajós (C Amazonian Brazil)

❏ *Atlantoxeros getulus* Barbary Ground Squirrel

Rocky slopes and stony fields of Morocco and NW Algeria;
feral Fuerteventura (Canary Is.)

❏ *Xerus erythropus* Striped Ground Squirrel

Open woods and dry savannas of SE Morocco, W & C Africa
from Gambia to N Tanzania

❏ *Xerus inauris* South African Ground Squirrel

Open hard-sand areas of S Angola, Botswana, W Zimbabwe
and C South Africa

❏ *Xerus princeps* Damara Ground Squirrel

Rocky habitats of Kaokoveld escarpments of extreme SW
Angola and W Namibia

❏ *Xerus rutilus* Unstriped Ground Squirrel

Semi-deserts, arid savannas and grasslands of SE Sudan and
Djibouti to NE Tanzania and Somalia

❏ *Spermophilopsis leptodactylus*
Long-clawed Ground Squirrel

Uzbekistan and SE Kazakhstan to NE Iran and SW Afghanistan

❏ *Marmota baibacina* Grey Marmot

SW Siberia and Kyrgyzstan to Mongolia and NW China

❏ *Marmota bobak* Bobak Marmot

Steppes of E Europe to Kazakhstan

❏ *Marmota broweri* Alaska Marmot

Boulder fields in the Brooks Range (N Alaska, USA)

❏ *Marmota caligata* Hoary Marmot

Treeless rocky and montane habitats of Alaska, W Canada and
NW USA

❏ *Marmota camtschatica* Black-capped Marmot

E Siberia

❏ *Marmota caudata* Long-tailed Marmot

Montane forests and meadows of Hindu Kush (Afghanistan)
and N Pakistan to Tien Shan (NW China)

❏ *Marmota flaviventris* Yellow-bellied Marmot

Rocky meadows of interior SW Canada and W USA

❏ *Marmota himalayana* Himalayan Marmot

High-altitude montane grasslands of N India, Nepal and W
China

❏ *Marmota marmota* Alpine Marmot

Montane meadows of C Europe from Pyrenees to Carpathian
Mtns

❏ *Marmota menzbieri* Menzbier's Marmot

W Tien Shan (S Kazakhstan and NW Kyrgyzstan)

❏ *Marmota monax* Woodchuck

Meadows of EC Alaska, Canada, NE & EC USA

❏ *Marmota olympus* Olympic Marmot

Rocky montane slopes and meadows of Olympic Peninsula (W Washington, USA)

❏ *Marmota sibirica* Tarbagan Marmot

S Siberia, Mongolia and NE China

❏ *Marmota vancouverensis*
Vancouver Island Marmot

Montane alpine and subalpine meadows of S Vancouver I. (SW British Columbia, Canada); Endangered

Spermophilus (Callospermophilus)

❏ *Spermophilus lateralis*
Golden-mantled Ground Squirrel

Mountains of SW Canada and W USA

❏ *Spermophilus madrensis*
Sierra Madre Ground Squirrel

SW Chihuahua (Mexico)

❏ *Spermophilus saturatus*
Cascade Golden-mantled Ground Squirrel

Cascade Mtns of SW British Columbia (Canada) and W Washington (USA)

Spermophilus (Colobotis)

❏ *Spermophilus erythrogenys*
Red-cheeked Ground Squirrel

E Kazakhstan and SW Siberia to Mongolia and N China

❏ *Spermophilus fulvus* Yellow Ground Squirrel

Kazakhstan and NW China to NE Iran and N Afghanistan

❏ *Spermophilus major* Russet Ground Squirrel

C Russia and N Kazakhstan

Spermophilus (Ictidomys)

❏ *Spermophilus mexicanus*
Mexican Ground Squirrel

Grasslands and desert brush of S New Mexico and W Texas (USA) to C Mexico

❏ *Spermophilus perotensis* Perote Ground Squirrel

Veracruz and Puebla (EC Mexico)

❏ *Spermophilus spilosoma* Spotted Ground Squirrel

Grasslands and desert brush of SC USA to C Mexico

❏ *Spermophilus tridecemlineatus*
Thirteen-lined Ground Squirrel

Roadsides and grasslands of SC Canada and C USA

Spermophilus (Notocitellus)

❏ *Spermophilus annulatus*
Ring-tailed Ground Squirrel

Nayarit to N Guerrero (WC & SW Mexico)

Spermophilus (Otospermophilus)

❏ *Spermophilus adocetus* Tropical Ground Squirrel

E Jalisco to N Guerrero (WC & SW Mexico)

❏ *Spermophilus atricapillus*
Baja California Ground Squirrel

Baja California (Mexico)

❏ *Spermophilus beecheyi*
California Ground Squirrel

Roadsides and grassy habitats of SW Washington (USA) to Baja California Norte (Mexico)

❏ *Spermophilus variegatus*
Variegated Ground Squirrel (Rock Squirrel)

Rocky slopes and cliffs of SW USA and N Mexico

Spermophilus (Poliocitellus)

❏ *Spermophilus franklinii*
Franklin's Ground Squirrel

Grassland and scrub habitats of SC Canada and NC USA

Spermophilus (Spermophilus)

❏ *Spermophilus alashanicus*
Ala Shan Ground Squirrel

SC Mongolia, Ala Shan and E Nan Shan (NC China)

❏ *Spermophilus armatus* Uinta Ground Squirrel

Sagebrush and grasslands of S Montana and W Wyoming to C Utah (USA)

❏ *Spermophilus beldingi* Belding's Ground Squirrel

Fields and shortgrass meadows of E Oregon and C California to SW Idaho, N Nevada and extreme NW Utah (USA)

❏ *Spermophilus brunneus* Idaho Ground Squirrel

Very locally in montane meadows of WC Idaho (USA); Endangered

❏ *Spermophilus canus* Merriam's Ground Squirrel

Sagebrush and grasslands of SE Oregon, SW Idaho, extreme NE California and NW Nevada (USA)

❏ *Spermophilus citellus* European Souslik

Shortgrass steppes of EC & SE Europe

❏ *Spermophilus dauricus* Daurian Ground Squirrel

E Siberia, Mongolia and N China

❏ *Spermophilus elegans* Wyoming Ground Squirrel

Montane meadows and rocky slopes of SE Oregon and NE Nevada to SW Montana and N Colorado (USA)

❏ *Spermophilus mollis* Piute Ground Squirrel

Fields and desert flats of SC Washington, SE Oregon, SW Idaho, E California and Nevada (USA)

❏ *Spermophilus musicus*
Caucasian Mountain Ground Squirrel

N Caucasus Mtns (Georgia)

❏ *Spermophilus pygmaeus* Little Ground Squirrel

SW Ukraine and W Russia to Georgia and NW Uzbekistan

❏ *Spermophilus relictus*
Tien Shan Ground Squirrel

Mountains of Tien Shan (Kyrgyzstan and SE Kazakhstan)

❏ *Spermophilus richardsonii*
Richardson's Ground Squirrel

SC Canada and NC USA

❏ *Spermophilus suslicus* Spotted Souslik

Steppes and fields of SE Poland to W Russia and Ukraine

❏ *Spermophilus townsendii*
Townsend's Ground Squirrel

Sagebrush and fields between Yakima and Columbia Rs. (SC Washington, USA)

❏ *Spermophilus washingtoni*
Washington Ground Squirrel

Locally in sagebrush and grasslands of SE Washington and NE Oregon (USA)

❏ *Spermophilus xanthoprymnus*
Asia Minor Ground Squirrel

Armenia, Turkey, Israel and Syria

Spermophilus (Urocitellus)

❏ *Spermophilus columbianus*
Columbian Ground Squirrel

Meadows and grasslands of SE British Columbia and SW Alberta (Canada) to interior NW USA

❏ *Spermophilus parryii* Arctic Ground Squirrel

Tundra of E Siberia, Alaska (USA) and NW Canada

❏ *Spermophilus undulatus*
Long-tailed Ground Squirrel

E Kazakhstan, S Siberia, N Mongolia and N China

Spermophilus (Xerospermophilus)

❏ *Spermophilus mohavensis*
Mojave Ground Squirrel

Desert flats of Mojave Desert (SC California, USA)

❏ *Spermophilus tereticaudus*
Round-tailed Ground Squirrel

Desert flats of SE California, S Nevada and SW Arizona (USA) to NW Mexico

❏ *Ammospermophilus harrisii*
Harris' Antelope Squirrel

Deserts of Arizona and SW New Mexico (USA) to Sonora (Mexico)

❏ *Ammospermophilus insularis*
Espíritu Santo Island Antelope Squirrel

Espíritu Santo Is. (Baja California, Mexico)

❏ *Ammospermophilus interpres*
Texas Antelope Squirrel

Rocky slopes of SC New Mexico and SW Texas (USA) to NC Mexico

❏ *Ammospermophilus leucurus*
White-tailed Antelope Squirrel

Desert scrub of SE Oregon to C New Mexico (USA) and Baja California Sur (Mexico)

❏ *Ammospermophilus nelsoni*
Nelson's Antelope Squirrel

Desert scrub of San Joaquin Valley (S California, USA); Endangered

❏ *Cynomys gunnisoni* Gunnison Prairie-Dog

Montane valleys and plains of SE Utah, SW Colorado, NE Arizona and NW New Mexico (USA)

❏ *Cynomys leucurus* **White-tailed Prairie-Dog**

Montane meadows of SC Montana to NE Utah and NW Colorado (USA)

❏ *Cynomys ludovicianus* **Black-tailed Prairie-Dog**

Prairies of SC Canada and C USA, S to NW Mexico

❏ *Cynomys mexicanus* **Mexican Prairie-Dog**

Coahuila and San Luis Potosi (Mexico); Endangered

❏ *Cynomys parvidens* **Utah Prairie-Dog**

Grasslands of SC Utah (USA)

Tamias (Eutamias)

❏ *Tamias sibiricus* **Siberian Chipmunk**

Forests of Siberia and Kazakhstan to N China and N Japan; feral Netherlands and NW Germany to C Italy

Tamias (Neotamias)

❏ *Tamias alpinus* **Alpine Chipmunk**

High-altitude rocky alpine meadows of Sierra Nevada range (California, USA)

❏ *Tamias amoenus* **Yellow-pine Chipmunk**

Rocky slopes and meadows of SW Canada to C California (USA)

❏ *Tamias bulleri* **Buller's Chipmunk**

S Sierra Madre Occidental (Mexico)

❏ *Tamias canipes* **Grey-footed Chipmunk**

Rocky brush slopes of SC New Mexico and extreme W Texas (USA)

❏ *Tamias cinereicollis* **Grey-collared Chipmunk**

Montane coniferous forests of C Arizona to SC New Mexico (USA)

❏ *Tamias dorsalis* **Cliff Chipmunk**

Rocky slopes of SW USA and N Mexico

❏ *Tamias durangae* **Durango Chipmunk**

NC Mexico

❏ *Tamias merriami* **Merriam's Chipmunk**

Forests of C California (USA) to N Baja California (Mexico)

❏ *Tamias minimus* **Least Chipmunk**

Canada and W & NC USA

❏ *Tamias obscurus* **California Chipmunk**

Locally in rocky scrub slopes of S California (USA) to C Baja California (Mexico)

❏ *Tamias ochrogenys* **Yellow-cheeked Chipmunk**

Dark redwood forests of coastal N California (USA)

❏ *Tamias palmeri* **Palmer's Chipmunk**

High-elevation rocky coniferous habitats of Spring Mtns (S Nevada, USA)

❏ *Tamias panamintinus* **Panamint Chipmunk**

High-elevation coniferous habitats of S California / S Nevada border region (USA)

❏ *Tamias quadrimaculatus*
Long-eared Chipmunk

Forests and scrub of Sierra Nevada range of EC California and extreme WC Nevada (USA)

❏ *Tamias quadrivittatus* **Colorado Chipmunk**

Rocky slopes in coniferous forests of E Utah and Colorado to NE Arizona and N New Mexico (USA)

❏ *Tamias ruficaudus* **Red-tailed Chipmunk**

Dense coniferous forests of SE British Columbia (Canada) and NE Washington to W Montana (USA)

❏ *Tamias rufus* **Hopi Chipmunk**

Rocky slopes in coniferous habitats of SE Utah, W Colorado and NE Arizona (USA)

❏ *Tamias senex* **Shadow Chipmunk**

Forests of C Oregon to EC California and extreme WC Nevada (USA)

❏ *Tamias siskiyou* **Siskiyou Chipmunk**

Coniferous forests of C Oregon to extreme NW California (USA)

❏ *Tamias sonomae* **Sonoma Chipmunk**

Brush habitats of NW California (USA)

❏ *Tamias speciosus* **Lodgepole Chipmunk**

Coniferous forests of Sierra Nevada of E California and extreme WC Nevada (USA)

❏ *Tamias townsendii* **Townsend's Chipmunk**

Forests of SW British Columbia (Canada), W Washington and Oregon (USA)

❏ *Tamias umbrinus* Uinta Chipmunk

Coniferous forests of interior WC USA

Tamias (Tamias)

❏ *Tamias striatus* Eastern Chipmunk

Deciduous forests of SC & SE Canada and E USA

❏ *Ratufa affinis* Pale Giant Squirrel

S Vietnam, Malay Peninsula, Sumatra and Borneo

❏ *Ratufa bicolor* Black Giant Squirrel

Forests of E Nepal, NE India and Bhutan to S China, Java and Bali

❏ *Ratufa indica* Indian Giant Squirrel

Forests of C & S India

❏ *Ratufa macroura* Sri Lankan Giant Squirrel

Locally in forests of S India and Sri Lanka

❏ *Dremomys everetti*
Bornean Mountain Ground Squirrel

Mountains of NW Borneo

❏ *Dremomys lokriah*
Orange-bellied Himalayan Squirrel

Forests of Nepal, NE India, Bhutan and Tibet to N Myanmar

❏ *Dremomys pernyi* Perny's Squirrel

Tibet and NE India to S China, Taiwan and N Vietnam

❏ *Dremomys pyrrhomerus* Red-hipped Squirrel

C & S China incl. Hainan, and N Vietnam

❏ *Dremomys rufigenis*
Asian Red-cheeked Squirrel

NE India and Myanmar to SW China and Malay Peninsula

❏ *Callosciurus adamsi* Ear-spot Squirrel

Borneo

❏ *Callosciurus albescens* Kloss' Squirrel

Sumatra

❏ *Callosciurus baluensis* Kinabalu Squirrel

Borneo

❏ *Callosciurus caniceps* Grey-bellied Squirrel

S Myanmar, Thailand and Malay Peninsula

❏ *Callosciurus erythraeus* Pallas' Squirrel

Forests, parks and gardens of India and Malay Peninsula to SE China and Taiwan; feral Cap d'Antibes (SE France)

❏ *Callosciurus finlaysonii* Finlayson's Squirrel

Forests and plantations of S Myanmar to Vietnam; feral Acqui Terme (Alessandria, Italy)

❏ *Callosciurus inornatus* Inornate Squirrel

S Yunnan (SW China), Laos and N Vietnam

❏ *Callosciurus melanogaster* Mentawai Squirrel

Mentawai Is. (Sumatra)

❏ *Callosciurus nigrovittatus* Black-striped Squirrel

S Vietnam and Malay Peninsula to Borneo and Java

❏ *Callosciurus notatus* Plantain Squirrel

Thailand to Borneo and W Lesser Sundas

❏ *Callosciurus orestes*
Bornean Black-banded Squirrel

Borneo

❏ *Callosciurus phayrei* Phayre's Squirrel

Myanmar

❏ *Callosciurus prevostii* Prevost's Squirrel

Malay Peninsula, Sumatra and Borneo

❏ *Callosciurus pygerythrus* Irrawaddy Squirrel

Nepal and NE India to SW China and N Vietnam

❏ *Callosciurus quinquestriatus* Anderson's Squirrel

NE Myanmar and Yunnan (SW China)

❏ *Tamiops macclellandi* Himalayan Striped Squirrel

Nepal and NE India to SW China and Malay Peninsula

❏ *Tamiops maritimus* Maritime Striped Squirrel

Laos, Vietnam, SE China, Hainan (S China) and Taiwan

❏ *Tamiops rodolphei* Cambodian Striped Squirrel

E Thailand, S Laos, Cambodia and S Vietnam

❏ *Tamiops swinhoei* Swinhoe's Striped Squirrel

Tibet and N Myanmar to C China and N Vietnam

❏ *Glyphotes simus* Sculptor Squirrel

Mountains of Borneo

❏ *Rhinosciurus laticaudatus* Shrew-faced Squirrel

Malay Peninsula, Sumatra and Borneo

❏ *Lariscus hosei* Four-striped Ground Squirrel

Mountains of Borneo

❏ *Lariscus insignis* Three-striped Ground Squirrel

Malay Peninsula, Sumatra, Java and Borneo

❏ *Lariscus niobe* Niobe Ground Squirrel

Mountains of Sumatra, Mentawai Is. (Sumatra) and Java

❏ *Lariscus obscurus*
Mentawai Three-striped Squirrel

Siberut and S Pagai Is. (Mentawai Is., Sumatra)

❏ *Menetes berdmorei* Indochinese Ground Squirrel

Myanmar and Thailand to SW China and S Vietnam

❏ *Prosciurillus abstrusus* Secretive Dwarf Squirrel

SE Sulawesi

❏ *Prosciurillus leucomus* Whitish Dwarf Squirrel

Sulawesi and Buton

❏ *Prosciurillus murinus* Sulawesi Dwarf Squirrel

C & NE Sulawesi

❏ *Prosciurillus rosenbergii*
Rosenberg's Dwarf Squirrel

Sangihe Is. (N Moluccas)

❏ *Prosciurillus weberi* Weber's Dwarf Squirrel

C Sulawesi

❏ *Exilisciurus concinnus* Philippine Pygmy Squirrel

SE Philippines

❏ *Exilisciurus exilis* Least Pygmy Squirrel

Borneo

❏ *Exilisciurus whiteheadi* Tufted Pygmy Squirrel

Mountains of Borneo

Sundasciurus (Aletesciurus)

❏ *Sundasciurus davensis* Davao Squirrel

Davao region (Mindanao, SE Philippines)

❏ *Sundasciurus hippurus* Horse-tailed Squirrel

S Vietnam, Malay Peninsula, Sumatra and Borneo

❏ *Sundasciurus hoogstraali* Busuanga Squirrel

Busuanga (WC Philippines)

❏ *Sundasciurus juvencus*
Northern Palawan Tree Squirrel

N Palawan (SW Philippines); Endangered

❏ *Sundasciurus mindanensis* Mindanao Squirrel

Mindanao and near islands (SE Philippines)

❏ *Sundasciurus moellendorffi* Culion Tree Squirrel

Culion I. (WC Philippines)

❏ *Sundasciurus philippinensis*
Philippine Tree Squirrel

Basilan, W & S Mindanao (S & SE Philippines)

❏ *Sundasciurus rabori* Palawan Montane Squirrel

Mountains of Palawan (SW Philippines)

❏ *Sundasciurus samarensis* Samar Squirrel

Samar and Leyte Is. (EC Philippines)

❏ *Sundasciurus steerii*
Southern Palawan Tree Squirrel

Balabac and near islands (SW Philippines)

Sundasciurus (Sundasciurus)

❏ *Sundasciurus brookei* Brooke's Squirrel

Mountains of Borneo

❏ *Sundasciurus fraterculus* Fraternal Squirrel

S Mentawai Is. (Sumatra)

❏ *Sundasciurus jentinki* Jentink's Squirrel

Mountains of N Borneo

❏ *Sundasciurus lowii* Low's Squirrel

Malay Peninsula, Sumatra and Borneo

❏ *Sundasciurus tenuis* Slender Squirrel

Malay Peninsula, Singapore, Sumatra and Borneo

❏ *Rubrisciurus rubriventer*
Sulawesi Giant Squirrel

Sulawesi

❏ *Nannosciurus melanotis* Black-eared Squirrel

Sumatra, Java, Borneo and near islands

❏ *Hyosciurus heinrichi*
Montane Long-nosed Squirrel

Mountains of C Sulawesi

❏ *Hyosciurus ileile* Lowland Long-nosed Squirrel

Mountains of N Sulawesi

❏ *Heliosciurus gambianus* Gambian Sun Squirrel

Woods and savannas of subsaharan Africa S to Angola and
Zimbabwe

❏ *Heliosciurus mutabilis* Mutable Sun Squirrel

Forests, riparian woods and thickets of S Tanzania, Malawi,
SE Zimbabwe and EC Mozambique

❏ *Heliosciurus punctatus* Small Sun Squirrel

Forests of E Liberia, S Ivory Coast and S Ghana

❏ *Heliosciurus rufobrachium*
Red-legged Sun Squirrel

Widespread from Senegal to Uganda, S to Zimbabwe and
Mozambique

❏ *Heliosciurus ruwenzorii*
Ruwenzori Sun Squirrel

Montane forests of E D.R. Congo, Rwanda, Burundi and SW
Uganda

❏ *Heliosciurus undulatus* Zanj Sun Squirrel

Forests, regrowth areas and thickets of SE Kenya and NE
Tanzania incl. Zanzibar and Mafia Is.

❏ *Protoxerus aubinnii* Slender-tailed Squirrel

Locally in palm groves in lowland primary rainforests of
Liberia, Ivory Coast and Ghana

❏ *Protoxerus stangeri* Forest Giant Squirrel

Rainforests of Sierra Leone to S Sudan, S to N Angola and
Tanzania

❏ *Epixerus ebii* Western Palm Squirrel

Palm groves in lowland primary rainforests of Sierra Leone,
Liberia, Ivory Coast and Ghana

❏ *Epixerus wilsoni* Biafran Palm Squirrel

Palm groves in lowland primary rainforests of Cameroon,
Equatorial Guinea and Gabon

❏ *Paraxerus alexandri* Alexander's Bush Squirrel

Lowland primary rainforests of NE D.R. Congo and Uganda

❏ *Paraxerus boehmi* Böhm's Bush Squirrel

Dense undergrowth in forests and woods of S Sudan and W
Kenya to S D.R. Congo and N Zambia

❏ *Paraxerus cepapi* Smith's Bush Squirrel

Miombo (*Brachystegia*) woods and thickets of SE D.R. Congo
and SW Tanzania to N Namibia, Botswana, N South Africa
and S Mozambique

❏ *Paraxerus cooperi* Cooper's Mountain Squirrel

Mid-elevation and montane forests of Cameroon

❏ *Paraxerus flavovittis* Striped Bush Squirrel

Forests, savannas, thickets and farmland of S Kenya, Tanzania
and N Mozambique

❏ *Paraxerus lucifer* Black-and-red Bush Squirrel

Montane forests of E Zambia, N Malawi and SW Tanzania

❏ *Paraxerus ochraceus* Ochre Bush Squirrel

Dry forests and riparian woods in arid areas of S Sudan, Kenya
and Tanzania

❏ *Paraxerus palliatus* Red Bush Squirrel

Coastal and riparian forests and thickets of S Somalia to
KwaZulu-Natal (SE South Africa)

❏ *Paraxerus poensis* Green Bush Squirrel

Forests, forest edges and plantations of Sierra Leone and
Guinea to Rep. Congo and W D.R. Congo

❏ *Paraxerus vexillarius*
Swynnerton's Bush Squirrel

Montane forests of W Usambara Mtns (C & E Tanzania)

❑ *Paraxerus vincenti* Vincent's Bush Squirrel

Montane forests of Mt Namuli (N Mozambique)

❑ *Funisciurus anerythrus* Thomas' Rope Squirrel

Riparian woods and plantations of SW Nigeria to Uganda and SW D.R. Congo

❑ *Funisciurus bayonii* Lunda Rope Squirrel

Forests and woods of NE Angola and SW D.R. Congo

❑ *Funisciurus carruthersi*
Carruthers' Mountain Squirrel

Montane forests of S Uganda, Rwanda and Burundi

❑ *Funisciurus congicus* Congo Rope Squirrel

Rainforests and forest remnants of D.R. Congo, Angola and NW Namibia

❑ *Funisciurus isabella* Lady Burton's Rope Squirrel

Locally in montane forests of Cameroon, Rep. Congo and Central African Republic

❑ *Funisciurus lemniscatus* Ribboned Rope Squirrel

Forests of S Cameroon, Central African Republic and D.R. Congo

❑ *Funisciurus leucogenys*
Red-cheeked Rope Squirrel

Primary rainforests of Ghana to Equatorial Guinea and Central African Republic

❑ *Funisciurus pyrropus* Fire-footed Rope Squirrel

Dense undergrowth in forests of Senegal to NW Angola, D.R. Congo and Uganda

❑ *Funisciurus substriatus* Kintampo Rope Squirrel

Woods and forest edges of Ivory Coast to SE Nigeria

❑ *Funambulus layardi* Layard's Palm Squirrel

Mountains of S India and Sri Lanka

❑ *Funambulus palmarum* Indian Palm Squirrel

Forests and scrub of S India and Sri Lanka

❑ *Funambulus pennantii* Northern Palm Squirrel

Woods, gardens and built areas of SE Iran and Pakistan to Nepal and N India; feral Perth and Sydney (Australia)

❑ *Funambulus sublineatus* Dusky Palm Squirrel

SW India and C Sri Lanka

❑ *Funambulus tristriatus* Jungle Palm Squirrel

Coastal W India

❑ *Myosciurus pumilio* African Pygmy Squirrel

Wet primary rainforests of SE Nigeria and Cameroon to Gabon

❑ *Tamiasciurus douglasii* Douglas' Squirrel

Forests of SW British Columbia (Canada) to C California (USA)

❑ *Tamiasciurus hudsonicus* American Red Squirrel

Coniferous and mixed forests of Alaska, Canada and interior W & NE USA

❑ *Tamiasciurus mearnsi* Mearns' Squirrel

Sierra San Pedro Martir Mtns (Baja California Norte, Mexico)

❑ *Pteromys momonga* Japanese Flying Squirrel

Honshu and Kyushu (C & S Japan)

❑ *Pteromys volans* Siberian Flying Squirrel

Mixed forests of NE Europe and Siberia to N China, Korea and Hokkaido (N Japan)

❑ *Hylopetes alboniger* Particoloured Flying Squirrel

Montane forests of Nepal and NE India to C China and Indochina; Endangered

❑ *Hylopetes baberi* Afghan Flying Squirrel

Mountains of C Afghanistan and Kashmir (NW India)

❑ *Hylopetes bartelsi* Bartels' Flying Squirrel

Java

❑ *Hylopetes lepidus* Grey-cheeked Flying Squirrel

S Vietnam, Malay Peninsula, Sumatra, Java and Borneo

❑ *Hylopetes nigripes* Palawan Flying Squirrel

Palawan (Philippines)

❑ *Hylopetes phayrei* Indochinese Flying Squirrel

Myanmar to S Vietnam and S China

❑ *Hylopetes sipora* Sipora Flying Squirrel

Sipora (Mentawai Is., Sumatra); Endangered

❑ *Hylopetes spadiceus*
Red-cheeked Flying Squirrel

Myanmar, S Vietnam, Malay Peninsula and Sumatra

❏ *Hylopetes winstoni* Sumatran Flying Squirrel

Mountains of N Sumatra; Critically Endangered

❏ *Eoglaucomys fimbriatus* Kashmir Flying Squirrel

Montane forests of N Pakistan and NW India

❏ *Petinomys crinitus* Mindanao Flying Squirrel

Mindanao and near islands (SE Philippines)

❏ *Petinomys fuscocapillus*
Travancore Flying Squirrel

S India and Sri Lanka

❏ *Petinomys genibarbis* Whiskered Flying Squirrel

Malay Peninsula, Sumatra, Java and Borneo

❏ *Petinomys hageni* Hagen's Flying Squirrel

Sumatra and Borneo

❏ *Petinomys lugens* Siberut Flying Squirrel

Siberut and Sipora (Mentawai Is., Sumatra)

❏ *Petinomys sagitta* Arrow Flying Squirrel

Java

❏ *Petinomys setosus* Temminck's Flying Squirrel

Myanmar, Malay Peninsula, Sumatra and Borneo

❏ *Petinomys vordermanni*
Vordermann's Flying Squirrel

S Myanmar, Malaya and Borneo

❏ *Petaurista alborufus*
Red-and-white Giant Flying Squirrel

C & S China, Taiwan

❏ *Petaurista elegans* Spotted Giant Flying Squirrel

Nepal and C China to Malay Peninsula, Sumatra, Java and Borneo

❏ *Petaurista leucogenys*
Japanese Giant Flying Squirrel

C China and S Japan

❏ *Petaurista magnificus*
Hodgson's Giant Flying Squirrel

Nepal, Tibet, Bhutan and Sikkim (NE India)

❏ *Petaurista nobilis* Bhutan Giant Flying Squirrel

C Nepal, Bhutan and Sikkim (NE India)

❏ *Petaurista petaurista* Red Giant Flying Squirrel

Forests of E Afghanistan to C China, Malay Peninsula, Sumatra, Java and Borneo

❏ *Petaurista philippensis*
Indian Giant Flying Squirrel

India and Sri Lanka to S China, Taiwan and Thailand

❏ *Petaurista xanthotis*
Chinese Giant Flying Squirrel

Mountains of Tibet and C China

❏ *Glaucomys sabrinus* Northern Flying Squirrel

Forests of E Alaska, Canada and W & NE USA

❏ *Glaucomys volans* Southern Flying Squirrel

Forests of SE Canada and E USA to Honduras

❏ *Aeretes melanopterus*
North Chinese Flying Squirrel

Sichuan and Hebei (C China)

❏ *Eupetaurus cinereus* Woolly Flying Squirrel

Mountains of N Pakistan and Tibet to NE India; Endangered

❏ *Petaurillus emiliae* Lesser Pygmy Flying Squirrel

Borneo

❏ *Petaurillus hosei* Hose's Pygmy Flying Squirrel

Borneo

❏ *Petaurillus kinlochii*
Selangor Pygmy Flying Squirrel

Selangor (Malaysia)

❏ *Belomys pearsonii* Hairy-footed Flying Squirrel

NE India and Myanmar to SC China and Indochina

❏ *Trogopterus xanthipes*
Complex-toothed Flying Squirrel

Montane forests of C & E China; Endangered

❏ *Aeromys tephromelas* Black Flying Squirrel

Malay Peninsula, Sumatra and Borneo

❏ *Aeromys thomasi* Thomas' Flying Squirrel

Borneo

❏ *Pteromyscus pulverulentus*
Smoky Flying Squirrel

Malay Peninsula, Sumatra and Borneo

❏ *Biswamoyopterus biswasi*
Namdapha Flying Squirrel

Namdapha region (Arunachal Pradesh, India); Critically
Endangered

❏ *Olisthomys morrisi* Morris' Flying Squirrel

N Myanmar

❏ *Iomys horsfieldii* Javan Flying Squirrel

Malay Peninsula, Sumatra, Java and Borneo

❏ *Iomys sipora* Mentawai Flying Squirrel

Mentawai Is. (Sumatra)

Family: CASTORIDAE (Beavers—2)

❏ *Castor canadensis* American Beaver

Wooded margins of lakes, swamps and rivers of Alaska,
Canada, USA and N Mexico; feral E Europe and Asia

❏ *Castor fiber* Eurasian Beaver

Wooded margins of lakes, swamps and rivers of C & N
Europe, Siberia and Mongolia

Family: DIPODIDAE
(Jerboas and Jumping Mice—51)

❏ *Sicista armenica* Armenian Birch Mouse

Subalpine zone of NW Armenia; Critically Endangered

❏ *Sicista betulina* Northern Birch Mouse

Boreal forest and alpine zones from C & N Europe to E Siberia

❏ *Sicista caucasica* Caucasian Birch Mouse

NW Caucasus Mtns (Russia)

❏ *Sicista caudata* Long-tailed Birch Mouse

E Siberia and NE China; Endangered

❏ *Sicista concolor* Chinese Birch Mouse

Pakistan and N India to NW & NC China

❏ *Sicista kazbegica* Kazbeg Birch Mouse

Kazbegi District (Georgia)

❏ *Sicista kluchorica* Kluchor Birch Mouse

NW Caucasus Mtns (Russia)

❏ *Sicista napaea* Altai Birch Mouse

NW Altai Mtns (E Kazakhstan and SC Siberia)

❏ *Sicista pseudonapaea* Grey Birch Mouse

Taiga of S Altia Mtns (E Kazakhstan)

❏ *Sicista severtzovi* Severtzov's Birch Mouse

S Russia

❏ *Sicista strandi* Strand's Birch Mouse

N Caucasus Mtns (Russia)

❏ *Sicista subtilis* Southern Birch Mouse

Steppes, grasslands and fields of EC Europe to SW Siberia, N
Kazakhstan, NW China and W Mongolia

❏ *Sicista tianshanica* Tien Shan Birch Mouse

Mountains of N Tien Shan (Kazakhstan and NW China)

❏ *Eozapus setchuanus* Chinese Jumping Mouse

C China

❏ *Zapus hudsonius* Meadow Jumping Mouse

Fields and meadows of S Alaska, S Canada and NC & NE
USA

❏ *Zapus princeps* Western Jumping Mouse

Montane meadows and streamsides of SW Canada and W USA

❏ *Zapus trinotatus* Pacific Jumping Mouse

Wet meadows of extreme SW British Columbia (Canada) and
coastal W USA, S to C California

❏ *Napaeozapus insignis*
Woodland Jumping Mouse

Woods of SE Canada and NC & NE USA, S to Appalachian
Mtns

Allactaga (Allactaga)

❏ *Allactaga elater* Small Five-toed Jerboa

NE Turkey and Transcaucasia to N China

❑ *Allactaga firouzi* Iranian Jerboa

Mountain steppes of C Iran; Critically Endangered

❑ *Allactaga hotsoni* Hotson's Jerboa

SE Iran, S Afghanistan and W Pakistan

❑ *Allactaga major* Great Jerboa

Transcaucasia and W Russia to Kazakhstan and Turkmenistan

❑ *Allactaga severtzovi* Severtzov's Jerboa

Kazakhstan, Uzbekistan, NE Turkmenistan and SW Tajikistan

❑ *Allactaga vinogradovi* Vinogradov's Jerboa

S Kazakhstan, E Uzbekistan, Kyrgyzstan and Tajikistan

Allactaga (Orientallactaga)

❑ *Allactaga balikunica* Balikun Jerboa

NE Xinxiang (NW China) and Mongolia

❑ *Allactaga bullata* Gobi Jerboa

Deserts of Mongolia and N China

❑ *Allactaga sibirica* Mongolian Five-toed Jerboa

Kazakhstan and Turkmenistan to Mongolia and N China

Allactaga (Paralactaga)

❑ *Allactaga euphratica* Euphrates Jerboa

Turkey, Caucasus Mtns and Middle East to E Afghanistan

Allactaga (Scarturus)

❑ *Allactaga tetradactyla* Four-toed Jerboa

Coastal gravel plains of E Libya and Egypt; Endangered

Pygeretmus (Alactagulus)

❑ *Pygeretmus pumilio* Dwarf Fat-tailed Jerboa

Russia and NE Iran to Mongolia and N China

Pygeretmus (Pygeretmus)

❑ *Pygeretmus platyurus* Lesser Fat-tailed Jerboa

Kazakhstan

❑ *Pygeretmus shitkovi* Greater Fat-tailed Jerboa

Lake Balkhash region (E Kazakhstan)

❑ *Allactodipus bobrinskii* Bobrinski's Jerboa

N Turkmenistan and W Uzbekistan

❑ *Jaculus blanfordi* Blanford's Jerboa

SE Iran, SW Afghanistan and W Pakistan

❑ *Jaculus jaculus* Lesser Egyptian Jerboa

Sand dunes of N Africa, SC Sahara Desert, Arabian Peninsula and SW Iran

❑ *Jaculus orientalis* Greater Egyptian Jerboa

Sandy areas of N Africa, S Israel and Arabian Peninsula

❑ *Jaculus turcmenicus* Turkmen Jerboa

Turkmenistan and C Uzbekistan

❑ *Eremodipus lichtensteini* Lichtenstein's Jerboa

Kazakhstan, Turkmenistan and Uzbekistan

❑ *Dipus sagitta* Northern Three-toed Jerboa

Russia and N Iran to Mongolia and N China

❑ *Stylodipus andrewsi* Andrews' Three-toed Jerboa

Mongolia and N China

❑ *Stylodipus sungorus* Mongolian Three-toed Jerboa

SW Mongolia

❑ *Stylodipus telum* Thick-tailed Three-toed Jerboa

E Ukraine and Caucasus Mtns to N China

❑ *Cardiocranius paradoxus* Five-toed Pygmy Jerboa

N China

Salpingotus (Anguistodontus)

❑ *Salpingotus crassicauda* Thick-tailed Pygmy Jerboa

E Kazakhstan, SW Mongolia and NW China

Salpingotus (Prosalpingotus)

❑ *Salpingotus heptneri* Heptner's Pygmy Jerboa

S Kazakhstan and Uzbekistan

❑ *Salpingotus pallidus* Pallid Pygmy Jerboa

Deserts of C & E Kazakhstan

❏ *Salpingotus thomasi* Thomas' Pygmy Jerboa

"Afghanistan" (type locality doubtful)

Salpingotus (Salpingotus)

❏ *Salpingotus kozlovi* Kozlov's Pygmy Jerboa

Deserts of S Mongolia and N China

Salpingotus (Salpingotulus)

❏ *Salpingotus michaelis* Baluchistan Pygmy Jerboa

W Pakistan

❏ *Paradipus ctenodactylus* Comb-toed Jerboa

C Kazakhstan, Uzbekistan and Turkmenistan

❏ *Euchoreutes naso* Long-eared Jerboa

S Mongolia and N China; Endangered

Family: MURIDAE
(Mice, Rats, Voles and Gerbils—1394)
Subfamily: SIGMODONTINAE
(Neotropical Mice—461)

❏ *Baiomys musculus* Southern Pygmy Mouse

Seasonally dry grasslands and fields of C Mexico to NW Nicaragua

❏ *Baiomys taylori* Northern Pygmy Mouse

SE Arizona, extreme SW New Mexico and E Texas (USA) to C Mexico

❏ *Scotinomys teguina* Alston's Singing Mouse

Montane forests and forest edge of S Mexico to W Panama

❏ *Scotinomys xerampelinus*
Chiriqui Singing Mouse

Wet high-elevation montane forests, forest edge and páramo of C Costa Rica to W Panama

❏ *Rhagomys rufescens* Brazilian Arboreal Mouse

Very few records from Minas Gerais and Rio de Janeiro (SE Brazil); Critically Endangered

❏ *Otonyctomys hatti* Yucatán Vesper Mouse

Very few records from lowland forests and regrowth areas of Yucatán Peninsula (SE Mexico, N Belize and N Guatemala)

❏ *Nyctomys sumichrasti* Vesper Rat

Forests and tall regrowth areas of C Mexico to C Panama

❏ *Tylomys bullaris* Chiapan Climbing Rat

Tuxtla Gutiérrez region (Chiapas, SE Mexico); only known from subadult holotype specimen, Critically Endangered

❏ *Tylomys fuliventer* Fulvous-bellied Climbing Rat

Tacarcuna region (Darién Prov., E Panama)

❏ *Tylomys mirae* Mira Climbing Rat

W Colombia and NW Ecuador

❏ *Tylomys nudicaudus* Northern Climbing Rat

Rocky areas in forests and regrowth areas of EC Mexico to N Nicaragua

❏ *Tylomys panamensis* Panamanian Climbing Rat

Cana and Boca de Río Paya regions (Darién Prov., E Panama)

❏ *Tylomys tumbalensis* Tumbalá Climbing Rat

Tumbalá region (Chiapas, SE Mexico); only known from subadult holotype specimen, Critically Endangered

❏ *Tylomys watsoni* Watson's Climbing Rat

Lowland forests of Costa Rica and Panama

❏ *Ototylomys phyllotis* Big-eared Climbing Rat

Rocky areas in forests and regrowth areas of S Mexico to C Costa Rica

Neotoma (Neotoma)

❏ *Neotoma albigula* White-throated Woodrat

SW & SC USA to C Mexico

❏ *Neotoma angustapalata* Tamaulipan Woodrat

NE Mexico

❏ *Neotoma anthonyi* Anthony's Woodrat

Todos Santos I. (Baja California Norte, Mexico); Endangered

❏ *Neotoma bryanti* Bryant's Woodrat

Cedros I. (Baja California Norte, Mexico); Endangered

❏ *Neotoma bunkeri* Bunker's Woodrat

Coronados I. (Baja California Sur, Mexico); Endangered

❏ *Neotoma chrysomelas* Nicaraguan Woodrat

Very few records from mountains of Honduras and NW Nicaragua

❏ *Neotoma devia* Arizona Woodrat

Rocky slopes and cliffs of W Arizona (USA)

❏ *Neotoma floridana* Eastern Woodrat

Woods and meadows of C & SE USA

❏ *Neotoma goldmani* Goldman's Woodrat

NE Mexico

❏ *Neotoma lepida* Desert Woodrat

Desert scrub and coastal sagebrush of SW USA and Baja California (Mexico)

❏ *Neotoma magister* Allegheny Woodrat

Rocky slopes and cliffs of Allegheny Mtns (E USA)

❏ *Neotoma martinensis* San Martin Island Woodrat

San Martin I. (Baja California Norte, Mexico); Endangered

❏ *Neotoma mexicana* Mexican Woodrat

Rocky slopes in woods and forests of interior SW USA to W Honduras and W El Salvador

❏ *Neotoma micropus* Southern Plains Woodrat

Rocky slopes and scrub flats of SC USA and NE Mexico

❏ *Neotoma nelsoni* Nelson's Woodrat

Perote region (Veracruz, Mexico); Endangered

❏ *Neotoma palatina* Bolaos Woodrat

EC Jalisco (Mexico)

❏ *Neotoma stephensi* Stephens' Woodrat

Semi-desert scrub of extreme SC Utah, N Arizona and NW New Mexico (USA)

❏ *Neotoma varia* Turner Island Woodrat

Turner I. (Sonora, Mexico); Endangered

Neotoma (*Teanopus*)

❏ *Neotoma phenax* Sonoran Woodrat

SW Sonora and NW Sinaloa (Mexico)

Neotoma (*Teonoma*)

❏ *Neotoma cinerea* Bushy-tailed Woodrat

Montane forests of SW Canada and NW USA

❏ *Neotoma fuscipes* Dusky-footed Woodrat

Woods and scrub of coastal W Oregon (USA) S to N Baja California (Mexico)

❏ *Xenomys nelsoni* Magdalena Rat

W Jalisco and Colima (Mexico)

❏ *Hodomys alleni* Allen's Woodrat

SW Mexico

❏ *Nelsonia goldmani*
Nelson and Goldman's Woodrat

N & C Mexico

❏ *Nelsonia neotomodon* Diminutive Woodrat

S Sierra Madre Occidental (WC Mexico)

❏ *Peromyscus attwateri* Texas Deer Mouse

Coniferous forests and woods of SE Kansas and SW Missouri to C Texas (USA)

❏ *Peromyscus aztecus* Aztec Deer Mouse

Wet montane forest edge and regrowth areas of C Mexico to SE Honduras

❏ *Peromyscus boylii* Brush Deer Mouse

Montane rocky slopes and brush of SW & SC USA to C Mexico incl. San Esteban I. (Sonora, Mexico)

❏ *Peromyscus bullatus* Perote Deer Mouse

Perote region (Veracruz, Mexico); Endangered

❏ *Peromyscus californicus* California Deer Mouse

Forests, woods and coastal scrub of C California (USA) to Baja California Norte (Mexico)

❏ *Peromyscus crinitus* Canyon Deer Mouse

Grasslands, scrub and bare-rock deserts of interior SW USA and extreme NW Mexico

❏ *Peromyscus difficilis* Zacatecan Deer Mouse

Mexico

❏ *Peromyscus eremicus* Cactus Deer Mouse

Rocky desert flats and foothills of SW & SC USA to C Mexico, incl. San Lorenzo island group and Monserrate I. (Baja California, Mexico)

❏ *Peromyscus eva* Eva's Deer Mouse

Carmen I. and S Baja California Sur (Mexico)

❏ *Peromyscus furvus* **Blackish Deer Mouse**

E slope of S Sierra Madre Oriental (Mexico)

❏ *Peromyscus gossypinus* **Cotton Deer Mouse**

Swamps and bottomland forests of SC & SE USA

❏ *Peromyscus grandis* **Giant Deer Mouse**

Cloud forests of C Guatemala

❏ *Peromyscus gratus* **Saxicolous Deer Mouse**

Rocky habitats of SW New Mexico (USA) to S Mexico

❏ *Peromyscus guardia* **Angel Island Deer Mouse**

Angel de la Guarda island group (N Gulf of California, Mexico)

❏ *Peromyscus guatemalensis* **Guatemalan Deer Mouse**

Wet montane oak forests of Sierra Madre (S Chiapas, SE Mexico) to SC Guatemala

❏ *Peromyscus gymnotis* **Naked-eared Deer Mouse**

Forests, plantations and scrub of Pacific slope of extreme SE Chiapas (SE Mexico) to SW Nicaragua

❏ *Peromyscus hooperi* **Hooper's Deer Mouse**

C Coahuila and NE Zacatecas (NC Mexico)

❏ *Peromyscus keeni* **Northwestern Deer Mouse**

Alexander Archipelago (SE Alaska) to coastal NW Washington (USA)

❏ *Peromyscus leucopus* **White-footed Deer Mouse**

Dry forests, scrub and roadsides of SC & SE Canada, C & E USA and Mexico

❏ *Peromyscus levipes* **Southern Deer Mouse**

Forest edge, scrub and clearings of Mexico to SE Honduras

❏ *Peromyscus madrensis* **Tres Marías Deer Mouse**

Tres Marías Is. (Nayarit, Mexico)

❏ *Peromyscus maniculatus* **North American Deer Mouse**

SE Alaska, Canada, USA and Mexico

❏ *Peromyscus mayensis* **Maya Deer Mouse**

Wet montane oak forests between Santa Eulalia and San Mateo Ixtatán (W Guatemala); Endangered

❏ *Peromyscus megalops* **Brown Deer Mouse**

Sierra Madre del Sur (S Mexico)

❏ *Peromyscus mekisturus* **Puebla Deer Mouse**

SE Puebla (Mexico)

❏ *Peromyscus melanocarpus* **Zempoaltepec Deer Mouse**

Cloud forest of NC Oaxaca (S Mexico)

❏ *Peromyscus melanophrys* **Plateau Deer Mouse**

Dry rocky deserts of C & S Mexico plateau

❏ *Peromyscus melanotis* **Black-eared Deer Mouse**

N & C Mexico

❏ *Peromyscus melanurus* **Black-tailed Deer Mouse**

Pacific slope of Sierra Madre del Sur (S Mexico)

❏ *Peromyscus merriami* **Merriam's Deer Mouse**

Mesquite and cactus thickets of SC Arizona (USA) to WC Mexico incl. Tortuga I. (Baja California Sur, Mexico)

❏ *Peromyscus mexicanus* **Mexican Deer Mouse**

Forests, plantations and scrub of NE Mexico to W Panama

❏ *Peromyscus nasutus* **Northern Rock Deer Mouse**

Boulder-strewn woods of SE Utah and NC Colorado (USA) to NW Coahuila (Mexico)

❏ *Peromyscus ochraventer* **El Carrizo Deer Mouse**

Forests of NE Mexico

❏ *Peromyscus pectoralis* **White-ankled Deer Mouse**

Arid upland rocky habitats of C Texas to C Mexico

❏ *†Peromyscus pembertoni* **Pemberton's Deer Mouse**

Formerly San Pedro Nolasco I. (Sonora, Mexico); Extinct

❏ *Peromyscus perfulvus* **Marsh Deer Mouse**

Coastal SW Mexico

❏ *Peromyscus polionotus* **Oldfield Deer Mouse (Beach Mouse)**

Open sandy habitats and coastal dunes of SE USA

❑ *Peromyscus polius* Chihuahuan Deer Mouse

WC Chihuahua (Mexico)

❑ *Peromyscus pseudocrinitus*
False Canyon Deer Mouse

Coronados I. (Baja California Sur, Mexico); Critically
Endangered

❑ *Peromyscus sejugis* Santa Cruz Deer Mouse

Santa Cruz and San Diego Is. (S Gulf of California, Mexico)

❑ *Peromyscus simulus* Nayarit Deer Mouse

Coastal Nayarit and S Sinaloa (NW Mexico)

❑ *Peromyscus slevini* Slevin's Deer Mouse

Santa Catalina I. (Baja California Sur, Mexico); Critically
Endangered

❑ *Peromyscus spicilegus* Gleaning Deer Mouse

S Sierra Madre Occidental (WC Mexico)

❑ *Peromyscus stirtoni* Stirton's Deer Mouse

Locally in dry rocky lowland forests and scrub from SE
Guatemala to W Nicaragua

❑ *Peromyscus truei* Piñon Deer Mouse

Rocky coniferous scrub habitats of SW & SC USA and N Mexico

❑ *Peromyscus winkelmanni*
Winkelmann's Deer Mouse

Michoacán and Guerrero (SW Mexico)

❑ *Peromyscus yucatanicus* Yucatán Deer Mouse

Lowland forests and regrowth areas of N Yucatán Peninsula
(SE Mexico)

❑ *Peromyscus zarhynchus* Chiapan Deer Mouse

Wet montane forests, regrowth areas and scrub of NC Chiapas
(Mexico)

Reithrodontomys (Aporodon)

❑ *Reithrodontomys brevirostris*
Short-nosed Harvest Mouse

Montane forests and plantations of Cordillera Dariense
(Nicaragua) and Cordillera Central (C Costa Rica)

❑ *Reithrodontomys creper* Chiriqui Harvest Mouse

Montane forest, clearings and forest edge of N Costa Rica to
W Panama

❑ *Reithrodontomys darienensis*
Darién Harvest Mouse

Lowland rainforests, forest edge and clearings of C & E Panama

❑ *Reithrodontomys gracilis* Slender Harvest Mouse

Forests, regrowth areas, scrub and clearings of S Mexico to
NW Costa Rica

❑ *Reithrodontomys mexicanus*
Mexican Harvest Mouse

Forests, scrub, clearings and farmland of NE Mexico to W
Panama; Andes of W Colombia and N Ecuador

❑ *Reithrodontomys microdon*
Small-toothed Harvest Mouse

Very locally in montane moss forests of C Mexico to W
Guatemala

❑ *Reithrodontomys paradoxus*
Nicaraguan Harvest Mouse

Very few records from mid-elevation deciduous forests of SW
Nicaragua and NW Costa Rica

❑ *Reithrodontomys rodriguezi*
Rodriguez's Harvest Mouse

Wet montane forests and forest edge of slopes of Volcán
Barva, Volcán Irazu and Cerro Asunción (C Costa Rica)

❑ *Reithrodontomys spectabilis*
Cozumel Harvest Mouse

Forest edge and dense regrowth areas of Cozumel I. (Quintana
Roo, Mexico); Endangered

❑ *Reithrodontomys tenuirostris*
Narrow-nosed Harvest Mouse

Wet montane moss forests of extreme SE Chiapas (SE Mexico)
to C Guatemala

Reithrodontomys (Reithrodontomys)

❑ *Reithrodontomys burti* Sonoran Harvest Mouse

WC Sonora to C Sinaloa (Mexico)

❑ *Reithrodontomys chrysopsis*
Volcano Harvest Mouse

Transverse volcanic range of C Mexico

❑ *Reithrodontomys fulvescens*
Fulvous Harvest Mouse

Forest clearings, scrub and grasslands of SC USA to W
Nicaragua

❏ *Reithrodontomys hirsutus* Hairy Harvest Mouse

SC Nayarit and NW Jalisco (Mexico)

❏ *Reithrodontomys humulis*
Eastern Harvest Mouse

Damp meadows and scrub of SC & SE USA

❏ *Reithrodontomys megalotis*
Western Harvest Mouse

Tallgrass prairies and meadows of SW Canada, W & C USA and Mexico

❏ *Reithrodontomys montanus*
Plains Harvest Mouse

Upland grasslands of Great Plains region of SC USA to NC Mexico

❏ *Reithrodontomys raviventris*
Saltmarsh Harvest Mouse

Saltmarshes of San Francisco Bay (California, USA)

❏ *Reithrodontomys sumichrasti*
Sumichrast's Harvest Mouse

Wet montane forest edge and clearings from C Mexico to W Panama

❏ *Reithrodontomys zacatecae*
Zacatecan Harvest Mouse

NW & WC Mexico

❏ *Onychomys arenicola*
Chihuahuan Grasshopper Mouse

Desert scrub of Chihuahuan Desert of SC USA and NC Mexico

❏ *Onychomys leucogaster*
Northern Grasshopper Mouse

Arid and semiarid grasslands of SC Canada, interior W & C USA, and NE Mexico

❏ *Onychomys torridus*
Southern Grasshopper Mouse

Arid deserts and chaparral of SW USA and NW Mexico

❏ *Neotomodon alstoni* Mexican Volcano Mouse

Transverse volcanic range of C Mexico

❏ *Ochrotomys nuttalli* Golden Mouse

Dense forests of SE USA

❏ *Podomys floridanus* Florida Mouse

Sandy uplands of peninsular Florida (USA)

❏ *Isthmomys flavidus* Yellow Isthmus Rat

Locally in primary montane forests of W Panama

❏ *Isthmomys pirrensis* Mount Pirre Isthmus Rat

Wet elfin forests of Cerro Malí, Tacarcuna and Cerro Pirre (Darién Prov., E Panama)

❏ *Megadontomys cryophilus*
Oaxaca Giant Deer Mouse

Sierra de Juarez (Oaxaca, Mexico)

❏ *Megadontomys nelsoni*
Nelson's Giant Deer Mouse

E slopes of S Sierra Madre Oriental (EC Mexico)

❏ *Megadontomys thomasi*
Thomas' Giant Deer Mouse

Mountains of Sierra Madre del Sur (S Mexico)

❏ *Habromys chinanteco* Chinanteco Deer Mouse

Sierra de Juarez (Oaxaca, Mexico)

❏ *Habromys lepturus* Slender-tailed Deer Mouse

Cloud forests of NC Oaxaca (Mexico)

❏ *Habromys lophurus*
Crested-tailed Deer Mouse

Wet montane oak forests of Chiapas (SE Mexico) to NW El Salvador

❏ *Habromys simulatus* Jico Deer Mouse

E slopes of S Sierra Madre Oriental (EC Mexico); Endangered

❏ *Osgoodomys banderanus*
Michoacán Deer Mouse

SW Mexico

❏ *†Megalomys desmarestii*
Antillean Giant Rice Rat

Formerly Martinique (Lesser Antilles); Extinct (still plentiful in 1890, almost certainly became extinct after devastating volcanic eruption of 8 May 1902)

❏ *†Megalomys lucia* St Lucia Giant Rice Rat

Formerly St Lucia (Lesser Antilles); Extinct (last recorded prior to 1900)

❑ *Oryzomys albigularis* Tomes' Rice Rat

Wet montane forests of E Panama and Venezuela to Ecuador and N Peru

❑ *Oryzomys alfaroi* Alfaro's Rice Rat

Forests, forest edge and scrub of NE Mexico to W Colombia and Ecuador

❑ *Oryzomys auriventer* Ecuadorian Rice Rat

Andes of Ecuador and N Peru

❑ *Oryzomys balneator* Peruvian Rice Rat

Andes of SE Ecuador and N Peru

❑ *Oryzomys bolivaris* Long-whiskered Rice Rat

Primary rainforests of Caribbean slope of E Honduras to W Colombia and NW Ecuador

❑ *Oryzomys buccinatus* Paraguayan Rice Rat

Forests and forest edge of S Bolivia, E Paraguay and N Argentina

❑ *Oryzomys caracolus* Caracol Rice Rat

Montane forests of N Venezuela

❑ *Oryzomys chapmani* Chapman's Rice Rat

Cloud forests of NE and S Mexico

❑ *Oryzomys couesi* Coues' Rice Rat

Wetlands, scrub and forest edge of extreme S Texas (USA) to NW Colombia; Jamaica

❑ *Oryzomys devius* Montane Rice Rat

Wet montane forests of Costa Rica and W Panama

❑ *Oryzomys dimidiatus* Nicaraguan Rice Rat

Very few records from riparian habitats of Zelaya Prov. (SE Nicaragua)

❑ *Oryzomys emmonsae* Emmons' Rice Rat

Rio Xingu (Pará, E Amazonian Brazil)

❑ *Oryzomys galapagoensis* Galápagos Rice Rat

San Cristóbal and Santa Fe Is. (Galápagos Is.)

❑ *Oryzomys gorgasi* Gorgas' Rice Rat

Sautata region (Antioquia Dept., NW Colombia); only known from holotype specimen, Critically Endangered

❑ *Oryzomys hammondi* Hammond's Rice Rat

Highlands of Mindo region (Pichincha Prov., N Ecuador)

❑ *Oryzomys intectus* Colombian Rice Rat

Santa Elena region, near Medellin (Antioquia Dept., NC Colombia); status uncertain

❑ *Oryzomys intermedius* Intermediate Rice Rat

N Argentina, E Paraguay and SE Brazil

❑ *Oryzomys keaysi* Keays' Rice Rat

Mid-elevation montane forests of C Peru to Bolivia

❑ *Oryzomys kelloggi* Kellogg's Rice Rat

Fazenda São Geraldo region (Minas Gerais, SE Brazil)

❑ *Oryzomys lamia* Monster Rice Rat

Rio Jordas region (Minas Gerais, SE Brazil)

❑ *Oryzomys levipes* Light-footed Rice Rat

High montane cloud forests of SE Peru and WC Bolivia

❑ *Oryzomys macconnelli* MacConnell's Rice Rat

Lowland rainforests of S Colombia to Suriname, S to E Peru and E Amazonian Brazil

❑ *Oryzomys megacephalus* Large-headed Rice Rat

Lowland forests of E Venezuela to French Guiana, S to Peru, Bolivia, Paraguay and E Brazil

❑ *Oryzomys melanotis* Black-eared Rice Rat

SW Mexico

❑ †*Oryzomys nelsoni* Nelson's Rice Rat

Formerly Maria Madre I. (Nayarit, Mexico); Extinct

❑ *Oryzomys nitidus* Elegant Rice Rat

Lowland forests and regrowth areas from E Ecuador to Bolivia, SW Brazil, Paraguay and NW Argentina

❑ *Oryzomys oniscus* Sowbug Rice Rat

São Torenzo region (Pernambuco, EC Brazil)

❑ *Oryzomys palustris* Marsh Rice Rat

Wetlands of SC & SE USA

❑ *Oryzomys polius* Grey Rice Rat

Mountains of Tambo Carrizal region (Amazonas, EC Peru)

❏ *Oryzomys ratticeps* Rat-headed Rice Rat

Forests and forest edge of NE Argentina, Paraguay and E Brazil

❏ *Oryzomys rhabdops* Striped Rice Rat

Mountains of S Mexico and C Guatemala

❏ *Oryzomys rostratus* Rusty Rice Rat

Forest clearings, plantations and forest edge from NE Mexico to W Nicaragua

❏ *Oryzomys russatus* Big-headed Rice Rat

E Andes of SC Bolivia and NW Argentina

❏ *Oryzomys saturatior* Cloud Forest Rice Rat

Cloud forests from S Mexico to NC Nicaragua

❏ *Oryzomys seuanezi* Seuánez's Rice Rat

SE Brazil

❏ *Oryzomys subflavus* Terraced Rice Rat

Dry forests of E Brazil

❏ *Oryzomys talamancae* Talamancan Rice Rat

Forests and forest edge of E Costa Rica to NW Venezuela, W Colombia and W Ecuador

❏ *Oryzomys tatei* Tate's Rice Rat

Tungurahua region (EC Ecuador)

❏ *Oryzomys xantheolus* Yellowish Rice Rat

SW Ecuador and W Peru

❏ *Oryzomys yunganus* Yungas Rice Rat

Lowland rainforests of C Colombia to Guianas, S to WC Bolivia

❏ †*Nesoryzomys darwini*
Darwin's Galápagos Mouse

Formerly Santa Cruz I. (Galápagos Is.); Extinct (last collected in 1930)

❏ *Nesoryzomys fernandinae*
Fernandina Galápagos Mouse

Fernandina I. (Galápagos Is.)

❏ †*Nesoryzomys indeffesus*
Indefatigable Galápagos Mouse

Formerly Santa Cruz, Baltra and Fernandina I. (Galápagos Is.); Extinct (probably since about 1945)

❏ *Nesoryzomys swarthi*
Santiago Galápagos Mouse

San Salvador I. (Galápagos Is.)

❏ *Melanomys caliginosus* Dusky Rice Rat

Forest edge and regrowth areas of E Honduras to extreme NW Venezuela, W Colombia and SW Ecuador

❏ *Melanomys robustulus* Robust Dark Rice Rat

Highlands of SE Ecuador

❏ *Melanomys zunigae* Zuniga's Dark Rice Rat

WC Peru

❏ *Sigmodontomys alfari* Cana Rice Rat

Lowland marshes and wet regrowth areas of SE Honduras to NW Venezuela, W Colombia and NW Ecuador

❏ *Sigmodontomys aphrastus* Long-tailed Rice Rat

Very few records from semideciduous forests of S Costa Rica and SW Panama; Critically Endangered

❏ *Nectomys melanius* Small-footed Water Rat

French Guiana; Critically Endangered

❏ *Nectomys palmipes* Trinidad Water Rat

NE Venezuela; Trinidad

❏ *Nectomys squamipes* South American Water Rat

Rainforests, riparian habitats and farmland of Colombia to Suriname, S to Paraguay, NE Argentina and SE Brazil

❏ *Amphinectomys savamis* Amphibious Rat

Forest streams of R. Ucayali region (Loreto Dept., NE Peru)

❏ *Oligoryzomys andinus* Andean Pygmy Rice Rat

Highlands of W Peru to WC Bolivia

❏ *Oligoryzomys arenalis* Sandy Pygmy Rice Rat

Arid desert coastal plains of NW Peru

❏ *Oligoryzomys chacoensis*
Chacoan Pygmy Rice Rat

Grasslands, thorn scrub and forests of Gran Chaco region of WC Brazil, Bolivia, W Paraguay and N Argentina

❏ *Oligoryzomys delticola* Delta Pygmy Rice Rat

Forests and thorn scrub of NE Argentina, Uruguay and S Brazil

❑ *Oligoryzomys destructor*
Destructive Pygmy Rice Rat

Andes from S Colombia to WC Bolivia

❑ *Oligoryzomys eliurus* **Brazilian Pygmy Rice Rat**

C & SE Brazil

❑ *Oligoryzomys flavescens* **Yellow Pygmy Rice Rat**

Marshes and riparian habitats of NE Argentina, Uruguay and
SE Brazil

❑ *Oligoryzomys fulvescens*
Northern Pygmy Rice Rat

Forest clearings and regrowth areas of NE Mexico to
Suriname, S to Ecuador and N Brazil

❑ *Oligoryzomys griseolus* **Greyish Pygmy Rice Rat**

Cordillera Occidental (E Colombia) and Táchira Andes (W
Venezuela)

❑ *Oligoryzomys longicaudatus*
Long-tailed Pygmy Rice Rat

Montane cloud forests and brush of Andes of Chile and W
Argentina

❑ *Oligoryzomys magellanicus*
Magellanic Pygmy Rice Rat

Forest clearings, brush and roadsides of S Chile and S
Argentina incl. Tierra del Fuego

❑ *Oligoryzomys microtis*
Small-eared Pygmy Rice Rat

Forest edge and regrowth areas of Amazonian Peru to C Brazil, S
to marshes and wet grasslands of Paraguay and NE Argentina

❑ *Oligoryzomys nigripes*
Black-footed Pygmy Rice Rat

Forests, regrowth areas and farmland of E Paraguay, N
Argentina, Uruguay and SE Brazil

❑ *Oligoryzomys stramineus*
Straw-coloured Pygmy Rice Rat

Terezina de Goiás (Goiás, Brazil)

❑ *Oligoryzomys vegetus* **Sprightly Pygmy Rice Rat**

Montane forest clearings, regrowth areas and scrub of Costa
Rica and W Panama

❑ *Oligoryzomys victus* **St Vincent Pygmy Rice Rat**

St. Vincent (Lesser Antilles); Endangered

❑ *Neacomys dubosti* **Dubost's Bristly Mouse**

Primary rainforests of SE Suriname, French Guiana and N
Amapá (NE Brazil)

❑ *Neacomys guianae* **Guianan Bristly Mouse**

Dense rainforests of S Venezuela to Suriname and N
Amazonian Brazil

❑ *Neacomys minutus* **Small Bristly Mouse**

Primary rainforests of Amazonian Brazil

❑ *Neacomys musseri* **Musser's Bristly Mouse**

Primary rainforests of Amazonian Brazil

❑ *Neacomys paracou* **Paracou Bristly Mouse**

Primary rainforests of Guyana, Suriname, French Guiana and
NE Amazonian Brazil

❑ *Neacomys pictus* **Painted Bristly Mouse**

Rainforest edge habitats of Darién Prov. (E Panama)

❑ *Neacomys spinosus* **Common Bristly Mouse**

Lowlands of Amazon basin from E Colombia to Peru, E
Bolivia and C Brazil

❑ *Neacomys tenuipes* **Narrow-footed Bristly Mouse**

Dense rainforests of Colombia, NW Venezuela, E Ecuador and
N Brazil

❑ *Zygodontomys brevicauda*
Common Cane Mouse

Lowland marshes, grassland and farmland of SE Costa Rica to
Suriname and N Amazonian Brazil; Trinidad & Tobago

❑ *Zygodontomys brunneus*
Colombian Cane Mouse

Intermontane valleys of N Colombia

❑ *Pseudoryzomys simplex*
Brazilian False Rice Rat

Lowland alluvial grasslands of SE Bolivia, Paraguay, NE
Argentina and S & E Brazil

❑ *Lundomys molitor* **Greater Marsh Rat**

Dense reedbeds and wet grasslands of Uruguay and Rio
Grande do Sul (SE Brazil)

❑ *Holochilus brasiliensis* **Web-footed Marsh Rat**

Lowland marshes of EC Argentina, Uruguay and SE Brazil

❏ *Holochilus chacarius* Chaco Marsh Rat

Rice fields and plantations of Paraguay and NE Argentina

❏ *Holochilus sciureus* Common Marsh Rat

Rice and sugarcane fields of Venezuela to Suriname, S to Bolivia and C Brazil

❏ *Microakodontomys transitorius*
Intermediate Lesser Grass Mouse

Brasília region (Distrito Federal, E Brazil)

❏ *Oecomys auyantepui*
North Amazonian Arboreal Rice Rat

Forests of SE Venezuela to French Guiana, S to R. Amazonas (C Amazonian Brazil)

❏ *Oecomys bicolor* Bicoloured Arboreal Rice Rat

Lowland forests and savannas of C Panama to French Guiana, S to Peru, E Bolivia and SC Brazil

❏ *Oecomys cleberi* Cleber's Arboreal Rice Rat

Brasília (Distrito Federal, E Brazil); only known from type locality, Endangered

❏ *Oecomys concolor* Unicoloured Arboreal Rice Rat

Lowland forests and clearings of E Colombia, S Venezuela, NW Brazil and N Bolivia

❏ *Oecomys flavicans* Yellow Arboreal Rice Rat

NE Colombia, W & N Venezuela

❏ *Oecomys mamorae* Mamore Arboreal Rice Rat

E Bolivia, N Paraguay and SW Brazil

❏ *Oecomys paricola*
South Amazonian Arboreal Rice Rat

Forests S of R. Amazonas (C Amazonian Brazil)

❏ *Oecomys phaeotis* Dusky Arboreal Rice Rat

E slope of Andes of C Peru

❏ *Oecomys rex* King Arboreal Rice Rat

E Venezuela, Guianas and N Brazil

❏ *Oecomys roberti* Robert's Arboreal Rice Rat

Amazon basin of S Venezuela and Guianas to E Peru and N Bolivia

❏ *Oecomys rutilus* Red Arboreal Rice Rat

Guyana, Suriname and French Guiana

❏ *Oecomys speciosus*
Venezuelan Arboreal Rice Rat

Savannas of NE Colombia, C & N Venezuela; Trinidad

❏ *Oecomys superans* Foothill Arboreal Rice Rat

Foothills of Andes of E Colombia, Ecuador and Peru

❏ *Oecomys trinitatis* Big Arboreal Rice Rat

Forest and forest clearings of S Costa Rica to Guianas, S to Peru and S Brazil; Trinidad & Tobago

❏ *Microryzomys altissimus*
Highland Small Rice Rat

High montane areas of Andes of Colombia, Ecuador and Peru

❏ *Microryzomys minutus* Forest Small Rice Rat

Montane cloud forests of Andes of Colombia and N Venezuela to WC Bolivia

❏ *Scolomys juruaense* Rio Juruá Spiny Mouse

Amazon basin of E Ecuador and W Brazil

❏ *Scolomys melanops* Ecuadorian Spiny Mouse

Foothills of Mera region (Pastaza Prov., EC Ecuador); Endangered

❏ *Scolomys ucayalensis* Ucayali Spiny Mouse

R. Ucayali region (Loreto Dept., NE Peru); Endangered

❏ *Chilomys instans* Colombian Forest Mouse

Montane cloud forests of Andes of N Colombia, W Venezuela and N Ecuador

❏ *Abrawayaomys ruschii* Ruschi's Rat

Very few records from forests of Espírito Santo and Minas Gerais (Brazil); Endangered

❏ *Delomys dorsalis* Striped Atlantic Forest Rat

Coastal forests of extreme NE Argentina and SE Brazil

❏ *Delomys sublineatus* Pallid Atlantic Forest Rat

Atlantic coastal forests of SE Brazil

❏ *Thomasomys apeco* Apeco Oldfield Mouse

Andes of San Martín Dept. (NC Peru)

❏ *Thomasomys aureus* Golden Oldfield Mouse

Montane cloud forests of Colombia, NW Venezuela, Ecuador and NE Peru

❏ *Thomasomys baeops* Beady-eyed Oldfield Mouse

Highlands of Chilla valley region (El Oro Prov., SW Ecuador)

❏ *Thomasomys bombycinus* Silky Oldfield Mouse

Highlands of Cordillera Occidental (W Colombia)

❏ *Thomasomys cinereiventer*
Ashy-bellied Oldfield Mouse

Highlands of NC Colombia and Ecuador

❏ *Thomasomys cinereus*
Ash-coloured Oldfield Mouse

Andes of SW Ecuador and N Peru

❏ *Thomasomys daphne* Daphne's Oldfield Mouse

Andes from S Peru to C Bolivia

❏ *Thomasomys eleusis* Peruvian Oldfield Mouse

Tambo Jemes region (Andes of NC Peru)

❏ *Thomasomys gracilis* Slender Oldfield Mouse

Andes of Ecuador to SE Peru

❏ *Thomasomys hylophilus*
Woodland Oldfield Mouse

Dense wet montane forests of N Colombia and NW Venezuela

❏ *Thomasomys incanus* Incan Oldfield Mouse

Junín region (Andes of C Peru)

❏ *Thomasomys ischyurus*
Strong-tailed Oldfield Mouse

Highlands of N & C Peru

❏ *Thomasomys kalinowskii*
Kalinowski's Oldfield Mouse

Highlands of Vitoc valley (Junín Dept, C Peru)

❏ *Thomasomys ladewi* Ladew's Oldfield Mouse

Highlands of R. Aceramarca region (La Paz Dept., NW Bolivia)

❏ *Thomasomys laniger* Butcher Oldfield Mouse

Montane forests of Cordillera Oriental (NC Colombia and NW Venezuela)

❏ *Thomasomys macrotis*
Large-eared Oldfield Mouse

Rio Abiseo National Park (W slope of Andes of N Peru)

❏ *Thomasomys monochromos*
Unicoloured Oldfield Mouse

Sierra Nevada de Santa Marta (Magdalena Dept., NE Colombia)

❏ *Thomasomys niveipes*
Snow-footed Oldfield Mouse

C Colombia

❏ *Thomasomys notatus*
Distinguished Oldfield Mouse

R. Urubamba and Alto Madre de Dios drainages in Andes of SE Peru

❏ *Thomasomys onkiro* Onkiro Oldfield Mouse

Between Rs. Ene and Urubamba, E slope of Andes (Cusco Dept., Peru)

❏ *Thomasomys oreas* Montane Oldfield Mouse

Highlands of Cocopunco region (La Paz Dept., WC Bolivia)

❏ *Thomasomys paramorum*
Páramo Oldfield Mouse

Páramo of high Andes of Ecuador

❏ *Thomasomys pyrrhonotus*
Thomas' Oldfield Mouse

Andes of S Ecuador and NW Peru

❏ *Thomasomys rhoadsi* Rhoads' Oldfield Mouse

High Andes of Ecuador

❏ *Thomasomys rosalinda*
Rosalinda's Oldfield Mouse

Andes of NC Peru

❏ *Thomasomys silvestris* Forest Oldfield Mouse

W Andes of Ecuador

❏ *Thomasomys taczanowskii*
Taczanowski's Oldfield Mouse

Andes of NW Peru

❏ *Thomasomys vestitus* Dressy Oldfield Mouse

Montane cloud forests of NW Venezuela

❏ *Wilfredomys oenax* Wilfred's Mouse

Dense forests of C Uruguay to SE Brazil

❏ *Juliomys pictipes* Short-haired Julio Mouse

Dense forests of Misiones Prov. (NE Argentina) and SE Brazil

❏ *Juliomys rimofrons* Long-haired Julio Mouse

Brejo da Lapa (Serra da Mantiqueira, Minas Gerais, E Brazil)

❏ *Aepeomys fuscatus* Dusky Montane Mouse

Highlands of Cordilleras Occidental and Central (W Colombia)

❏ *Aepeomys lugens* Olive Montane Mouse

Highlands of Cordillera Oriental of NW Venezuela, Colombia and Ecuador

❏ *Aepeomys reigi* Reig's Montane Mouse

Andes of Venezuela

❏ *Phaenomys ferrugineus*
Rio de Janeiro Arboreal Rat

Rio de Janeiro region (Brazil); known only from holotype specimen collected in 19th Century, Endangered

❏ *Rhipidomys austrinus*
Southern Climbing Mouse

E Andean slopes of SC Bolivia and NW Argentina

❏ *Rhipidomys caucensis* Cauca Climbing Mouse

W Andes of Colombia

❏ *Rhipidomys couesi* Coues' Climbing Mouse

Colombia, Ecuador, Peru and Venezuela; Trinidad

❏ *Rhipidomys fulviventer*
Buff-bellied Climbing Mouse

Andes of Colombia and W Venezuela

❏ *Rhipidomys latimanus*
Broad-footed Climbing Mouse

Montane forests of Colombia and Ecuador

❏ *Rhipidomys leucodactylus*
White-footed Climbing Mouse

Lowland rainforests of S Venezuela to French Guiana, S to N Ecuador, Peru and N Brazil

❏ *Rhipidomys macconnelli*
MacConnell's Climbing Mouse

Submontane forests of SE Venezuela and W Guyana

❏ *Rhipidomys mastacalis*
Long-tailed Climbing Mouse

C & E Brazil

❏ *Rhipidomys nitela* Splendid Climbing Mouse

S Venezuela to French Guiana and NC Brazil

❏ *Rhipidomys ochrogaster*
Yellow-bellied Climbing Mouse

SE Peru

❏ *Rhipidomys scandens*
Mount Pirre Climbing Mouse

Very few records from montane rainforests of Cerro Malí, Tacarcuna and Cerro Pirre (E Panama)

❏ *Rhipidomys venezuelae*
Venezuelan Climbing Mouse

Mountains of E Colombia and N Venezuela

❏ *Rhipidomys venustus*
Charming Climbing Mouse

Montane forests of NW Venezuela

❏ *Rhipidomys wetzeli* Wetzel's Climbing Mouse

Highlands of S Venezuela

❏ *Wiedomys pyrrhorhinos* Red-nosed Mouse

Dry deciduous forests and scrub in cerrado and caatinga of Ceara to Rio Grande do Sul (E & SE Brazil)

Akodon (Akodon)

❏ *Akodon affinis* Colombian Grass Mouse

Cordillera Occidental (W Colombia)

❏ *Akodon albiventer* White-bellied Grass Mouse

Open grasslands and dry slopes of SE Peru, WC Bolivia, N Chile and NW Argentina

❏ *Akodon aliquantulus* Tucumán Grass Mouse

NW Argentina

❏ *Akodon azarai* Azara's Grass Mouse

Thorn scrub, grassland and marshes of S Bolivia, Paraguay, N Argentina, Uruguay and S Brazil

❏ *Akodon boliviensis* Bolivian Grass Mouse

Grasslands of altiplano of Andes of SE Peru and NC Bolivia

❏ *Akodon cursor* Cursorial Grass Mouse

Forests and savannas of Paraguay, NE Argentina, Uruguay and C & SE Brazil

❏ *Akodon dayi* Day's Grass Mouse

Lowlands of C & SC Bolivia

❏ *Akodon dolores* Dolorous Grass Mouse

Sierra de Córdoba (Córdoba Prov., C Argentina)

❏ *Akodon fumeus* Smoky Grass Mouse

E slope of Andes of SE Peru and W Bolivia

❏ *Akodon hershkovitzi* Hershkovitz's Grass Mouse

Steppe and coastal forests of islands of S Chilean Archipelago (Chile)

❏ *Akodon iniscatus* Intelligent Grass Mouse

SC Argentina

❏ *Akodon juninensis* Junín Grass Mouse

Highlands of Andes of C Peru

❏ *Akodon kofordi* Koford's Grass Mouse

Highlands of Cusco and Puno Depts. (SE Peru)

❏ *Akodon lindberghi* Lindbergh's Grass Mouse

Brasília region (Distrito Federal, Brazil)

❏ *Akodon markhami* Markham's Grass Mouse

Isla Wellington (Magallanes Prov., S Chile)

❏ *Akodon molinae* Molina's Grass Mouse

Riparian grasslands of C Argentina

❏ *Akodon mollis* Soft Grass Mouse

High Andes of Ecuador and NC Peru

❏ *Akodon mystax* Caparao Grass Mouse

Mt Caparao (Minas Gerais, Brazil)

❏ *Akodon neocenus* Neuquen Grass Mouse

C & S Argentina

❏ *Akodon nigrita* Blackish Grass Mouse

Rainforests of SE Paraguay, NE Argentina and SE Brazil

❏ *Akodon oenos* Wine Grass Mouse

Mendoza Prov. (Argentina)

❏ *Akodon olivaceus* Olive Grass Mouse

Brush and grasslands of N & C Chile

❏ *Akodon orophilus* El Dorado Grass Mouse

Andes of N Peru

❏ *Akodon paranaensis* Paraná Grass Mouse

SE Brazil and NE Misiones Prov. (NE Argentina)

❏ *Akodon pervalens* Robust Grass Mouse

NW Argentina

❏ *Akodon puer* Altiplano Grass Mouse

Altiplano of Andes of C Peru, W Bolivia and NW Argentina

❏ *Akodon sanctipaulensis* São Paulo Grass Mouse

N bank of Rs. Juquia and Etá (São Paulo, SE Brazil)

❏ *Akodon serrensis* Serra do Mar Grass Mouse

Forests of Misiones Prov. (NE Argentina) and SE Brazil

❏ *Akodon simulator* Grey-bellied Grass Mouse

E Andes of SC Bolivia and NW Argentina

❏ *Akodon spegazzinii* Spegazzini's Grass Mouse

E Andes of NW Argentina

❏ *Akodon subfuscus* Puno Grass Mouse

Bunchgrass habitats of Andes of SC Peru and NW Bolivia

❏ *Akodon surdus* Silent Grass Mouse

Andes of SE Peru; status uncertain

❏ *Akodon sylvanus* Forest Grass Mouse

NW Argentina

❏ *Akodon toba* Chaco Grass Mouse

Gran Chaco region of E Bolivia, W Paraguay and NW Argentina

❏ *Akodon torques* Cloud Forest Grass Mouse

Cloud forests of SE Peru

❑ *Akodon urichi* Northern Grass Mouse

Rainforests, regrowth areas and clearings of E Colombia, N Brazil and Venezuela; Trinidad & Tobago

❑ *Akodon varius* Variable Grass Mouse

E slope of Andes of W Bolivia

Akodon (*Chalcomys*)

❑ *Akodon aerosus* Highland Grass Mouse

Mid-elevation montane forests of Andes of SE Ecuador, E Peru and NW Bolivia

Akodon (*Deltamys*)

❑ *Akodon kempi* Kemp's Grass Mouse

Wet grasslands and riparian habitats of R. Paraná estuary (EC Argentina), S Uruguay and Rio Grande do Sul (S Brazil)

Akodon (*Hypsimys*)

❑ *Akodon budini* Budin's Grass Mouse

Few records from mountains of NW Argentina

❑ *Akodon siberiae* Cochabamba Grass Mouse

Cloud forests of Comarapa region (E Cochabamba Dept., C Bolivia)

Akodon (*Microxus*)

❑ *Akodon bogotensis* Bogotá Grass Mouse

Forest edge in montane cloud forests of C Colombia to W Venezuela

❑ *Akodon latebricola* Ecuadorian Grass Mouse

R. Cusutagua region in Andes of Ecuador; only known from holotype specimen

❑ *Akodon mimus* Thespian Grass Mouse

E slope of Andes of SE Peru and WC Bolivia

❑ *Bolomys amoenus* Pleasant Bolo Mouse

Highlands of Andes of SE Peru and WC Bolivia

❑ *Bolomys lactens* Rufous-bellied Bolo Mouse

Mountains of SC Bolivia and NW Argentina

❑ *Bolomys lasiurus* Hairy-tailed Bolo Mouse

Forest edge, grasslands and cerrado of C & S Brazil, E Bolivia, Paraguay and Misiones Prov. (NE Argentina)

❑ *Bolomys obscurus* Dark Bolo Mouse

Lowland wet grasslands of EC Argentina and S Uruguay

❑ *Bolomys punctulatus* Spotted Bolo Mouse

"Ecuador" (type locality doubtful); status unknown

❑ *Bolomys temchuki* Temchuk's Bolo Mouse

NE Argentina

❑ *Podoxymys roraimae* Roraima Mouse

Summit of Mt Roraima (WC Guyana)

❑ *Thalpomys cerradensis* Cerrado Mouse

Cerrado of C Brazil

❑ *Thalpomys lasiotis* Hairy-eared Cerrado Mouse

Cerrado of C Brazil

❑ *Blarinomys breviceps* Brazilian Shrew-Mouse

Deep litter in montane rainforests of Bahia to Rio de Janeiro (EC & SE Brazil)

❑ *Brucepattersonius albinasus* White-nosed Brucie

SE Brazil

❑ *Brucepattersonius griserufescens* Grey-bellied Brucie

SE Brazil

❑ *Brucepattersonius guarani* Guaraní Brucie

Misiones Prov. (NE Argentina)

❑ *Brucepattersonius igniventris* Red-bellied Brucie

SE Brazil

❑ *Brucepattersonius iheringi* Ihering's Brucie

Forests of Misiones Prov. (NE Argentina) and Rio Grande do Sul (SE Brazil)

❑ *Brucepattersonius misionensis* Misiones Brucie

Misiones Prov. (NE Argentina)

❑ *Brucepattersonius paradisus* Beautiful Brucie

Misiones Prov. (NE Argentina)

❑ *Brucepattersonius soricinus* Soricine Brucie

SE Brazil

❏ *Oxymycterus akodontius* Argentine Hocicudo

NW Argentina

❏ *Oxymycterus amazonicus* Amazon Hocicudo

Lower R. Tapajós (E Amazonian Brazil)

❏ *Oxymycterus angularis* Angular Hocicudo

Pernambuco (E Brazil)

❏ *Oxymycterus caparaoe* Caparao Hocicudo

Caparao region (Minas Gerais, Brazil)

❏ *Oxymycterus delator* Spy Hocicudo

Lowland marshes of E Paraguay

❏ *Oxymycterus hiska* Small Hocicudo

Yanahuaya region (Puna Dept., SC Peru)

❏ *Oxymycterus hispidus* Hispid Hocicudo

Misiones Prov. (NE Argentina) and E Brazil

❏ *Oxymycterus hucucha* Quechuan Hocicudo

Compara region (Cochabamba Dept., C Bolivia)

❏ *Oxymycterus inca* Incan Hocicudo

Lower montane forests of C & SE Peru and WC Bolivia

❏ *Oxymycterus josei* José's Hocicudo

Uruguay

❏ *Oxymycterus nasutus* Long-nosed Hocicudo

Uruguay and SE Brazil

❏ *Oxymycterus paramensis* Páramo Hocicudo

Mid- and high-elevation forests and brush of E Andes of SE Peru, WC Bolivia and NW Argentina

❏ *Oxymycterus roberti* Robert's Hocicudo

Rio Jordao region (Minas Gerias, Brazil)

❏ *Oxymycterus rufus* Red Hocicudo

Wet grasslands, riparian habitats and scrub of SE Paraguay, NE & EC Argentina, Uruguay and SE Brazil

❏ *Juscelinomys candango*
Brasília Burrowing Mouse

Brasília region (Distrito Federal, E Brazil)

❏ *Juscelinomys guaporensis*
Rio Guaporé Burrowing Mouse

Open savanna woodland of Parque Nacional Noel Kempff Mercado (E Bolivia)

❏ *Juscelinomys huanchacae*
Huanchaca Burrowing Mouse

Dense grassland of Parque Nacional Noel Kempff Mercado (E Bolivia)

❏ *Lenoxus apicalis* Andean Rat

Montane forests of E slope of Andes of SE Peru and W Bolivia

❏ *Abrothrix illuteus* Grey Grass Mouse

Streamside brush of Jujuy and Tucumán Provs. (NW Argentina)

❏ *Abrothrix lanosus* Woolly Grass Mouse

Cool damp forests of extreme S Chile and S Argentina

❏ *Abrothrix longipilis* Long-haired Grass Mouse

Forests, brush and grasslands of C & S Chile and SW Argentina

❏ *Abrothrix mansoensis* Manso Grass Mouse

Río Negro Prov. (WC Argentina)

❏ *Abrothrix sanborni* Sanborn's Grass Mouse

Forests of Los Lagos region (SC Chile) and adjacent SW Argentina

❏ *Abrothrix xanthorhinus*
Yellow-nosed Grass Mouse

Forests, forest edge and wet grasslands of S Chile and S Argentina

❏ *Chroeomys andinus* Andean Altiplano Mouse

Rocky slopes of Andes of Peru, Bolivia, N Chile and N Argentina

❏ *Chroeomys jelskii* Jelski's Altiplano Mouse

High altiplano of Andes of S Peru, WC Bolivia and NW Argentina

❏ *Chelemys macronyx*
Andean Long-clawed Mouse

Forests and scrub of Andes of S Chile and S Argentina

❏ *Chelemys megalonyx* Large Long-clawed Mouse

Humid soils of C & S Chile

❏ *Geoxus valdivianus* Long-clawed Mole-Mouse

Rainforests, marshes and meadows of SC Chile incl. Chiloé I., and SW Argentina

❏ *Notiomys edwardsii*
Milne-Edwards' Long-clawed Mouse

Sapium brush of S Argentina

❏ *Pearsonomys annectens*
Pearson's Long-clawed Mouse

Chile

❏ *Calomys boliviae* Bolivian Vesper Mouse

W Bolivia

❏ *Calomys callidus* Crafty Vesper Mouse

E Paraguay and EC Argentina

❏ *Calomys callosus* Large Vesper Mouse

Forest clearings, regrowth areas, farmland and riparian habitats of E Bolivia, Paraguay, N Argentina and S & E Brazil

❏ *Calomys expulsus* Rejected Vesper Mouse

Minas Gerais (EC Brazil)

❏ *Calomys hummelincki*
Hummelinck's Vesper Mouse

Llanos and semiarid habitats of NE Colombia and N Venezuela; Aruba and Curaçao Is.

❏ *Calomys laucha* Small Vesper Mouse

Savannas and grasslands of SE Bolivia, WC & SE Brazil, Paraguay, NE & EC Argentina and Uruguay

❏ *Calomys lepidus* Andean Vesper Mouse

Grasslands of altiplano of Andes of C Peru to N Chile and NW Argentina

❏ *Calomys musculinus* Drylands Vesper Mouse

Pampas of NW & C Argentina (and E Paraguay?)

❏ *Calomys sorellus* Peruvian Vesper Mouse

Highlands of Andes of SC Peru

❏ *Calomys tener* Delicate Vesper Mouse

EC Brazil

❏ *Calomys venustus* Pretty Vesper Mouse

NW Argentina

❏ *Eligmodontia moreni* Monte Gerbil-Mouse

Andes of NW Argentina

❏ *Eligmodontia morgani* Morgan's Gerbil-Mouse

S Chile and S Argentina

❏ *Eligmodontia puerulus* Andean Gerbil-Mouse

Altiplano of S Peru to NE Chile and NW Argentina

❏ *Eligmodontia typus* Highland Gerbil-Mouse

C Chile and C Argentina

Graomys (Andalgalomys)

❏ *Graomys olrogi* Olrog's Chaco Mouse

Semi-deserts of Andalgala region (Catamarca Prov., NW Argentina)

❏ *Graomys pearsoni* Pearson's Chaco Mouse

Dry grasslands of Gran Chaco region of SE Bolivia and NW Paraguay

❏ *Graomys roigi* Roig's Chaco Mouse

Argentina

Graomys (Graomys)

❏ *Graomys domorum* Pale Leaf-eared Mouse

Forests and regrowth areas of Andes of SC Bolivia and NW Argentina

❏ *Graomys edithae* Edith's Leaf-eared Mouse

La Rioja and Catamarca Provs. (NW Argentina)

❏ *Graomys griseoflavus* Grey Leaf-eared Mouse

Dry thorn scrub, riparian habitats and farmland of S Bolivia, WC Brazil, Paraguay and N & C Argentina

❏ *Tapecomys primus* Tapecua Rat

Tapecua region (Tarija Dept., Bolivia)

❏ *Salinomys delicatus* Delicate Salt-flat Mouse

Argentina

❏ *Phyllotis amicus* Friendly Leaf-eared Mouse

Lowlands of WC Peru

❏ *Phyllotis andium* Andean Leaf-eared Mouse

Andes from C Ecuador to C Peru

❏ *Phyllotis bonaeriensis*
Buenos Aires Leaf-eared Mouse

Buenos Aires Prov. (Argentina)

❏ *Phyllotis caprinus* **Capricorn Leaf-eared Mouse**

Thorn scrub of E slope of Andes of S Bolivia and NW Argentina

❏ *Phyllotis darwini* **Darwin's Leaf-eared Mouse**

Forests and brush of coastal C Chile

❏ *Phyllotis definitus* **Definitive Leaf-eared Mouse**

Mountains of Macate region (Ancash Dept., C Peru)

❏ *Phyllotis gerbillus* **Gerbil Leaf-eared Mouse**

Sechura Desert (NW Peru)

❏ *Phyllotis haggardi* **Haggard's Leaf-eared Mouse**

Andes of NC Ecuador

❏ *Phyllotis limatus*
Narrow-toothed Leaf-eared Mouse

W slope of Andes from C Peru to N Chile

❏ *Phyllotis magister* **Master Leaf-eared Mouse**

Brush and rocky areas of W slope of Andes from C Peru to N Chile

❏ *Phyllotis osgoodi* **Osgood's Leaf-eared Mouse**

Altiplano of Andes of Arica Prov. (NE Chile)

❏ *Phyllotis osilae* **Bunchgrass Leaf-eared Mouse**

High Andean bunchgrass habitats from SC Peru to NW Argentina

❏ *Phyllotis wolffsohni* **Wolffsohn's Leaf-eared Mouse**

E slope of Andes of C Bolivia

❏ *Phyllotis xanthopygus*
Yellow-rumped Leaf-eared Mouse

Forests and brush of C Peru to S Chile and Argentina

❏ *Loxodontomys micropus*
Southern Big-eared Mouse

Dense forests and scrub of S Chile and SW Argentina

❏ *Auliscomys boliviensis* **Bolivian Big-eared Mouse**

Open rocky slopes of high altiplano of Andes of S Peru, W Bolivia and extreme N Chile

❏ *Auliscomys pictus* **Painted Big-eared Mouse**

High Andes from C Peru to NW Bolivia

❏ *Auliscomys sublimis* **Andean Big-eared Mouse**

Open rocky slopes of high altiplano of Andes of S Peru, SW Bolivia, N Chile and NW Argentina

❏ *Galenomys garleppi* **Garlepp's Mouse**

Arid areas of high altiplano of Andes of S Peru, SW Bolivia and N Chile

❏ *Chinchillula sahamae*
Altiplano Chinchilla-Mouse

Rocky areas of high altiplano of Andes of S Peru, SW Bolivia, N Chile and NW Argentina

❏ *Punomys kofordi* **Koford's Puna Mouse**

Highlands E of Lake Titicaca, in regions of Abra Aricema and Limbani valley in Cordillera de Carabaya (Puno Dept., S Peru)

❏ *Punomys lemminus* **Puna Mouse**

Rocky treeless areas in high altiplano W of Lake Titicaca (Puno Dept., S Peru)

❏ *Andinomys edax* **Andean Mouse**

Dense streamside vegetation and rocky areas of altiplano of Andes of S Peru, SW Bolivia, N Chile and NW Argentina

❏ *Irenomys tarsalis* **Chilean Climbing Mouse**

Dense wet forests with bamboo thickets of SC Chile and SW Argentina

❏ *Euneomys chinchilloides*
Patagonian Chinchilla-Mouse

S Chile

❏ *Euneomys fossor* **Burrowing Chinchilla-Mouse**

"Salta Prov." (Argentina) (type locality doubtful)

❏ *Euneomys mordax* **Biting Chinchilla-Mouse**

C Chile and WC Argentina

❏ *Euneomys petersoni*
Peterson's Chinchilla-Mouse

S Chile and S Argentina

❏ *Neotomys ebriosus* **Andean Swamp Rat**

Dense grasslands, streamsides and marshes of altiplano of Andes of C & S Peru, W Bolivia, N Chile and NW Argentina

❏ *Reithrodon auritus* Bunny Rat

Open grasslands and brush of S Chile, Argentina and Uruguay

❏ *Sigmodon alleni* Allen's Cotton Rat

SW & S Mexico

❏ *Sigmodon alstoni* Alston's Cotton Rat

Lowland grasslands of NE Colombia to Suriname and N Brazil

❏ *Sigmodon arizonae* Arizona Cotton Rat

Damp grasslands and fields of S Arizona (USA) to WC Mexico

❏ *Sigmodon fuliventer* Tawny-bellied Cotton Rat

Grasslands and scrub of SE Arizona and W New Mexico (USA) to C Mexico

❏ *Sigmodon hispidus* Hispid Cotton Rat

Grasslands and farmland of SC & SE USA to N Colombia and Venezuela

❏ *Sigmodon inopinatus* Unexpected Cotton Rat

High Andean summits of C Ecuador

❏ *Sigmodon leucotis* White-eared Cotton Rat

Mexico

❏ *Sigmodon mascotensis* Jaliscan Cotton Rat

SW & S Mexico

❏ *Sigmodon ochrognathus* Yellow-nosed Cotton Rat

Arid upland rocky habitats of extreme SC USA and NC Mexico

❏ *Sigmodon peruanus* Peruvian Cotton Rat

Coastal plains of W Ecuador and NW Peru

❏ *Scapteromys tumidus* Argentine Swamp Rat

Flooded grasslands and saltmarshes of Paraguay, NE & EC Argentina, Uruguay and SE Brazil

❏ *Kunsia fronto* Fossorial Giant Rat

Marshes of NE Argentina and EC Brazil

❏ *Kunsia tomentosus* Woolly Giant Rat

Deep soils of NE Bolivia and WC Brazil

❏ *Bibimys chacoensis* Chaco Crimson-nosed Rat

Few records from wet grasslands of NE Argentina

❏ *Bibimys torresi* Torres' Crimson-nosed Rat

Few records from NE & EC Argentina

❏ *Neusticomys monticolus* Montane Fish-eating Rat

Mountains of Colombia and Ecuador

❏ *Neusticomys mussoi* Musso's Fish-eating Rat

Pregonero region (Táchira state, Venezuela); Endangered

❏ *Neusticomys oyapocki* Oyapock Fish-eating Rat

Riparian habitats of French Guiana; Endangered

❏ *Neusticomys peruviensis* Peruvian Fish-eating Rat

Balta region (Ucayali Dept., Peru); only known from holotype specimen, Endangered

❏ *Neusticomys venezuelae*
Venezuelan Fish-eating Rat

Montane streams of S Venezuela; Endangered

❏ *Rheomys mexicanus* Mexican Water Mouse

Oaxaca (S Mexico)

❏ *Rheomys raptor* Goldman's Water Mouse

Fast-flowing forest streams in mountains of Costa Rica and W & E Panama

❏ *Rheomys thomasi* Thomas' Water Mouse

Streams in forests and regrowth areas of Chiapas (SE Mexico), W Guatemala and El Salvador

❏ *Rheomys underwoodi*
Underwood's Water Mouse

Clear montane streams of C Costa Rica and W Panama

❏ *Anotomys leander* Ecuador Fish-eating Rat

High Andes of Pichincha Prov. (C Ecuador); Endangered

❏ *Chibchanomys orcesi*
Orces' Chibchan Water Mouse

Lake Luspa region, Las Cajos plateau (Azuay Prov., Ecuador)

❏ *Chibchanomys trichotis* Chibchan Water Mouse

Streams in montane cloud forests of NE Colombia and W Venezuela (and Peru?)

❏ *Ichthyomys hydrobates* Common Crab-eating Rat

Fast-flowing forest streams in mountains of Colombia, W Venezuela and Ecuador

❏ *Ichthyomys pittieri* Pittier's Crab-eating Rat

Fast-flowing montane forest streams of NC Venezuela

❏ *Ichthyomys stolzmanni*
Stolzmann's Crab-eating Rat

Fast-flowing montane forest streams of Ecuador and Peru

❏ *Ichthyomys tweedii* Northern Crab-eating Rat

Fast-flowing forest streams of C Panama and W Ecuador

Subfamily: CALOMYSCINAE
(Mouse-like Hamsters—6)

❏ *Calomyscus bailwardi* Iranian Mouse-like Hamster

Iran

❏ *Calomyscus baluchi*
Baluchistan Mouse-like Hamster

NC & E Afghanistan, W Pakistan

❏ *Calomyscus hotsoni*
Hotson's Mouse-like Hamster

Gwambuk Kaul region (Baluchistan, Pakistan); Endangered

❏ *Calomyscus mystax* Afghan Mouse-like Hamster

NC & NE Iran, S Turkmenistan and NW Afghanistan

❏ *Calomyscus tsolovi* Tsolov's Mouse-like Hamster

SW Syria

❏ *Calomyscus urartensis*
Urartsk Mouse-like Hamster

Azerbaijan and NW Iran

Subfamily: DENDROMURINAE
(Climbing Mice—23)

❏ *Dendromus insignis*
Remarkable Climbing Mouse

Highlands of Ethiopia to W Uganda, Rwanda and NE Tanzania

❏ *Dendromus kahuziensis*
Mount Kahuzi Climbing Mouse

Kivu region (E D.R. Congo)

❏ *Dendromus kivu* Kivu Climbing Mouse

E D.R. Congo and W Uganda

❏ *Dendromus lovati* Lovat's Climbing Mouse

Highlands of C Ethiopia

❏ *Dendromus melanotis* Grey Climbing Mouse

Grasslands of Guinea to Ethiopia, S to South Africa

❏ *Dendromus mesomelas* Brants' Climbing Mouse

Grasslands of SE D.R. Congo and EC Tanzania to NE Namibia, NW Botswana and S & E South Africa

❏ *Dendromus messorius* Banana Climbing Mouse

Benin and Nigeria to Sudan and Kenya

❏ *Dendromus mystacalis*
Chestnut Climbing Mouse

Grasslands of S Sudan and Ethiopia to Zimbabwe and E South Africa

❏ *Dendromus nyikae* Nyika Climbing Mouse

Locally in grasslands of Angola, SC D.R. Congo and SW Tanzania to NE South Africa

❏ *Dendromus oreas* Cameroon Climbing Mouse

Mountains of W Cameroon

❏ *Dendromus vernayi* Vernay's Climbing Mouse

Chitau region (EC Angola); Critically Endangered

❏ *Steatomys caurinus* Northwestern Fat Mouse

Senegal to C Nigeria

❏ *Steatomys cuppedius* Dainty Fat Mouse

Senegal to SC Niger and NC Nigeria

❏ *Steatomys jacksoni* Jackson's Fat Mouse

Ghana and Nigeria

❏ *Steatomys krebsii* Krebs' Fat Mouse

Locally in sandy habitats of S Angola and W Zambia to N Botswana and SW & NC South Africa

❏ *Steatomys parvus* Tiny Fat Mouse

Sandy habitats of S Sudan, S Ethiopia and Somalia to N Namibia, NW Botswana and SE South Africa

❏ *Steatomys pratensis* Common Fat Mouse

Sandy habitats of Cameroon, D.R. Congo and SW Sudan to E South Africa

❏ *Malacothrix typica* Long-eared Mouse

Shortgrass meadows and karoo scrub of SW Angola, Namibia, S Botswana and W & C South Africa

❏ *Megadendromus nikolausi* Nikolaus' Mouse

Montane grasslands and sagebrush in Bale Mtns (C Ethiopia)

❏ *Dendroprionomys rousseloti*
Velvet Climbing Mouse

Only recorded from zoological gardens on banks of R. D.R. Congo at Brazzaville (SE Rep. Congo)

❏ *Prionomys batesi* Dollman's Tree Mouse

Rainforests of SW Cameroon and S Central African Republic

❏ *Deomys ferrugineus* Congo Forest Mouse

Seasonally flooded forests of Cameroon and Rep. Congo to Uganda and Rwanda

❏ *Leimacomys buettneri*
Groove-toothed Forest Mouse

Forest-savannas (?) of Bismarckburg region (Togo); possibly extinct

Subfamily: LOPHIOMYINAE (Crested Rat—1)

❏ *Lophiomys imhausi* Crested Rat

Rocky habitats in woods of E Sudan, Ethiopia and Somalia to Tanzania

Subfamily: CRICETOMYINAE (Pouched Rats—6)

❏ *Saccostomus campestris* Pouched Mouse

C Angola and SW Tanzania to South Africa

❏ *Saccostomus mearnsi* Mearns' Pouched Mouse

S Ethiopia and S Somalia to NE Tanzania

❏ *Beamys hindei* Long-tailed Pouched Rat

Sandy riparian habitats in forests of S Kenya and NE Tanzania

❏ *Beamys major* Greater Long-tailed Pouched Rat

Sandy riparian habitats in forests of NE Zambia, Malawi and S Tanzania

❏ *Cricetomys emini* Giant Pouched Rat

Lowland rainforests from Sierra Leone to Gabon and S Uganda

❏ *Cricetomys gambianus*
Gambian Giant Pouched Rat

Savannas, woods and gardens of Senegal and S Sudan to E South Africa

Subfamily: CRICETINAE (Hamsters—18)

❏ *Cricetus cricetus* Black-bellied Hamster

Farmland of EC Europe, W Siberia, N Kazakhstan and NW China

❏ *Cricetulus alticola* Tibetan Dwarf Hamster

NW India, W Tibet and W Nepal

❏ *Cricetulus barabensis* Striped Dwarf Hamster

S Siberia, Mongolia, N China and Korea

❏ *Cricetulus kamensis* Kam Dwarf Hamster

Tibet

❏ *Cricetulus longicaudatus*
Long-tailed Dwarf Hamster

C Siberia, Kazakhstan, Mongolia and W China

❏ *Cricetulus migratorius* Grey Dwarf Hamster

Steppes and deserts of SE Europe and Middle East to N China and Pakistan

❏ *Cricetulus sokolovi* Sokolov's Dwarf Hamster

SW Mongolia and Nei Mongol (N China)

❏ *Mesocricetus auratus* Golden Hamster

Aleppo region (Syria); Endangered

❏ *Mesocricetus brandti* Brandt's Hamster

Turkey, Transcaucasia and Middle East

❏ *Mesocricetus newtoni* Romanian Hamster

Locally in dry steppes and fields of E Romania and N Bulgaria

❏ *Mesocricetus raddei* Ciscaucasian Hamster

Steppes of S Russia N of Caucasus Mtns

❏ *Phodopus campbelli* Campbell's Hamster

E Siberia, Mongolia and N China

❏ *Phodopus roborovskii* Desert Hamster

E Kazakhstan, C Siberia, Mongolia and N China

❏ *Phodopus sungorus* Dzhungarian Hamster

E Kazakhstan and SW Siberia

❏ *Tscherskia triton* Greater Long-tailed Hamster

E Siberia, NE China and Korea

❏ *Cansumys canus* Gansu Hamster

Gansu and Shaanxi (China)

❏ *Allocricetulus curtatus* Mongolian Hamster

Steppes of Mongolia and N China

❏ *Allocricetulus eversmanni* Eversmann's Hamster

Steppes of N Kazakhstan

Subfamily: ARVICOLINAE
(Voles, Lemmings and Muskrats—140)

❏ *Arvicola sapidus* Southwestern Water Vole

Streams and marshes of Portugal, Spain and France

❏ *Arvicola terrestris* European Water Vole

Wetlands of Europe to E Siberia, S to Mediterranean Sea, Middle East and NW China

Microtus (Agricola)

❏ *Microtus agrestis* Field Vole

Marshes, damp grasslands, meadows and woods of Europe to C Siberia and NW China

❏ *Microtus cabrerae* Cabrera's Vole

Damp pastures and fields of Portugal and Spain

Microtus (Alexandromys)

❏ *Microtus evoronensis* Evorsk Vole

Lake Evoron region (Khabarovsk Krai, Russia); Critically Endangered

❏ *Microtus fortis* Reed Vole

E Siberia to E China

❏ *Microtus hyperboreus* North Siberian Vole

NE Siberia

❏ *Microtus maximowiczii* Maximowicz's Vole

E Siberia, E Mongolia and NE China

❏ *Microtus middendorffii* Middendorff's Vole

NC Siberia

❏ *Microtus mongolicus* Mongolian Vole

E Siberia, Mongolia and NE China

❏ *Microtus mujanensis* Muisk Vole

Vitim R. region (Buryat, Russia); Critically Endangered

❏ *Microtus sachalinensis* Sakhalin Vole

Sakhalin I. (Russia)

Microtus (Aulacomys)

❏ *Microtus chrotorrhinus* Rock Vole

Damp rocky areas in forests of SE Canada and Appalachian Mtns from Maine to W North Carolina (USA)

❏ *Microtus longicaudus* Long-tailed Vole

Montane forests and brush of E Alaska, W Canada and W USA

❏ *Microtus richardsoni*
Richardson's Vole (Water Vole)

High-elevation waterside meadows of SW Canada and NW USA

❏ *Microtus xanthognathus* Taiga Vole

Taiga zone of EC Alaska (USA) and NW & NC Canada

Microtus (Herpetomys)

❏ *Microtus guatemalensis* Guatemalan Vole

Montane forests and clearings of Cerro Tzontehuitz (Chiapas, SE Mexico) and S Guatemala

Microtus (Lasiopodomys)

❏ *Microtus brandtii* Brandt's Vole

E Siberia, Mongolia and NE China

❏ *Microtus fuscus* Plateau Vole

Qinghai (NC China)

❏ *Microtus mandarinus* Mandarin Vole

E Siberia, N Mongolia, C & NE China, and Korea

Microtus (Microtus)

❏ *Microtus arvalis* Common Vole

Meadows, pastures and fields of C Europe and Russia E to Ural Mtns

❏ *Microtus dogramacii* Dogramaci's Vole

C Anatolia (Turkey)

❏ *Microtus guentheri* Günther's Vole

Dry meadows and pastures of S Yugoslavia, Macedonia, E Greece, SE Bulgaria and W Turkey

❏ *Microtus irani* Persian Vole

E Libya and Middle East to E Turkey and Turkmenistan

❏ *Microtus kermanensis* Baluchistan Vole

SE Iran; Endangered

❏ *Microtus kirgisorum* Tien Shan Vole

SE Turkmenistan, Tajikistan, Kyrgyzstan and S Kazakhstan

❏ *Microtus obscurus* Altai Vole

Russia and the Caucasus Mtns to C Siberia and NW China

❏ *Microtus rossiaemeridionalis* Sibling Vole

Open forests, meadows and fields of E Europe to Ural Mtns (Russia)

❏ *Microtus socialis* Social Vole

Ukraine and Middle East to Kazakhstan and NW China

❏ *Microtus transcaspicus* Transcaspian Vole

N Iran, S Turkmenistan and N Afghanistan

Microtus (Mynomes)

❏ *Microtus breweri* Beach Vole

Beach grass stands of Muskeget I. (Massachusetts, USA)

❏ *Microtus californicus* California Vole

Grasslands and damp meadows of W Oregon and California (USA) and Baja California Norte (Mexico)

❏ *Microtus canicaudus* Grey-tailed Vole

Grasslands and fields of Willamette Valley (extreme SW Washington and NW Oregon, USA)

❏ *Microtus mexicanus* Mexican Vole

Mexico

❏ *Microtus mogollonensis* Mogollon Vole

Coniferous forest meadows of EC Arizona and C New Mexico (USA)

❏ *Microtus montanus* Montane Vole

Woods and damp meadows of SC British Columbia (Canada) and W USA

❏ *Microtus oaxacensis* Tarabundi Vole

Sierra de Juarez (NC Oaxaca, Mexico)

❏ *Microtus oregoni* Creeping Vole

Coniferous forest clearings of extreme SW British Columbia (Canada) to NW California (USA)

❏ *Microtus pennsylvanicus* Meadow Vole

Grasslands of Alaska, Canada and C & E USA

❏ *Microtus townsendii* Townsend's Vole

Damp meadows and marshes of SW British Columbia (Canada) to NW California (USA)

Microtus (Neodon)

❏ *Microtus irene* Chinese Scrub Vole

Mountains of N Myanmar and SW China

❏ *Microtus juldaschi* Juniper Vole

Mountains of Hindu Kush, Pamirs, Tien Shan and NW Tibet (Afghanistan to W China)

❏ *Microtus sikimensis* Sikkim Vole

Himalayas from W Nepal to Bhutan and E Tibet

Microtus (Orthriomys)

❏ *Microtus umbrosus* Zempoaltepec Vole

Mt Zempoaltepec (Oaxaca, Mexico)

Microtus (Pallasiinus)

❏ *Microtus limnophilus* Lacustrine Vole

W Mongolia to Qinghai (NC China)

❏ *Microtus montebelli* Japanese Grass Vole

Shikotan I. (Kuril Is., Russia) and Japan

❏ *Microtus oeconomus* Root Vole (Tundra Vole)

Wet meadows, riparian habitats and tundra of C & N Europe to E Siberia; Alaska (USA) and NW Canada

Microtus (Pedomys)

❏ *Microtus ochrogaster* Prairie Vole

Prairies and fields of SC Canada and C USA

Microtus (Phaiomys)

❏ *Microtus leucurus* Blyth's Vole

Himalayas, Tibet and NC China

Microtus (Pitymys)

❏ *Microtus pinetorum* Woodland Vole

Forests of S Ontario (Canada) and E USA

❏ *Microtus quasiater* Jalapan Pine Vole

C & S Mexico

Microtus (Stenocranius)

❏ *Microtus abbreviatus* Insular Vole

Low-elevation grasslands of Hall and St Matthew Is. (Alaska, USA)

❏ *Microtus gregalis* Narrow-headed Vole

Siberia, Kazakhstan, Kyrgyzstan, N Mongolia, NW & NE China

❏ *Microtus miurus* Singing Vole

Willow scrub in tundra of Alaska (USA) and NW Canada

Microtus (Terricola)

❏ *Microtus bavaricus* Bavarian Pine Vole

Austria; formerly S Germany (extinct)

❏ *Microtus daghestanicus* Daghestan Pine Vole

S Daghestan (Caucasus Mtns, Russia)

❏ *Microtus duodecimcostatus*
Mediterranean Pine Vole

Soils in fields and meadows of S Portugal, Spain and S France

❏ *Microtus felteni* Balkan Pine Vole

Montane forests and fields of S Yugoslavia, Macedonia, Albania and N Greece

❏ *Microtus gerbei* Pyrenean Pine Vole

Pastures and fields of NE Spain and SW France

❏ *Microtus lusitanicus* Lusitanian Pine Vole

Soils in pastures and fields of Portugal, NW Spain and extreme SW France

❏ *Microtus majori* Major's Pine Vole

S Yugoslavia and Greece to Caucasus Mtns

❏ *Microtus multiplex* Alpine Pine Vole

Forest clearings and damp meadows of SE France, S Alps and N Italy to W Yugoslavia

❏ *Microtus nasarovi* Nasarov's Pine Vole

NE Caucasus Mtns (Russia)

❏ *Microtus savii* Savi's Pine Vole

Widespread in extreme SE France, extreme S Switzerland and Italy incl. Sicily

❏ *Microtus schelkovnikovi*
Schelkovnikov's Pine Vole

S Azerbaijan

❏ *Microtus subterraneus* Common Pine Vole

Forests, meadows and rocky areas of C & SE Europe to NE Russia

❏ *Microtus tatricus* Tatra Pine Vole

Montane coniferous forests and damp meadows of Carpathian Mtns (S Poland, Slovakia, W Ukraine and N Romania)

❏ *Microtus thomasi* Thomas' Pine Vole

Soils of meadows and fields of S Bosnia-Herzegovina to S Greece

❏ *Lemmiscus curtatus* Sagebrush Vole

Sagebrush and arid grasslands of interior SW Canada and W USA

❏ *Blanfordimys afghanus* Afghan Vole

Turkmenistan, Uzbekistan, Tajikistan and Afghanistan

❏ *Blanfordimys bucharicus* Bucharian Vole

Mountains of SW Tajikistan

❏ *Chionomys gud* Caucasian Snow Vole

NE Turkey and Caucasus Mtns

❏ *Chionomys nivalis* European Snow Vole

Alpine and rocky areas in mountains of S Europe to Caucasus Mtns and Iran

❏ *Chionomys roberti* Robert's Snow Vole

Forests of NE Turkey and W Caucasus Mtns

❏ *Proedromys bedfordi* Duke of Bedford's Vole

Gansu and Sichuan (C China)

❑ *Volemys clarkei* Clarke's Vole

Mountains of N Myanmar and Yunnan (China)

❑ *Volemys kikuchii* Taiwan Vole

Highlands of Taiwan

❑ *Volemys millicens* Sichuan Vole

Mountains of E Tibet and Sichuan (C China)

❑ *Volemys musseri* Marie's Vole

Mountains of Qionglai Shan (W Sichuan, C China)

❑ *Lagurus lagurus* Common Steppe Lemming

Steppes from Ukraine to W Mongolia and NW China

❑ *Eolagurus luteus* Yellow Steppe Lemming

W Mongolia and NW China

❑ *Eolagurus przewalskii*
Przewalski's Steppe Lemming

S Mongolia and N China

❑ *Ondatra zibethicus* Muskrat

Wetlands of Alaska, Canada, USA and extreme N Mexico; feral C & NE Europe and Argentina

Phenacomys (Arborimus)

❑ *Phenacomys albipes* White-footed Vole

Riparian alder forests of coastal W Oregon and NW California (USA)

❑ *Phenacomys longicaudus* Red Tree Vole

Tree-tops in coniferous forests of coastal W Oregon and extreme NW California (USA)

❑ *Phenacomys pomo* Sonoma Tree Vole

Tree-tops in coniferous forests of coastal NW California (USA)

Phenacomys (Phenacomys)

❑ *Phenacomys intermedius* Western Heather Vole

Boreal forests and meadows of SW Canada and mountains of W USA

❑ *Phenacomys ungava* Eastern Heather Vole

Boreal forests and scrub of Canada

❑ *Dinaromys bogdanovi* Balkan Snow Vole

Rocky areas in mountains of Croatia, Bosnia-Herzegovina, S Yugoslavia, ?Albania and Macedonia

❑ *Hyperacrius fertilis* True's Vole

N Pakistan and Kashmir (NW India)

❑ *Hyperacrius wynnei* Murree Vole

N Pakistan

❑ *Alticola albicauda* White-tailed Mountain Vole

Himalayas of NW India

❑ *Alticola argentatus* Silver Mountain Vole

Mountains of Hindu Kush (Afghanistan), Pamirs (Tajikistan), Tien Shan and NW China

❑ *Alticola barakshin* Gobi Altai Mountain Vole

SC Siberia and Mongolia

❑ *Alticola macrotis* Large-eared Vole

C Siberia and Mongolia

❑ *Alticola montosa* Central Kashmir Vole

Kashmir (NW India)

❑ *Alticola roylei* Royle's Mountain Vole

Himalayas of N India

❑ *Alticola semicanus* Mongolian Silver Vole

C Siberia and N Mongolia

❑ *Alticola stoliczkanus* Stoliczka's Mountain Vole

NW India and Nepal to Tibet and N China

❑ *Alticola stracheyi* Strachey's Mountain Vole

Himalayas from E Kashmir and N Nepal to Tibet and NE India

❑ *Alticola tuvinicus* Tuva Silver Vole

C Siberia, NW China and NW Mongolia

❑ *Alticola lemminus* Lemming Vole

NE Siberia

❑ *Alticola strelzowi* Flat-headed Vole

Kazakhstan, NW China and NW Mongolia

❏ *Eothenomys chinensis* Pratt's Vole

Sichuan and Yunnan (C & SW China)

❏ *Eothenomys custos* Southwest China Vole

Sichuan and Yunnan (C & SW China)

❏ *Eothenomys eva* Gansu Vole

Mountains of C China

❏ *Eothenomys inez* Kolan Vole

Shaanxi and Shanxi (C China)

❏ *Eothenomys melanogaster* Père David's Vole

N Myanmar and N Thailand to NC China and Taiwan

❏ *Eothenomys olitor* Chaotung Vole

Mountains of Yunnan (China)

❏ *Eothenomys proditor* Yulung Shan Vole

Mountains of Sichuan and Yunnan (C & SW China)

❏ *Eothenomys regulus* Royal Vole

Korea

❏ *Eothenomys shanseius* Shansei Vole

Shanxi and Hebei (EC China)

❏ *Clethrionomys californicus*
Western Red-backed Vole

Coniferous forests of W Oregon and NW California (USA)

❏ *Clethrionomys centralis*
Tien Shan Red-backed Vole

Tien Shan Mtns (Kyrgyzstan, SE Kazakhstan and NW China)

❏ *Clethrionomys gapperi* Southern Red-backed Vole

Deep litter in forests of Canada, W, NC & NE USA

❏ *Clethrionomys glareolus* Bank Vole

Europe to C Siberia, S to W Turkey and N Kazakhstan

❏ *Clethrionomys rufocanus* Grey-sided Vole

Forests and tundra of Scandinavia to E Siberia and N Japan

❏ *Clethrionomys rutilus* Ruddy Vole

Taiga and tundra zones of NE Scandinavia to E Siberia, NE China and Japan; Alaska and NW Canada

❏ *Clethrionomys sikotanensis* Shikotan Vole

Kuril Is. (Russia)

❏ *Phaulomys andersoni*
Japanese Red-backed Vole

Honshu (C Japan)

❏ *Phaulomys smithii* Smith's Vole

Japan

❏ *Prometheomys schaposchnikowi*
Long-clawed Mole-Vole

SW Caucasus Mtns (NE Turkey and Georgia)

Ellobius (Afganomys)

❏ *Ellobius fuscocapillus* Southern Mole-Vole

E Iran and S Turkmenistan to Afghanistan and W Pakistan

❏ *Ellobius lutescens* Transcaucasian Mole-Vole

E Turkey, S Caucasus Mtns and NW Iran

Ellobius (Ellobius)

❏ *Ellobius alaicus* Alai Mole-Vole

Alai Mtns (S Kyrgyzstan); Endangered

❏ *Ellobius talpinus* Northern Mole-Vole

Steppes of S Ukraine and S Russia to Kazakhstan and Turkmenistan

❏ *Ellobius tancrei* Zaisan Mole-Vole

NE Turkmenistan and Uzbekistan to Mongolia and NE China

❏ *Synaptomys borealis* Northern Bog Lemming

Wet meadows, bogs and tundra zone of Alaska, Canada and extreme NW & NE USA

❏ *Synaptomys cooperi* Southern Bog Lemming

Meadows and forest clearings of SE Canada and NC & NE USA

❏ *Lemmus amurensis* Amur Lemming

E Siberia

❏ *Lemmus lemmus* Norway Lemming

Tundra and forests of N Scandinavia and NW Russia

❏ *Lemmus sibiricus* Siberian Brown Lemming

Tundra of N Europe to E Siberia

❏ *Lemmus trimucronatus*
North American Brown Lemming

Tundra zone of Alaska (USA) and W, NC & NE Canada

❏ *Myopus schisticolor* Wood Lemming

Coniferous mossy forests of Scandinavia to E Siberia, N
Mongolia and NE China

❏ *Neofiber alleni* Round-tailed Muskrat

Freshwater marshes of SE Georgia and peninsular Florida
(USA)

❏ *Dicrostonyx groenlandicus*
Northern Collared Lemming

Arctic tundra of W & N Alaska (USA), far N Canada and NW
Greenland

❏ *Dicrostonyx hudsonius*
Ungava Collared Lemming

Tundra of N Québec and Labrador (E Canada)

❏ *Dicrostonyx richardsoni*
Richardson's Collared Lemming

Tundra of SE Northwest Territories, S Nunavut and N
Manitoba (C Canada)

❏ *Dicrostonyx torquatus* Arctic Lemming

Tundra of N Siberia incl. Novaya Zemlya and New Siberian Is.

❏ *Dicrostonyx vinogradovi* Wrangel Lemming

Wrangel I. (SE Siberia); Critically Endangered

Subfamily: NESOMYINAE
(Malagasy Mice and Rats—19)

❏ *Brachytarsomys albicauda* White-tailed Rat

Rainforests of E Madagascar

❏ *Macrotarsomys bastardi*
Bastard Big-footed Mouse

W & S Madagascar

❏ *Macrotarsomys ingens*
Greater Big-footed Mouse

Ankarafantsika region (Mahajanga Prov., Madagascar);
Critically Endangered

❏ *Monticolomys koopmani*
Malagasy Mountain Mouse

Massifs of Ankaratra, Andringitra and Andohahela
(Madagascar)

❏ *Nesomys rufus* Island Mouse

WC, N & E Madagascar

❏ *Gymnuromys roberti* Voalavoanala

Rainforests of E Madagascar

❏ *Voalavo gymnocaudus* Naked-tailed Voalavo

Marojejy-Anjanaharibe-Sud massifs (Madagascar)

❏ *Eliurus antsingy* Antsingy Tufted-tailed Rat

Tsingy forests of WC Madagascar

❏ *Eliurus ellermani*
Ellerman's Tufted-tailed Rat

Madagascar

❏ *Eliurus majori* Major's Tufted-tailed Rat

Highlands of Madagascar; Endangered

❏ *Eliurus minor* Lesser Tufted-tailed Rat

E Madagascar

❏ *Eliurus myoxinus* Dormouse Tufted-tailed Rat

W, SW & S Madagascar

❏ *Eliurus penicillatus*
White-tipped Tufted-tailed Rat

Ampitambe region (Fianarantsoa Prov., Madagascar);
Critically Endangered

❏ *Eliurus petteri* Petter's Tufted-tailed Rat

Madagascar

❏ *Eliurus tanala* Tanala Tufted-tailed Rat

Rainforests of E Madagascar

❏ *Eliurus webbi* Webb's Tufted-tailed Rat

Rainforests of E Madagascar

❏ *Brachyuromys betsileoensis*
Betsileo Short-tailed Rat

Highlands of C Madagascar

❏ *Brachyuromys ramirohitra*
Gregarious Short-tailed Rat

Highlands of SC Madagascar

❏ *Hypogeomys antimena* **Malagasy Giant Rat**

Coastal WC Madagascar; Endangered

Subfamily: DELANYMYINAE
(Delany's Swamp Mouse—1)

❏ *Delanymys brooksi* **Delany's Swamp Mouse**

Upland marshes of EC D.R. Congo, Rwanda and SW Uganda

Subfamily: MYSTROMYINAE
(White-tailed Mouse—1)

❏ *Mystromys albicaudatus* **White-tailed Mouse**

Savanna grasslands, heath and karoo scrub of S & E South
Africa, Lesotho and Swaziland; Endangered

Subfamily: PETROMYSCINAE
(Rock Mice—4)

❏ *Petromyscus barbouri* **Barbour's Rock Mouse**

NW Cape Prov. (South Africa); Endangered

❏ *Petromyscus collinus* **Pygmy Rock Mouse**

Arid upland rocky habitats of SW Angola, Namibia and W
Northern Cape Prov. (South Africa)

❏ *Petromyscus monticularis*
Brukkaros Pygmy Rock Mouse

Brukkaros Mtns. (S Namibia and extreme N Cape Prov., South
Africa)

❏ *Petromyscus shortridgei* **Shortridge's Rock Mouse**

SW Angola and N Namibia

Subfamily: ACOMYINAE
(Spiny Mice and Brush-furred Rats—35)

❏ *Acomys cahirinus* **Cairo Spiny Mouse**

N Africa S to N Nigeria and Ethiopia, Middle East to Pakistan

❏ *Acomys cilicicus* **Asia Minor Spiny Mouse**

Vil Mersin region (Turkey); Critically Endangered

❏ *Acomys cineraceus* **Grey Spiny Mouse**

Burkina Faso and N Ghana to Ethiopia and N Uganda

❏ *Acomys ignitus* **Fiery Spiny Mouse**

Usambara Mtns (SW Kenya and NE Tanzania)

❏ *Acomys kempi* **Kemp's Spiny Mouse**

S Somalia, Kenya and NE Tanzania

❏ *Acomys minous* **Crete Spiny Mouse**

Lowland dry rocky slopes of Crete (Greece)

❏ *Acomys mullah* **Mullah Spiny Mouse**

Ethiopia and Somalia

❏ *Acomys nesiotes* **Cyprus Spiny Mouse**

Cyprus

❏ *Acomys percivali* **Percival's Spiny Mouse**

S Sudan, Ethiopia, Uganda, Kenya and S Somalia

❏ *Acomys russatus* **Golden Spiny Mouse**

E Egypt, Israel, Jordan and Saudi Arabia

❏ *Acomys spinosissimus* **Common Spiny Mouse**

Rocky habitats and woods of SE D.R. Congo and Tanzania to
NE South Africa

❏ *Acomys subspinosus* **Cape Spiny Mouse**

Rocky habitats of Western Cape Prov. (South Africa)

❏ *Acomys wilsoni* **Wilson's Spiny Mouse**

S Sudan, S Ethiopia and S Somalia to Kenya and EC
Tanzania

❏ *Acomys louisae* **Louise's Spiny Mouse**

Somalia

❏ *Uranomys ruddi* **Rudd's Mouse**

Savannas (often with borassus palms) of Senegal and Guinea
to Kenya; E Zimbabwe, Malawi and C Mozambique

Lophuromys (*Kivumys*)

❏ *Lophuromys luteogaster*
Yellow-bellied Brush-furred Rat

NE D.R. Congo

❏ *Lophuromys medicaudatus*
Medium-tailed Brush-furred Rat

Highland forests of E D.R. Congo and Rwanda

❏ *Lophuromys woosnami*
Woosnam's Brush-furred Rat

Highland forests of E D.R. Congo, W Uganda, Rwanda and Burundi

Lophuromys (Lophuromys)

❏ *Lophuromys angolensis* Angolan Brush-furred Rat

Mbwambala region (Angola)

❏ *Lophuromys aquilus* Blackish Brush-furred Rat

NW D.R. Congo to Kenya, S to N Mozambique

❏ *Lophuromys brevicaudus*
Short-tailed Brush-furred Rat

Highlands of S Ethiopia

❏ *Lophuromys brunneus* Brown Brush-furred Rat

Highlands of SW Ethiopia

❏ *Lophuromys chrysopus*
Gold-footed Brush-furred Rat

Highlands of S Ethiopia

❏ *Lophuromys cinereus* Grey Brush-furred Rat

Marais Mukaba region (D.R. Congo)

❏ *Lophuromys dieterleni*
Dieterlen's Brush-furred Rat

Mt Oku (Cameroon)

❏ *Lophuromys dudui* Dudu's Brush-furred Rat

Rainforests of NE D.R. Congo

❏ *Lophuromys huttereri* Hutterer's Brush-furred Rat

D.R. Congo

❏ *Lophuromys flavopunctatus*
Yellow-spotted Brush-furred Rat

Highlands of C Ethiopia

❏ *Lophuromys melanonyx*
Black-clawed Brush-furred Rat

C & S Ethiopia

❏ *Lophuromys nudicaudus*
Fire-bellied Brush-furred Rat

W Cameroon, Equatorial Guinea and Gabon

❏ *Lophuromys rahmi* Rahm's Brush-furred Rat

Highland forests of E D.R. Congo and Rwanda

❏ *Lophuromys roseveari*
Rosevear's Brush-furred Rat

Mt Cameroun (Cameroon)

❏ *Lophuromys sikapusi*
Rusty-bellied Brush-furred Rat

Sierra Leone to N Angola and W Kenya

❏ *Lophuromys verhageni*
Verhagen's Brush-furred Rat

Clearings in montane forests of Mt Meru region (Tanzania)

❏ *Lophuromys zena* Zena Brush-furred Rat

Aberdare Range and Mt Kenya (Kenya)

Subfamily: GERBILLINAE
(Gerbils and Jirds—111)

❏ *Gerbillus acticola* Berbera Gerbil

Somalia

❏ *Gerbillus agag* Agag Gerbil

Mali, Niger and N Nigeria to Sudan and Kenya

❏ *Gerbillus allenbyi* Allenby's Gerbil

Coastal dunes of Israel

❏ *Gerbillus amoenus* Pleasant Gerbil

Libya and Egypt

❏ *Gerbillus andersoni* Anderson's Gerbil

NC Egypt

❏ *Gerbillus aquilus* Swarthy Gerbil

SE Iran, S Afghanistan and W Pakistan

❏ *Gerbillus bilensis* Bilen Gerbil

Bilen region (Ethiopia); Critically Endangered

❏ *Gerbillus bonhotei* Bonhote's Gerbil

NE Sinai Peninsula (Egypt)

❏ *Gerbillus bottai* Botta's Gerbil

Sudan and Kenya

❏ *Gerbillus brockmani* Brockman's Gerbil

Somalia

❏ *Gerbillus burtoni* Burton's Gerbil

Darfur Plains region (Sudan); Critically Endangered

❏ *Gerbillus campestris* North African Gerbil

Morocco to Egypt and Sudan

❏ *Gerbillus cheesmani* Cheesman's Gerbil

C Iraq and SW Iran to Yemen and Oman

❏ *Gerbillus cosensi* Cosens' Gerbil

Kozibiri River region (Kenya); Critically Endangered

❏ *Gerbillus dalloni* Dallon's Gerbil

Tibesti region (N Chad); Critically Endangered

❏ *Gerbillus dasyurus* Wagner's Gerbil

Syria and Iraq to Arabian Peninsula

❏ *Gerbillus diminutus* Diminutive Gerbil

Kenya

❏ *Gerbillus dongolanus* Dongola Gerbil

Dongola region (Sudan)

❏ *Gerbillus dunni* Dunn's Gerbil

Ethiopia, Djibouti and Somalia

❏ *Gerbillus famulus* Black-tufted Gerbil

Yemen

❏ *Gerbillus floweri* Flower's Gerbil

Sinai (Egypt); Critically Endangered

❏ *Gerbillus garamantis* Algerian Gerbil

Algeria

❏ *Gerbillus gerbillus* Lesser Egyptian Gerbil

Morocco and N Mali to Israel and N Sudan

❏ *Gerbillus gleadowi* Indian Hairy-footed Gerbil

Dunes of E Pakistan and NW India

❏ *Gerbillus grobbeni* Grobben's Gerbil

Darnah region (NE Libya); Critically Endangered

❏ *Gerbillus harwoodi* Harwood's Gerbil

Kenya

❏ *Gerbillus henleyi* Pygmy Gerbil

Algeria to Jordan and Arabian Peninsula; N Senegal and Burkina Faso

❏ *Gerbillus hesperinus* Western Gerbil

Coastal Morocco

❏ *Gerbillus hoogstraali* Hoogstraal's Gerbil

Taroudannt region (Morocco); Critically Endangered

❏ *Gerbillus jamesi* James' Gerbil

Tunisia

❏ *Gerbillus juliani* Julian's Gerbil

Somalia

❏ *Gerbillus latastei* Lataste's Gerbil

Tunisia and Libya

❏ *Gerbillus lowei* Lowe's Gerbil

Jebel Marra region (Sudan); Critically Endangered

❏ *Gerbillus mackillingini* Mackillingin's Gerbil

Deserts of SE Egypt

❏ *Gerbillus maghrebi*
Greater Short-tailed Gerbil

Taounate region (N Morocco)

❏ *Gerbillus mauritaniae* Mauritanian Gerbil

Archane Titarek region (Aouker Region, Mauritania); Critically Endangered

❏ *Gerbillus mesopotamiae* Mesopotamian Gerbil

River valleys of E Iraq and SW Iran

❏ *Gerbillus muriculus* Barfur Gerbil

Sudan

❏ *Gerbillus nancillus* Sudan Gerbil

El Fasher region (Darfur Plains, WC Sudan)

❏ *Gerbillus nanus* Baluchistan Gerbil

Morocco to Middle East, Arabian Peninsula and NW India

❏ *Gerbillus nigeriae* Nigerian Gerbil

Burkina Faso and N Nigeria

❏ *Gerbillus occiduus* Occidental Gerbil

Aoreora region (SW Morocco); Critically Endangered

❏ *Gerbillus percivali* Percival's Gerbil

Kenya

❏ *Gerbillus perpallidus* Pale Gerbil

N Egypt

❏ *Gerbillus poecilops* Large Aden Gerbil

SW Saudi Arabia and Yemen

❏ *Gerbillus principulus* Principal Gerbil

Jebel Meidob region (Sudan); Critically Endangered

❏ *Gerbillus pulvinatus* Cushioned Gerbil

Ethiopia

❏ *Gerbillus pusillus* Least Gerbil

S Sudan, Ethiopia and Kenya

❏ *Gerbillus pyramidum* Greater Egyptian Gerbil

Egypt and N Sudan

❏ *Gerbillus quadrimaculatus* Four-spotted Gerbil

Nubian Desert (NE Sudan); Critically Endangered

❏ *Gerbillus riggenbachi* Riggenbach's Gerbil

Western Sahara and N Senegal

❏ *Gerbillus rosalinda* Rosalinda Gerbil

Sudan

❏ *Gerbillus ruberrimus* Little Red Gerbil

E Ethiopia, Somalia and Kenya

❏ *Gerbillus rupicola* Rock-loving Gerbil

Emnal'here region (Mali)

❏ *Gerbillus simoni* Lesser Short-tailed Gerbil

Algeria, Tunisia, Libya and NW Egypt

❏ *Gerbillus somalicus* Somali Gerbil

Somalia

❏ *Gerbillus stigmonyx* Khartoum Gerbil

Sudan

❏ *Gerbillus syrticus* Sand Gerbil

Nofilia region (Libya); Critically Endangered

❏ *Gerbillus tarabuli* Tarabul's Gerbil

Libya

❏ *Gerbillus vivax* Vivacious Gerbil

Libya

❏ *Gerbillus watersi* Waters' Gerbil

Sudan and Somalia

❏ *Microdillus peeli* Somali Pygmy Gerbil

Somalia

Meriones (Cheliones)

❏ *Meriones hurrianae* Indian Desert Jird

SE Iran, Pakistan and NW India

Meriones (Meriones)

❏ *Meriones tamariscinus* Tamarisk Jird

N Caucasus Mtns and Kazakhstan to NC China

Meriones (Pallasiomys)

❏ *Meriones arimalius* Arabian Jird

Sand deserts of Saudi Arabia and Oman; Endangered

❏ *Meriones chengi* Cheng's Jird

Da-Ho-Yien region (N Xinjiang, NW China); Critically
Endangered

❏ *Meriones crassus* Sundevall's Jird

Morocco and Niger to Sudan, Middle East and Afghanistan

❏ *Meriones dahli* Dahl's Jird

Armenia; Endangered

❏ *Meriones libycus* Libyan Jird

Western Sahara to Saudi Arabia, Afghanistan and W China

❏ *Meriones meridianus* Mid-day Jird

Caucasus Mtns and E Iran to Mongolia and N China

❏ *Meriones sacramenti* Buxton's Jird

Israel; Endangered

❏ *Meriones shawi* Shaw's Jird

Coastal Morocco to N Sinai (Egypt)

❏ *Meriones tristrami* Tristram's Jird

Deserts and dry steppes of Kos (Greece), Israel and Jordan to E Turkey and NW Iran

❏ *Meriones unguiculatus* Mongolian Jird

C Siberia, Mongolia and N China

❏ *Meriones vinogradovi* Vinogradov's Jird

SE Turkey and Transcaucasia to N Syria and N Iran

❏ *Meriones zarudnyi* Zarudny's Jird

NE Iran, S Turkmenistan and N Afghanistan; Endangered

Meriones (Parameriones)

❏ *Meriones persicus* Persian Jird

Turkey and Transcaucasia to Afghanistan and W Pakistan

❏ *Meriones rex* King Jird

SW Arabian Peninsula

❏ *Rhombomys opimus* Great Gerbil

Kazakhstan, Iran and SW Pakistan to Mongolia and N China

❏ *Psammomys obesus* Fat Sand Rat

Hard-sand areas of Algeria to Middle East and Sudan

❏ *Psammomys vexillaris* Thin Sand Rat

Hard-sand areas of Algeria, Tunisia and Libya

❏ *Sekeetamys calurus* Bushy-tailed Jird

Rocky habitats of E Egypt and S Israel to C Saudi Arabia

❏ *Brachiones przewalskii* Przewalski's Gerbil

Deserts of NW China

❏ *Desmodilliscus braueri* Pouched Gerbil

Sahelian steppes of W Mauritania and Senegal to N Sudan and N Cameroon

❏ *Pachyuromys duprasi* Fat-tailed Gerbil

Hard-sand areas in deserts from W Morocco to N Egypt

Tatera (Gerbilliscus)

❏ *Tatera boehmi* Böhm's Gerbil

Angola, Zambia and Malawi to Uganda and Kenya

Tatera (Tatera)

❏ *Tatera indica* Indian Gerbil

Syria and Kuwait to India, S Nepal and Sri Lanka

Tatera (Taterona)

❏ *Tatera afra* Cape Gerbil

Sandy habitats of Western Cape Prov. (South Africa)

❏ *Tatera brantsii* Highveld Gerbil

Sandy habitats of S Angola and SW Zambia to C & E South Africa

❏ *Tatera guineae* Guinea Gerbil

Senegal to Burkina Faso and Ghana

❏ *Tatera inclusa* Gorongoza Gerbil

Sandy habitats of NE Tanzania to E Zimbabwe and C Mozambique

❏ *Tatera kempi* Kemp's Gerbil

Senegal and Guinea to Cameroon

❏ *Tatera leucogaster* Bushveld Gerbil

Sandy habitats of S D.R. Congo and SW Tanzania to N South Africa

❏ *Tatera nigricauda* Black-tailed Gerbil

Kenya, Somalia and Tanzania

❏ *Tatera phillipsi* Phillips' Gerbil

Ethiopia, Somalia ad Kenya

❏ *Tatera robusta* Fringe-tailed Gerbil

Burkina Faso to Somalia and Tanzania

❏ *Tatera valida* Savanna Gerbil

Chad and Sudan to Angola, Zambia and SW Tanzania

❏ *Taterillus arenarius* Sahel Gerbil

Sahel region of Mauritania, Mali and W Niger

❏ *Taterillus congicus* Congo Gerbil

Cameroon and Chad to Sudan and Uganda

❏ *Taterillus emini* Emin's Gerbil

Sudan and W Ethiopia to NE D.R. Congo, Uganda and NW Kenya

❏ *Taterillus gracilis* Slender Gerbil

Senegal to Niger and N Nigeria

❏ *Taterillus harringtoni* Harrington's Gerbil

Central African Republic to Somalia and Tanzania

❏ *Taterillus lacustris* Lake Chad Gerbil

NE Nigeria and Cameroon

❏ *Taterillus petteri* Petter's Gerbil

Sahel region of E Burkina Faso and W Niger

❏ *Taterillus pygargus* Senegal Gerbil

S Mauritania, Gambia, Senegal and W Mali

❏ *Desmodillus auricularis*
Cape Short-eared Gerbil

Shortgrass hard-soil habitats of Namibia, S Botswana and W & C South Africa

Gerbillurus (Gerbillurus)

❏ *Gerbillurus setzeri* Setzer's Hairy-footed Gerbil

Arid sandy habitats of Namib Desert (extreme SW Angola and coastal W Namibia)

❏ *Gerbillurus vallinus*
Bushy-tailed Hairy-footed Gerbil

Arid sandy habitats of S Namibia and NW Northern Cape Prov. (South Africa)

Gerbillurus (Paratatera)

❏ *Gerbillurus tytonis* Dune Hairy-footed Gerbil

Deserts of Namibia

Gerbillurus (Progerbillurus)

❏ *Gerbillurus paeba* Common Hairy-footed Gerbil

Sandy habitats of SW Angola, Namibia, Botswana, SE Zimbabwe, SW Mozambique and W South Africa

❏ *Ammodillus imbellis* Ammodile

Sandy deserts of E Ethiopia and Somalia

Subfamily: MURINAE
(True Mice and Rats—535)

❏ *Millardia gleadowi*
Sand-coloured Soft-furred Rat

Afghanistan, Pakistan and NW India

❏ *Millardia kathleenae*
Miss Ryley's Soft-furred Rat

C Myanmar

❏ *Millardia kondana* Kondana Soft-furred Rat

Sinhgarh Plateau (Maharashtra, India); Endangered

❏ *Millardia meltada* Common Soft-furred Rat

E Pakistan, Nepal, India and Sri Lanka

❏ *Cremnomys blanfordi* Blanford's Rat

India and Sri Lanka

❏ *Cremnomys cutchicus* Cutch Rat

India

❏ *Cremnomys elvira* Elvira Rat

SE India

❏ *Diomys crumpi* Crump's Mouse

W Nepal and NE India

❏ *Micromys minutus* Eurasian Harvest Mouse

Wetlands, damp meadows and fields of C & E Europe to NE India, E Siberia, Korea, China, Taiwan and Japan

❏ *Vandeleuria nolthenii*
Nolthenius' Long-tailed Climbing Mouse

Highlands of Sri Lanka

❏ *Vandeleuria oleracea*
Asiatic Long-tailed Climbing Mouse

Nepal, India and Sri Lanka to Yunnan (China) and Vietnam

❏ *Vernaya fulva* Red Climbing Mouse

N Myanmar and SW China

❏ *Chiropodomys calamianensis*
Palawan Pencil-tailed Tree Mouse

Busuanga, Palawan and Balabac (WC & SW Philippines)

❏ *Chiropodomys gliroides*
Common Pencil-tailed Tree Mouse

Assam (NE India) to SW China, S to Malay Peninsula,
Sumatra, Java and Bali

❏ *Chiropodomys karlkoopmani*
Koopman's Pencil-tailed Tree Mouse

Siberut and Pagai Is. (Mentawai Is., Sumatra); Endangered

❏ *Chiropodomys major*
Large Pencil-tailed Tree Mouse

Borneo

❏ *Chiropodomys muroides*
Grey-bellied Pencil-tailed Tree Mouse

N Borneo

❏ *Chiropodomys pusillus*
Small Pencil-tailed Tree Mouse

Borneo

❏ *Hapalomys delacouri* **Delacour's Marmoset-Rat**

Hainan (S China), N Laos and S Vietnam

❏ *Hapalomys longicaudatus* **Marmoset-Rat**

SE Myanmar, SW Thailand and Malay Peninsula

Apodemus (Alsomys)

❏ *Apodemus argenteus*
Small Japanese Field Mouse

Japan

❏ *Apodemus draco* **South China Field Mouse**

Assam (NE India), Myanmar and China

❏ *Apodemus gurkha* **Himalayan Field Mouse**

Nepal

❏ *Apodemus latronum* **Sichuan Field Mouse**

E Tibet and SW China

❏ *Apodemus peninsulae* **Korean Field Mouse**

E Siberia with Sakhalin (Russia), S to SW China, Korea and
Hokkaido (N Japan)

❏ *Apodemus semotus* **Taiwan Field Mouse**

Taiwan

❏ *Apodemus speciosus*
Large Japanese Field Mouse

Japan

Apodemus (Apodemus)

❏ *Apodemus agrarius* **Striped Field Mouse**

Forest edge, parks, grasslands and gardens of E Europe to
Caucasus Mtns, C Siberia, Korea, S China and Taiwan

❏ *Apodemus chevrieri* **Chevrier's Field Mouse**

C & SW China

Apodemus (Karstomys)

❏ *Apodemus mystacinus*
Broad-toothed Field Mouse (Rock Mouse)

Rocky areas of SE Europe and Caucasus Mtns to Middle East
and N Saudi Arabia

Apodemus (Sylvaemus)

❏ *Apodemus alpicola* **Alpine Field Mouse**

Montane forests of Alps from SE France to Austria

❏ *Apodemus arianus* **Persian Field Mouse**

Lebanon and N Israel to Iraq and N Iran

❏ *Apodemus flavicollis*
Yellow-necked Field Mouse

Forests of Europe to Russia and Middle East

❏ *Apodemus fulvipectus*
Yellow-breasted Field Mouse

Ukraine and Russia to Transcaucasia

❏ *Apodemus hermonensis*
Mount Hermon Field Mouse

Mt Hermon (Israel); Endangered

❏ *Apodemus hyrcanicus* **Caucasian Field Mouse**

E Caucasus Mtns (Azerbaijan)

❏ *Apodemus ponticus* **Black Sea Field Mouse**

SE Ukraine and S Russia to E Turkey and Iraq

❏ *Apodemus rusiges* **Kashmir Field Mouse**

N India

❏ *Apodemus sylvaticus*
Long-tailed Field Mouse (Wood Mouse)

Widespread in Europe (except NE) incl. Mediterranean islands, Belarus, NW Ukraine, Morocco, Algeria and Tunisia

❏ *Apodemus uralensis* Pygmy Field Mouse

Forest streams and farmland of EC Europe, Turkey and Caucasus Mtns to Altai Mtns and NW China

❏ *Apodemus wardi* Ward's Field Mouse

NW Iran and Afghanistan to N Pakistan, Kashmir and NC Nepal

❏ *Tokudaia muenninki* Muennink's Spiny Rat

Okinawa (Ryukyu Is., Japan); Critically Endangered

❏ *Tokudaia osimensis* Ryukyu Spiny Rat

Amami-oshima I. (Ryukyu Is., S Japan); Endangered

Mus (Coelomys)

❏ *Mus crociduroides* Sumatran Shrew-like Mouse

Highland rainforests of W Sumatra

❏ *Mus famulus* Servant Mouse

Nilgiri Hills (S India); Endangered

❏ *Mus mayori* Mayor's Mouse

Forests of Sri Lanka

❏ *Mus pahari* Gairdner's Shrew-Mouse

NE India and Myanmar to Yunnan (China) and Vietnam

❏ *Mus vulcani* Volcano Mouse

Mountain forests of W Java

Mus (Mus)

❏ *Mus booduga* Little Indian Field Mouse

S Nepal, India, Sri Lanka and C Myanmar

❏ *Mus caroli* Ryukyu Mouse

S China and S Japan to Indochina; feral Malay Peninsula to Flores

❏ *Mus cervicolor* Fawn-coloured Mouse

Nepal and NE India to Indochina; feral Sumatra and Java

❏ *Mus cookii* Cook's Mouse

Nepal and India to SW Yunnan (China) and Indochina

❏ *Mus domesticus* Western House Mouse

Commensal with humans in W & S Europe, N Africa and Middle East; introduced N & S America, subsaharan Africa, N Australia and oceanic islands

❏ *Mus macedonicus* Macedonian Mouse

Fields and riparian habitats of Balkan Peninsula and Transcaucasia to Israel, Jordan and Iran

❏ *Mus musculus* Eastern House Mouse

Commensal with humans in N & EC Europe to E China, S to Caucasus Mtns

❏ *Mus spicilegus* Mound-building Mouse

Forest edge, steppes and fields of E Austria and S Slovakia to W Russia and Greece

❏ *Mus spretus* Algerian Mouse

Field edge, grasslands and maquis of SW Europe and Morocco to Libya

❏ *Mus terricolor* Earth-coloured Mouse

Pakistan, Nepal and India

Mus (Nannomys)

❏ *Mus baoulei* Baoule's Mouse

E Guinea and Ivory Coast

❏ *Mus bufo* Toad Mouse

E D.R. Congo, Uganda, Rwanda and Burundi

❏ *Mus callewaerti* Callewaert's Mouse

SW D.R. Congo and NC Angola

❏ *Mus goundae* Gounda Mouse

Gounda R. region (Central African Republic)

❏ *Mus haussa* Haussa Mouse

S Mauritania and Senegal to S Niger and N Nigeria

❏ *Mus indutus* Desert Pygmy Mouse

Savannas and scrub habitats of E Namibia, Botswana, W Zimbabwe and NC South Africa

❏ *Mus kasaicus* Kasai Mouse

Kasai region (D.R. Congo); Critically Endangered

❏ *Mus mahomet* Mahomet Mouse

C Ethiopia, SW Uganda and SW Kenya

❏ *Mus mattheyi* Matthey's Mouse

Accra region (Ghana)

❏ *Mus minutoides* South African Pygmy Mouse

Savannas and woods of Zimbabwe, Mozambique, South Africa, Lesotho and Swaziland

❏ *Mus musculoides* Temminck's Mouse

Subsaharan Africa

❏ *Mus neavei* Neave's Mouse

S D.R. Congo and S Tanzania to E South Africa and Mozambique

❏ *Mus orangiae* Orange Mouse

Free State (South Africa)

❏ *Mus oubanguii* Oubangui Mouse

Oubangui R. region (Central African Republic)

❏ *Mus setulosus* Peters' Mouse

Sierra Leone and Guinea to Ethiopia, W Kenya and N Uganda

❏ *Mus setzeri* Setzer's Pygmy Mouse

NE Namibia, NW Botswana and W Zambia

❏ *Mus sorella* Thomas' Pygmy Mouse

E Cameroon to Kenya, S to EC Angola, Zimbabwe, Mozambique and NE South Africa

❏ *Mus tenellus* Tender Mouse

Sudan and S Ethiopia to C Tanzania

❏ *Mus triton* Grey-bellied Pygmy Mouse

N D.R. Congo and Kenya to Angola and NW Mozambique

Mus (Pyromys)

❏ *Mus fernandoni* Sri Lankan Spiny Mouse

Sri Lanka

❏ *Mus phillipsi* Phillips' Mouse

India

❏ *Mus platythrix* Flat-haired Mouse

India

❏ *Mus saxicola* Rock-loving Mouse

S Pakistan, S Nepal and India

❏ *Mus shortridgei* Shortridge's Mouse

Myanmar, Thailand, Cambodia and NW Vietnam

❏ *Muriculus imberbis* Stripe-backed Mouse

Subalpine heaths and moorlands of Ethiopia

❏ *Hadromys humei* Manipur Bush Rat

NE India and W Yunnan (China)

❏ *Golunda ellioti* Indian Bush Rat

SE Iran, Pakistan, Nepal, India and Sri Lanka

❏ *Pelomys campanae*
Bell Groove-toothed Swamp Rat

Marshes and riparian habitats of W D.R. Congo and NW Angola

❏ *Pelomys fallax*
Creek Groove-toothed Swamp Rat

Marshes and riparian habitats of SW Uganda and S Kenya to NE Namibia, N Botswana, E Zimbabwe and Mozambique

❏ *Pelomys hopkinsi*
Hopkins' Groove-toothed Swamp Rat

Marshes and riparian habitats of Uganda, Rwanda and SW Kenya

❏ *Pelomys isseli* Issel's Groove-toothed Swamp Rat

Riparian habitats of Lake Victoria islands (Uganda)

❏ *Pelomys minor* Least Groove-toothed Swamp Rat

Marshes and riparian habitats of SE D.R. Congo, N Angola, W Tanzania and NW Zambia

❏ *Desmomys harringtoni* Harrington's Rat

Montane savannas and scrub in highlands of Ethiopia

❏ *Mylomys dybowskii* African Groove-toothed Rat

Savannas and grasslands of Guinea to S Sudan, S to Rep. Congo, D.R. Congo, Rwanda and Tanzania

❏ *Arvicanthis abyssinicus* Abyssinian Grass Rat

Grasslands of highlands of Ethiopia

❏ *Arvicanthis blicki* Blick's Grass Rat

Grasslands of highlands of Ethiopia

❏ *Arvicanthis nairobae* Nairobi Grass Rat

Grasslands of C Kenya to EC Tanzania

❏ *Arvicanthis neumanni* Neumann's Grass Rat

Grasslands of SE Sudan to Somalia, S to C Tanzania

❏ *Arvicanthis niloticus* African Grass Rat

Grasslands of subsaharan Africa S to E Zambia and SW Tanzania; SW Saudi Arabia

❏ *Lemniscomys barbarus*
Barbary Striped Grass Mouse

Grasslands of NW and subsaharan Africa, S to E D.R. Congo and Tanzania

❏ *Lemniscomys bellieri*
Bellier's Striped Grass Mouse

Grasslands of Ivory Coast

❏ *Lemniscomys griselda*
Griselda's Striped Grass Mouse

Grasslands of Angola

❏ *Lemniscomys hoogstraali*
Hoogstraal's Striped Grass Mouse

Grasslands of Paloich region (Sudan)

❏ *Lemniscomys linulus*
Senegal One-striped Grass Mouse

Grasslands of Senegal and Ivory Coast

❏ *Lemniscomys macculus*
Buffoon Striped Grass Mouse

Grasslands of S Sudan and Ethiopia to NE D.R. Congo, Uganda and Kenya

❏ *Lemniscomys mittendorfi*
Mittendorf's Striped Grass Mouse

Grasslands of Lake Oku region (Cameroon); Endangered

❏ *Lemniscomys rosalia* Single-striped Grass Mouse

Savannas, scrub, grasslands and farmland of Angola to S Kenya, S to N Namibia and E South Africa

❏ *Lemniscomys roseveari*
Rosevear's Striped Grass Mouse

Grasslands of Zambia

❏ *Lemniscomys striatus*
Typical Striped Grass Mouse

Grasslands of Sierra Leone to Ethiopia, S to Angola, NE Zambia and N Malawi

❏ *Rhabdomys pumilio* Four-striped Grass Mouse

Grasslands and heaths of C Angola, SE D.R. Congo and Kenya, S to South Africa

❏ *Otomys anchietae* Angolan Vlei Rat

C Angola, N Malawi, SW Tanzania and W Kenya

❏ *Otomys angoniensis* Angoni Vlei Rat

Savannas and marshes of S Kenya to South Africa

❏ *Otomys denti* Dent's Vlei Rat

Uganda, EC Tanzania, Zambia and N Malawi

❏ *Otomys irroratus* Common Vlei Rat

Grasslands and marshes of E Zimbabwe, Mozambique and South Africa

❏ *Otomys laminatus* Laminate Vlei Rat

SE South Africa and Swaziland

❏ *Otomys maximus* Large Vlei Rat

S Angola, SW Zambia, NE Namibia and NW Botswana

❏ *Otomys occidentalis* Western Vlei Rat

SE Nigeria and W Cameroon; Endangered

❏ *Otomys sloggetti* Sloggett's Vlei Rat

High rocky habitats of SE South Africa and Lesotho

❏ *Otomys tropicalis* Tropical Vlei Rat

Highlands of Cameroon and E D.R. Congo to W Kenya

❏ *Otomys typus* Typical Vlei Rat

Highlands of Ethiopia, S to E Zambia and N Malawi

❏ *Otomys unisulcatus* Bush Vlei Rat

Arid karoo scrub of Cape Prov. (South Africa)

❏ *Parotomys brantsii* Brants' Whistling Rat

Arid sandy habitats of SE Namibia, extreme SW Botswana and W South Africa

❏ *Parotomys littledalei* Littledale's Whistling Rat

Arid sandy habitats of SW Namibia and Northern Cape Prov. (South Africa)

❏ *Dasymys foxi* Fox's Shaggy Rat

Marshes and wet grasslands of Jos Plateau (Nigeria)

❏ *Dasymys incomtus* **African Marsh Rat**

Marshes and wet grasslands of S Sudan and Ethiopia, S to S & E South Africa

❏ *Dasymys montanus* **Montane Shaggy Rat**

Montane swamps of Ruwenzori Mtns (Uganda)

❏ *Dasymys nudipes* **Angolan Marsh Rat**

Marshes and wet grasslands of S Angola, SW Zambia, NE Namibia and N Botswana

❏ *Dasymys rufulus* **West African Shaggy Rat**

Marshes and wet grasslands of Sierra Leone to Cameroon

❏ *Lamottemys okuensis* **Mount Oku Rat**

Montane regrowth areas of Mt Oku (W Cameroon); Endangered

❏ *Oenomys hypoxanthus* **Rufous-nosed Rat**

Forest clearings and regrowth areas of S Nigeria to S Sudan, SW Ethiopia, N Angola and W Tanzania

❏ *Oenomys ornatus* **Ghana Rufous-nosed Rat**

Forest clearings and regrowth areas of SE Guinea to Ghana

❏ *Thamnomys kempi* **Kemp's Thicket Rat**

Montane forests of E D.R. Congo, W Uganda and NW Burundi

❏ *Thamnomys venustus* **Charming Thicket Rat**

Montane forests of E D.R. Congo, Uganda and Rwanda

❏ *Thallomys loringi* **Loring's Rat**

Savannas of E Kenya, N & E Tanzania

❏ *Thallomys nigricauda* **Black-tailed Tree Rat**

Savannas of Angola and SE Zambia to NW South Africa

❏ *Thallomys paedulcus* **Acacia Rat**

Savannas of S Ethiopia and S Somalia, S to E South Africa and Swaziland

❏ *Thallomys shortridgei* **Shortridge's Rat**

Savannas of Orange R. region (Northern Cape Prov., South Africa)

❏ *Grammomys aridulus* **Arid Thicket Rat**

Arid montane thickets of WC Sudan

❏ *Grammomys buntingi* **Bunting's Thicket Rat**

Forests of Guinea, Sierra Leone, Liberia and Ivory Coast

❏ *Grammomys caniceps* **Grey-headed Thicket Rat**

Dense scrub of S Somalia and N Kenya

❏ *Grammomys cometes* **Mozambique Thicket Rat**

Forests and woods of E Zimbabwe, S Mozambique and SE South Africa

❏ *Grammomys dolichurus* **Woodland Thicket Rat**

Forests and woods of Nigeria to SW Ethiopia, S to Angola and S & E South Africa

❏ *Grammomys dryas* **Forest Thicket Rat**

Forests of E D.R. Congo, W Uganda and NW Burundi

❏ *Grammomys gigas* **Giant Thicket Rat**

Dense montane thickets of Mt Kenya (Kenya); Endangered

❏ *Grammomys ibeanus* **Ruwenzori Thicket Rat**

Dense scrub of S Sudan to NE Zambia

❏ *Grammomys macmillani* **Macmillan's Thicket Rat**

Forests of Sierra Leone to S Ethiopia, S to E Zimbabwe and Mozambique

❏ *Grammomys minnae* **Ethiopian Thicket Rat**

Dense scrub of S Ethiopia

❏ *Grammomys rutilans* **Shining Thicket Rat**

Dense scrub of Guinea to Central African Republic and W Uganda, S to N Angola

❏ *Aethomys bocagei* **Bocage's Rock Rat**

Rocky habitats of W & C Angola

❏ *Aethomys chrysophilus* **Red Rock Rat**

Savannas, grassland and rocky habitats of S Kenya, Tanzania, S Angola and Malawi to N South Africa

❏ *Aethomys granti* **Grant's Rock Rat**

Rocky habitats of C karoo region of Northern Cape Prov. (South Africa)

❏ *Aethomys hindei* **Hinde's Rock Rat**

Rocky habitats of N Cameroon and N D.R. Congo to SW Ethiopia and Tanzania

❏ *Aethomys kaiseri* **Kaiser's Rock Rat**

Rocky habitats of SW Uganda and S Kenya to S Angola and Malawi

❏ *Aethomys namaquensis* Namaqua Rock Rat

Rocky habitats of S Angola and Malawi, S to South Africa

❏ *Aethomys nyikae* Nyika Rock Rat

Rocky habitats of N Angola, S D.R. Congo, N Zambia, Malawi and extreme E Zimbabwe

❏ *Aethomys silindensis* Silinda Rock Rat

Rocky habitats of Mt Silinda region (extreme E Zimbabwe)

❏ *Aethomys stannarius* Tinfields Rock Rat

Rocky habitats of N Nigeria to W Cameroon

❏ *Aethomys thomasi* Thomas' Rock Rat

Rocky habitats of W & C Angola

❏ *Stochomys longicaudatus* Target Rat

Lowland forests, riparian habitats and regrowth areas of Togo to Central African Republic, S to Gabon, D.R. Congo and Uganda

❏ *Dephomys defua* Defua Rat

Regrowth areas and scrub of Guinea and Sierra Leone to Ghana

❏ *Dephomys eburnea* Ivory Coast Rat

Forests and scrub of Liberia and Ivory Coast

❏ *Hybomys basilii* Father Basilio's Striped Mouse

Forests of Bioko (Equatorial Guinea)

❏ *Hybomys eisentrauti* Eisentraut's Striped Mouse

Forests of Mt Lefo and Mt Oku (W Cameroon); Endangered

❏ *Hybomys lunaris* Moon Striped Mouse

Forests of NE D.R. Congo, W Uganda and Rwanda

❏ *Hybomys planifrons* Miller's Striped Mouse

Forests of NE Sierra Leone, SE Guinea, Liberia and W Ivory Coast

❏ *Hybomys trivirgatus* Temminck's Striped Mouse

Forests of E Sierra Leone to SW Nigeria

❏ *Hybomys univittatus* Peters' Striped Mouse

Forests of SE Nigeria to S Uganda, S to Rep. Congo and NW Zambia

❏ *Nilopegamys plumbeus* Ethiopian Water Mouse

Little Abbai R. region (NW Ethiopia); Critically Endangered

❏ *Colomys goslingi* African Water Rat

Riparian habitats in forests of Liberia; Cameroon to W Ethiopia, S to NE Angola and NW Zambia

❏ *Zelotomys hildegardeae* Hildegarde's Broad-headed Mouse

Forests and woods of Central African Republic and S Sudan to Angola, Zambia and Malawi

❏ *Zelotomys woosnami* Woosnam's Broad-headed Mouse

Sparsely vegetated arid and semiarid habitats of E Namibia, Botswana and N Northern Cape Prov. (South Africa)

❏ *Hylomyscus aeta* Beaded Wood Mouse

Cameroon to W Uganda, S to Gabon, Rep. Congo, D.R. Congo and NW Burundi

❏ *Hylomyscus alleni* Allen's Wood Mouse

Guinea to Cameroon and Gabon

❏ *Hylomyscus baeri* Baer's Wood Mouse

Ivory Coast and Ghana

❏ *Hylomyscus carillus* Angolan Wood Mouse

N Angola

❏ *Hylomyscus denniae* Montane Wood Mouse

Subalpine moorlands of mountains of E D.R. Congo, Uganda and Kenya to NE Zambia and Tanzania

❏ *Hylomyscus parvus* Little Wood Mouse

S Cameroon, S Central African Republic, N Gabon and NE D.R. Congo

❏ *Hylomyscus stella* Stella Wood Mouse

S Nigeria to S Sudan, S to N Angola, D.R. Congo, Burundi and EC Tanzania

❏ *Myomys albipes* Ethiopian White-footed Mouse

Highlands of Ethiopia

❏ *Myomys daltoni* Dalton's Mouse

Gambia to Central African Republic and SW Sudan

❏ *Myomys derooi* Deroo's Mouse

Ghana, Togo, Benin and W Nigeria

❏ *Myomys fumatus* African Rock Mouse

S Sudan, Ethiopia and Somalia to N Uganda and EC Tanzania

❏ *Myomys ruppi* Rupp's Mouse

Gamo Gofa region (SW Ethiopia)

❏ *Myomys verreauxii* Verreaux's Mouse

Dense wetland habitats of Western Cape Prov. (South Africa)

❏ *Myomys yemeni* Yemeni Mouse

SW Saudi Arabia and N Yemen

❏ *Heimyscus fumosus* African Smoky Mouse

Very locally in rainforests of S Cameroon, Central African Republic and Gabon

❏ *Praomys degraafi* De Graaf's Soft-furred Mouse

Nyamugari region (Burundi)

❏ *Praomys delectorum*
Delectable Soft-furred Mouse

Highlands of SE Kenya, Tanzania, NE Zambia and Malawi

❏ *Praomys hartwigi* Hartwig's Soft-furred Mouse

Gotel Mtns (Nigeria) and Lake Oku region (Cameroon); Endangered

❏ *Praomys jacksoni* African Soft-furred Mouse

C Nigeria to S Sudan, S to N Angola and Zambia

❏ *Praomys minor* Least Soft-furred Mouse

Lukolela region (C D.R. Congo)

❏ *Praomys misonnei* Misonne's Soft-furred Mouse

N & E D.R. Congo

❏ *Praomys morio* Cameroon Soft-furred Mouse

Mt Cameroun (Cameroon)

❏ *Praomys mutoni* Muton's Soft-furred Mouse

Batiabongena region (Masako Forest, N D.R. Congo)

❏ *Praomys rostratus* Forest Soft-furred Mouse

Guinea, Liberia and Ivory Coast

❏ *Praomys tullbergi* Tullberg's Soft-furred Mouse

Gambia to E D.R. Congo and Equatorial Guinea; NW Angola

❏ *Stenocephalemys albocaudata*
Ethiopian Narrow-headed Rat

High-altitude meadows and moorlands of Ethiopia

❏ *Stenocephalemys griseicauda*
Grey-tailed Narrow-headed Rat

High-altitude meadows and moorlands of Ethiopia

❏ *Mastomys angolensis*
Angolan Multimammate Mouse

Angola and S D.R. Congo

❏ *Mastomys awashensis*
Awash Multimammate Mouse

Awash region (Ethiopia)

❏ *Mastomys coucha*
Southern Multimammate Mouse

C Namibia, SW Zimbabwe and South Africa

❏ *Mastomys erythroleucus*
Guinea Multimammate Mouse

Morocco; Gambia to Somalia, S to E D.R. Congo and Burundi

❏ *Mastomys hildebrandtii*
Hildebrandt's Multimammate Mouse

Gambia to Somalia, S to N D.R. Congo, Burundi and Kenya

❏ *Mastomys natalensis* Natal Multimammate Mouse

Senegal; Namibia, EC Tanzania, Zimbabwe and South Africa

❏ *Mastomys pernanus* Dwarf Multimammate Mouse

SW Kenya, Rwanda and NW Tanzania

❏ *Mastomys shortridgei*
Shortridge's Multimammate Mouse

Dense wetland habitats of Okavango-Cuito R. confluence and Caprivi Strip (extreme NE Namibia and NW Botswana)

❏ *Mastomys verheyeni*
Verheyen's Multimammate Mouse

Lake Chad region (extreme NE Nigeria and N Cameroon)

❏ *Malacomys cansdalei* Cansdale's Swamp Rat

Riparian habitats in forests of E Liberia, S Ivory Coast and S Ghana

❏ *Malacomys edwardsi* Milne-Edwards' Swamp Rat

Riparian habitats in forests of Guinea and Sierra Leone to Ghana and Nigeria

❏ *Malacomys longipes* Big-eared Swamp Rat

Riparian habitats in forests of Guinea to S Sudan, S to NE Angola and Zambia

❏ *Malacomys lukolelae* Lukolela Swamp Rat

Riparian habitats in forests of Lukolela region (D.R. Congo)

❏ *Malacomys verschureni*
Verschuren's Swamp Rat

Riparian habitats in forests of NE D.R. Congo

❏ *Rattus adustus* Sunburned Rat

Enggano (Sumatra)

❏ *Rattus annandalei* Annandale's Rat

Malay Peninsula, Sumatra and near islands

❏ *Rattus argentiventer* Rice-field Rat

Indochina and Philippines to Java, Sulawesi, Lesser Sundas and New Guinea

❏ *Rattus baluensis* Summit Rat

Gunung Kinabalu (N Borneo); Endangered

❏ *Rattus bontanus* Bonthain Rat

Gunung Lompobatang (SW Sulawesi)

❏ *Rattus burrus* Nonsense Rat

Nicobar Is. (India)

❏ *Rattus colletti* Dusky Rat

Coastal Northern Territory (Australia)

❏ *Rattus elaphinus* Sula Rat

Sula Is. (Indonesia)

❏ *Rattus enganus* Enggano Rat

Enggano (Sumatra); Critically Endangered

❏ *Rattus everetti* Philippine Forest Rat

Philippines

❏ *Rattus exulans* Polynesian Rat

Southeast Asia; probably introduced Philippines, Indonesia, New Guinea, New Zealand and Pacific Ocean islands

❏ *Rattus feliceus* Spiny Seram Rat

Seram (S Moluccas)

❏ *Rattus foramineus* Hole Rat

SW Sulawesi

❏ *Rattus fuscipes* Australian Bush Rat

Coastal SW, S and E Australia

❏ *Rattus giluwensis* Giluwe Rat

Subalpine grasslands and moss forests on upper slopes of Mt Giluwe (EC New Guinea)

❏ *Rattus hainaldi* Hainald's Rat

Flores (Lesser Sundas)

❏ *Rattus hoffmanni* Hoffmann's Rat

Sulawesi

❏ *Rattus hoogerwerfi* Hoogerwerf's Rat

Gunung Leuser (Sumatra)

❏ *Rattus jobiensis* Japen Rat

Biak, Japen and Owi (Geelvinck Is., New Guinea)

❏ *Rattus koopmani* Koopman's Rat

Peleng I. (W Moluccas)

❏ *Rattus korinchi* Mount Kerinci Rat

Gunung Kerinci and Gunung Talakmau (W Sumatra)

❏ *Rattus leucopus* Cape York Rat

Rainforests of Aru Is., New Guinea and coastal NE Queensland (Australia)

❏ *Rattus losea* Lesser Rice-field Rat

S China, Taiwan, N Thailand, S Laos and C Vietnam

❏ *Rattus lugens* Mentawai Rat

Mentawai Is. (Sumatra)

❏ *Rattus lutreolus* Australian Swamp Rat

E & SE Australia incl. Tasmania

❏ †*Rattus macleari* MacLear's Rat

Formerly Christmas I. (Australia); Extinct (probably by 1908)

❏ *Rattus marmosurus* Opossum Rat

NE Sulawesi

❏ *Rattus mindorensis* Mindoro Black Rat

Highlands of Mindoro (WC Philippines)

❑ *Rattus mollicomulus* Mount Lompobatang Rat

Gunung Lompobatang (SW Sulawesi)

❑ *Rattus montanus* Nillu Rat

Mountains of C Sri Lanka; Critically Endangered

❑ *Rattus mordax* Eastern Rat

E New Guinea and D'Entrecasteaux Is.

❑ *Rattus morotaiensis*
Moluccan Prehensile-tailed Rat

Morotai, Halmahera and Bacan (N Moluccas)

❑ †*Rattus nativitatis* Bulldog Rat

Formerly Christmas I. (Australia); Extinct (probably by 1908)

❑ *Rattus nitidus* Himalayan Field Rat

Nepal and NE India to S China, Hainan and Vietnam; feral E to New Guinea

❑ *Rattus norvegicus* Brown Rat

Originally SE Siberia and N China; introduced worldwide with human settlements

❑ *Rattus novaeguineae* New Guinea Rat

Hill forests and grassland of E New Guinea

❑ *Rattus osgoodi* Osgood's Rat

Highlands of S Vietnam

❑ *Rattus palmarum* Palm Rat

Nicobar Is. (India)

❑ *Rattus pelurus* Peleng Rat

Peleng I. (W Moluccas)

❑ *Rattus praetor* Large Spiny Rat

Gardens and disturbed habitats of W & NC New Guinea, Admiralty Is., Bismarck Archipelago and Solomon Is.

❑ *Rattus ranjiniae* Kerala Rat

Kerala (SW India)

❑ *Rattus rattus* House Rat (Black Rat)

Originally India; introduced worldwide with human settlements

❑ *Rattus sikkimensis* Sikkim Rat

NE India to S China and Indochina

❑ *Rattus simalurensis* Simeulue Rat

Simeulue I. and near islands (off N Sumatra)

❑ *Rattus sordidus* Canefield Rat

Savanna grasslands and forest margins of SC & SE New Guinea and coastal NE Queensland (Australia)

❑ *Rattus steini* New Guinean Small Spiny Rat

Grasslands and gardens of New Guinea

❑ *Rattus stoicus* Andaman Rat

Andaman Is. (India)

❑ *Rattus tanezumi* Tanezumi Rat

E Afghanistan to Korea and Indochina; introduced Moluccas to Fiji with human settlements

❑ *Rattus tawitawiensis* Tawi-tawi Forest Rat

Tawi-tawi (S Philippines)

❑ *Rattus timorensis* Timor Rat

Timor (Lesser Sundas)

❑ *Rattus tiomanicus* Malaysian Field Rat

Malay Peninsula, E to Palawan (SW Philippines) and Bali

❑ *Rattus tunneyi* Pale Field Rat

W, N & E Australia

❑ *Rattus turkestanicus* Turkestan Rat

NE Iran, Afghanistan and Kyrgyzstan to N India, Nepal and S China

❑ *Rattus villosissimus* Long-haired Rat

Australia

❑ *Rattus xanthurus* Yellow-tailed Rat

N, C & SE Sulawesi

❑ *Nesokia bunnii* Bunn's Short-tailed Bandicoot-Rat

Marshes of SE Iraq

❑ *Nesokia indica* Short-tailed Bandicoot-Rat

NE Egypt and Middle East to Uzbekistan, N India and NW China

❑ *Bandicota bengalensis* Lesser Bandicoot-Rat

Pakistan to Myanmar and Sri Lanka; feral Sumatra, Java and Saudi Arabia

❏ *Bandicota indica* Greater Bandicoot-Rat

India and Nepal to S China, Taiwan and Indochina; feral Java

❏ *Bandicota maxima* Giant Bandicoot-Rat

Gardens and built areas of Nepal, India and Bangladesh

❏ *Bandicota savilei* Savile's Bandicoot-Rat

E Myanmar, Thailand and Vietnam

❏ *Berylmys berdmorei* Small White-toothed Rat

S Myanmar, Thailand, N Laos, Cambodia and S Vietnam

❏ *Berylmys bowersi* Bowers' White-toothed Rat

NE India to S China, Indochina, Malay Peninsula and NW Sumatra

❏ *Berylmys mackenziei*
Kenneth's White-toothed Rat

Assam (NE India), Myanmar, Sichuan (C China) and S Vietnam

❏ *Berylmys manipulus* Manipur White-toothed Rat

Assam (NE India), Myanmar and Yunnan (China)

❏ *Diplothrix legatus* Ryukyu Rat

Ryukyu Is. (S Japan); Endangered

❏ *Sundamys infraluteus* Mountain Giant Sunda Rat

Mountains of W Sumatra and N Borneo

❏ *Sundamys maxi* Bartels' Giant Sunda Rat

Highlands of W Java; Endangered

❏ *Sundamys muelleri* Müller's Giant Sunda Rat

S Myanmar and S Thailand to Sumatra, Borneo and Palawan (SW Philippines)

❏ *Palawanomys furvus*
Palawan Soft-furred Mountain Rat

Mt Mantalingajan (Palawan, Philippines); Endangered

❏ *Kadarsanomys sodyi* Sody's Tree Rat

Gunung Pangrango-Gede (W Java)

❏ *Tryphomys adustus* Luzon Short-nosed Rat

Luzon (N Philippines)

❏ *Abditomys latidens* Luzon Broad-toothed Rat

Luzon (N Philippines)

❏ *Bullimus bagobus* Bagobo Rat

EC & S Philippines

❏ *Bullimus gamay* Camiguin Forest Rat

Camiguin (Philippines)

❏ *Bullimus luzonicus* Luzon Forest Rat

Luzon (N Philippines)

❏ *Tarsomys apoensis* Long-footed Rat

Mountains of Mindanao (SE Philippines)

❏ *Tarsomys echinatus* Spiny Long-footed Rat

Mountains of Mindanao (SE Philippines)

❏ *Limnomys sibuanus* Mindanao Mountain Rat

Mt Apo and Mt Malindang (SE Mindanao, SE Philippines)

❏ *Taeromys arcuatus* Salokko Rat

Gunung Tanke Salokko (SE Sulawesi)

❏ *Taeromys callitrichus* Lovely-haired Rat

NE, C & SE Sulawesi

❏ *Taeromys celebensis* Sulawesi Rat

Sulawesi

❏ *Taeromys hamatus* Sulawesi Montane Rat

Mountains of C Sulawesi

❏ *Taeromys punicans* Sulawesi Forest Rat

C & SW Sulawesi

❏ *Taeromys taerae* Tondano Rat

Highlands of NE Sulawesi

❏ *Paruromys dominator* Sulawesi Giant Rat

Sulawesi

❏ *Paruromys ursinus* Sulawesi Bear Rat

Gunung Lompobatang (SW Sulawesi); Endangered

❏ *Papagomys armandvillei* Flores Giant Tree Rat

Flores (Lesser Sundas)

❏ *Komodomys rintjanus* Komodo Rat

Rintja, Padar and Flores (Lesser Sundas)

❏ *Bunomys andrewsi* Andrews' Hill Rat

C & SE Sulawesi

❏ *Bunomys chrysocomus* Yellow-haired Hill Rat

Sulawesi

❏ *Bunomys coelestis* Heavenly Hill Rat

Gunung Lompobatang (SW Sulawesi); Endangered

❏ *Bunomys fratrorum* Fraternal Hill Rat

NE Sulawesi

❏ *Bunomys heinrichi* Heinrich's Hill Rat

Gunung Lompobatang (SW Sulawesi)

❏ *Bunomys penitus* Inland Hill Rat

C & SE Sulawesi

❏ *Bunomys prolatus* Long-headed Hill Rat

Gunung Tambusisi (C Sulawesi); Endangered

❏ *†Paulamys naso* Flores Long-nosed Rat

Formerly Flores (Lesser Sundas); Extinct (known from subfossil remains plus a specimen collected in 1991 that has been assigned to this species)

❏ *Stenomys ceramicus* Seram Rat

Montane forests of Seram (S Moluccas)

❏ *Stenomys niobe* Moss-forest Rat

Hill and montane forests of New Guinea

❏ *Stenomys omichlodes* Arianus' Rat

Alpine heath and scrub in mountains of W New Guinea

❏ *Stenomys richardsoni* Glacier Rat

Alpine grassland and periglacial tundra in high mountains of W New Guinea

❏ *Stenomys vandeuseni* Van Deusen's Rat

Primary montane forests in far SE New Guinea; Endangered

❏ *Stenomys verecundus* Slender Rat

Forests and gardens of New Guinea

❏ *Maxomys alticola* Mountain Spiny Rat

Gunung Kinabalu and Gunung Trus Madi (N Borneo); Endangered

❏ *Maxomys baeodon* Bornean Small Spiny Rat

N Borneo; Endangered

❏ *Maxomys bartelsii* Bartels' Spiny Rat

Mountains of W & C Java

❏ *Maxomys dollmani* Dollmann's Spiny Rat

Highland forests of Rantekaroa and Gunung Tanke Salokko (Sulawesi)

❏ *Maxomys hellwaldii* Hellwald's Spiny Rat

Lowland rainforests of Sulawesi

❏ *Maxomys hylomyoides* Sumatran Spiny Rat

Highland forests of W Sumatra

❏ *Maxomys inas* Malayan Mountain Spiny Rat

Highland forests of Malay Peninsula

❏ *Maxomys inflatus* Fat-nosed Spiny Rat

Highland forests of W Sumatra

❏ *Maxomys moi* Mo's Spiny Rat

S Laos and S Vietnam

❏ *Maxomys musschenbroekii*
Musschenbroek's Spiny Rat

Sulawesi

❏ *Maxomys ochraceiventer*
Chestnut-bellied Spiny Rat

Highlands of N & E Borneo

❏ *Maxomys pagensis* Pagai Spiny Rat

Siberut, Sipora and Pagai Is. (Mentawai Is., Sumatra)

❏ *Maxomys panglima* Palawan Spiny Rat

Busuanga, Culion, Palawan and Balabac (WC & SW Philippines)

❏ *Maxomys rajah* Rajah Spiny Rat

S Thailand, Malay Peninsula, Riau Is., Sumatra and Borneo

❏ *Maxomys surifer* Red Spiny Rat

Indochina and Malay Peninsula to Borneo, Sumatra and Java

❏ *Maxomys wattsi* Watts' Spiny Rat

Gunung Tambusisi (C Sulawesi); Endangered

❏ *Maxomys whiteheadi* Whitehead's Spiny Rat

S Thailand, Malay Peninsula, Sumatra and Borneo

❏ *Leopoldamys edwardsi*
Milne-Edwards' Long-tailed Giant Rat

N India to C China, Indochina, Malay Peninsula and W Sumatra

❏ *Leopoldamys neilli* Neill's Long-tailed Giant Rat

W & C Thailand; Endangered

❏ *Leopoldamys sabanus*
Common Long-tailed Giant Rat

Bangladesh and Indochina to Sumatra, Borneo and Java

❏ *Leopoldamys siporanus*
Mentawai Long-tailed Giant Rat

Siberut, Sipora and Pagai Is. (Mentawai Is., Sumatra)

❏ *Niviventer andersoni*
Anderson's White-bellied Rat

SE Tibet, SW & C China

❏ *Niviventer brahma* Brahma White-bellied Rat

N Assam (NE India) and N Myanmar

❏ *Niviventer confucianus*
Chinese White-bellied Rat

N Myanmar, N Thailand and China

❏ *Niviventer coxingi* Coxing's White-bellied Rat

Taiwan

❏ *Niviventer cremoriventer* Dark-tailed Tree Rat

S Thailand, Malay Peninsula, Sumatra, Borneo, Java and Bali

❏ *Niviventer culturatus* Oldfield White-bellied Rat

Mountains of Taiwan

❏ *Niviventer eha* Smoke-bellied Rat

Nepal, NE India, N Myanmar and N Yunnan (China)

❏ *Niviventer excelsior* Large White-bellied Rat

Sichuan (C China)

❏ *Niviventer fulvescens* Chestnut White-bellied Rat

Nepal and N India to S China, Hainan, Indochina, Sumatra, Java and Bali

❏ *Niviventer hinpoon* Limestone Rat

Korat Plateau (Thailand)

❏ *Niviventer langbianis* Lang Bian White-bellied Rat

Assam (NE India), Myanmar, N Thailand, Laos and Vietnam

❏ *Niviventer lepturus*
Narrow-tailed White-bellied Rat

Highland forests of W & C Java

❏ *Niviventer niviventer*
Hodgson's White-bellied Rat

NE Pakistan, Nepal and N India

❏ *Niviventer rapit* Long-tailed Mountain Rat

Highlands of Malay Peninsula, Sumatra and N Borneo

❏ *Niviventer tenaster* Tenasserim White-bellied Rat

Mountains of Assam (NE India), S Myanmar and Vietnam

❏ *Chiromyscus chiropus* Fea's Tree Rat

E Myanmar, N Thailand, C Laos and Vietnam

❏ *Dacnomys millardi* Millard's Rat

E Nepal, NE India, N Laos and S Yunnan (China)

❏ *Srilankamys ohiensis* Ohiya Rat

Mountain forests of Sri Lanka

❏ *Lenothrix canus* Grey Tree Rat

Malay Peninsula and Borneo

❏ *Pithecheir melanurus* Red Tree Rat

Java

❏ *Pithecheir parvus* Malayan Tree Rat

Malay Peninsula

❏ *Eropeplus canus* Sulawesi Grey Rat

Montane forests of C Sulawesi; Endangered

❏ *Lenomys meyeri* Trefoil-toothed Giant Rat

N, C & SW Sulawesi and Sangihe I. (N Moluccas)

❏ *Margaretamys beccarii* Beccari's Margareta Rat

Lowland rainforests of NE & C Sulawesi

❏ *Margaretamys elegans* Elegant Margareta Rat

Gunung Nokilalaki (C Sulawesi)

❏ *Margaretamys parvus* Little Margareta Rat

Gunung Nokilalaki (C Sulawesi)

❏ *Melasmothrix naso* Sulawesi Shrew-Rat

Montane rainforests of C Sulawesi; Endangered

❏ *Tateomys macrocercus* Long-tailed Shrew-Rat

Gunung Nokilalaki (C Sulawesi)

❏ *Tateomys rhinogradoides* Tate's Shrew-Rat

Montane rainforests of C Sulawesi

❏ *Haeromys margarettae* Greater Ranee Mouse

Borneo

❏ *Haeromys minahassae* Minahassa Ranee Mouse

NE & C Sulawesi

❏ *Haeromys pusillus* Lesser Ranee Mouse

Palawan (SW Philippines) and Borneo

❏ *Anonymomys mindorensis* Mindoro Rat

Ilong Peak (Mindoro, WC Philippines)

❏ *Echiothrix leucura* Sulawesi Spiny Rat

Lowland rainforests of N & C Sulawesi

❏ *Phloeomys cumingi*
Southern Luzon Giant Cloud Rat

S Luzon, Marinduque and Catanduanes (N Philippines)

❏ *Phloeomys pallidus*
Northern Luzon Giant Cloud Rat

N Luzon (N Philippines)

❏ *Crateromys australis*
Dinagat Bushy-tailed Cloud Rat

Dinagat (EC Philippines); Endangered

❏ *Crateromys heaneyi* Panay Bushy-tailed Cloud Rat

Panay (C Philippines); Endangered

❏ *Crateromys paulus* Ilin Bushy-tailed Cloud Rat

Ilin (WC Philippines); Critically Endangered

❏ *Crateromys schadenbergi*
Luzon Bushy-tailed Cloud Rat

Mountains of N Luzon (N Philippines)

❏ *Batomys dentatus* Large-toothed Hairy-tailed Rat

Benguet region (N Luzon, N Philippines)

❏ *Batomys granti* Luzon Hairy-tailed Rat

Mt Data and Mt Isarog (Luzon, N Philippines)

❏ *Batomys russatus* Dinagat Hairy-tailed Rat

Dinagat (EC Philippines)

❏ *Batomys salomonseni* Mindanao Hairy-tailed Rat

Biliran, Leyte and Mindanao (EC & SE Philippines)

❏ *Carpomys melanurus*
Short-footed Luzon Tree Rat

Mt Data (N Luzon, N Philippines)

❏ *Carpomys phaeurus*
White-bellied Luzon Tree Rat

Mt Data and Mt Kapilingan (N Luzon, N Philippines)

❏ *Apomys abrae* Luzon Cordillera Forest Mouse

Luzon (N Philippines)

❏ *Apomys datae* Luzon Montane Forest Mouse

Highlands of N Luzon (N Philippines)

❏ *Apomys gracilirostris*
Large Mindoro Forest Mouse

Primary montane forests of Mindoro (WC Philippines)

❏ *Apomys hylocoetes* Mount Apo Forest Mouse

Highlands of Mindanao (SE Philippines)

❏ *Apomys insignis*
Mindanao Montane Forest Mouse

Highlands of Mindanao (SE Philippines)

❏ *Apomys littoralis*
Mindanao Lowland Forest Mouse

Biliran, Leyte, Bohol, Dinagat and Mindanao (EC & SE Philippines)

❏ *Apomys microdon* Small Luzon Forest Mouse

S Luzon and Catanduanes (N Philippines)

❏ *Apomys musculus* Least Forest Mouse

Luzon and Mindoro (N & WC Philippines)

❏ *Apomys sacobianus*
Long-nosed Luzon Forest Mouse

Sacobia R. region (Luzon, N Philippines)

❏ *Rhynchomys isarogensis* Mount Isarog Shrew-Rat

Mt Isarog (SE Luzon, N Philippines)

❏ *Rhynchomys soricoides* Mount Data Shrew-Rat

Mt Data (N Luzon, N Philippines)

❏ *Crunomys celebensis* Sulawesi Shrew-Mouse

C Sulawesi; Endangered

❏ *Crunomys fallax* North Luzon Shrew-Mouse

Isabella Prov. region (NC Luzon, N Philippines); Critically Endangered

❏ *Crunomys melanius*
Southern Philippine Shrew-Mouse

Leyte and Mindanao (EC & SE Philippines)

❏ *Crunomys suncoides*
Mount Katanglad Shrew-Mouse

Primary moss forest on Mt Katanglad (Mindanao, SE Philippines)

❏ *Sommeromys macrorhinos*
Long-nosed Shrew-Mouse

Moss forest of summit of Gunung Tokala (C Sulawesi)

❏ *Archboldomys luzonensis*
Mount Isarog Shrew-Mouse

Mt Isarog (SE Luzon, N Philippines); Endangered

❏ *Archboldomys musseri* Palanan Shrew-Mouse

Mossy forest on Mt Cetaceo (Luzon, N Philippines)

❏ *Chrotomys gonzalesi* Mount Isarog Striped Rat

Mt Isarog (SE Luzon, N Philippines); Critically Endangered

❏ *Chrotomys mindorensis* Mindoro Striped Rat

Lowlands of Luzon and Mindoro (N & WC Philippines)

❏ *Chrotomys whiteheadi* Luzon Striped Rat

Luzon (N Philippines)

❏ *Celaenomys silaceus* Blazed Luzon Shrew-Rat

N Luzon (N Philippines)

❏ *Pseudomys albocinereus* Ash-grey Mouse

SW Western Australia and near islands (Australia)

❏ *Pseudomys apodemoides* Silky Mouse

SE South Australia and W Victoria (Australia)

❏ *Pseudomys australis* Plains Mouse

S Northern Territory, South Australia, S Queensland and New South Wales (Australia)

❏ *Pseudomys bolami* Bolam's Mouse

S Western Australia and S South Australia (Australia)

❏ *Pseudomys calabyi*
Calaby's Pebble-mound Mouse

Headwaters of South Alligator and Mary Rs. (NW Northern Territory, Australia)

❏ *Pseudomys chapmani*
Western Pebble-mound Mouse

Pilbara Dist. region (NW Western Australia, Australia)

❏ *Pseudomys delicatulus* Delicate Mouse

Mixed savannas of trans-Fly region (SC New Guinea) and N Australia

❏ *Pseudomys desertor* Brown Desert Mouse

Arid deserts of interior C Australia

❏ *Pseudomys fieldi* Djoongari (Alice Springs Mouse)

Formerly widespread in W & C Australia, now confined to Bernier I. and Doole I. (Western Australia); Critically Endangered

❏ *Pseudomys fumeus* Smoky Mouse

Victoria (Australia)

❏ *Pseudomys fuscus* Broad-toothed Mouse

E New South Wales, S Victoria and Tasmania (Australia)

❏ *Pseudomys glaucus* Blue-grey Mouse

Murray-Darling basin (S Queensland and N New South Wales, Australia); last recorded 1956, Critically Endangered

❏ †*Pseudomys gouldii* Gould's Mouse

Formerly New South Wales (Australia); Extinct (last collected 1857)

❏ *Pseudomys gracilicaudatus*
Eastern Chestnut Mouse

Coastal E Australia

❏ *Pseudomys hermannsburgensis*
Sandy Inland Mouse

Arid deserts of interior W & C Australia

❏ *Pseudomys higginsi* **Long-tailed Mouse**

Tasmania (Australia)

❏ *Pseudomys johnsoni*
Central Pebble-mound Mouse

C Northern Territory (Australia)

❏ *Pseudomys laborifex* **Kimberley Mouse**

Kimberley Plateau (NE Western Australia and NW Northern Territory, Australia)

❏ *Pseudomys nanus* **Western Chestnut Mouse**

W & NE Western Australia, N Northern Territory and NW Queensland (Australia)

❏ *Pseudomys novaehollandiae* **New Holland Mouse**

Coastal E New South Wales, S Victoria and N Tasmania (Australia)

❏ *Pseudomys occidentalis* **Western Mouse**

SW Western Australia (Australia); Endangered

❏ *Pseudomys oralis* **Hastings River Mouse**

SE Queensland and NE New South Wales (Australia); Endangered

❏ *Pseudomys patrius* **Country Mouse**

Mt Inkerman region (Queensland, Australia)

❏ *Pseudomys pilligaensis* **Pilliga Mouse**

Pilliga Scrub region (N New South Wales, Australia)

❏ *Pseudomys praeconis* **Shark Bay Mouse**

Shark Bay region and Bernier I. (Western Australia, Australia)

❏ *Pseudomys shortridgei* **Heath Rat**

SW Western Australia and SW Victoria (Australia)

❏ *Notomys alexis* **Spinifex Hopping Mouse**

Western Australia, Northern Territory, South Australia and W Queensland (Australia)

❏ *†Notomys amplus* **Short-tailed Hopping Mouse**

Formerly SE Northern Territory and SE South Australia (Australia); Extinct (last collected in 1895)

❏ *Notomys aquilo* **Northern Hopping Mouse**

N Northern Territory and N Queensland (Australia)

❏ *Notomys cervinus* **Fawn Hopping Mouse**

S Northern Territory, South Australia and SW Queensland (Australia)

❏ *Notomys fuscus* **Dusky Hopping Mouse**

C & SE Australia

❏ *†Notomys longicaudatus*
Long-tailed Hopping Mouse

Formerly Western Australia and Northern Territory (Australia); Extinct (last seen in 1901)

❏ *†Notomys macrotis* **Big-eared Hopping Mouse**

Formerly Moore R. region (SW Western Australia, Australia); Extinct (known only from 2 specimens collected prior to 1844)

❏ *Notomys mitchellii* **Mitchell's Hopping Mouse**

S Western Australia, S South Australia and W Victoria (Australia)

❏ *†Notomys mordax* **Darling Downs Hopping Mouse**

Formerly Darling Downs region (SE Queensland, Australia); Extinct (known only from holotype specimen collected in 1840s)

❏ *†Conilurus albipes* **White-footed Rabbit Rat**

Formerly eucalypt forests of SE Australia; Extinct (last recorded in 1862)

❏ *Conilurus penicillatus* **Brush-tailed Rabbit Rat**

Mixed savannas of trans-Fly plains region (SC New Guinea) and coastal NC Australia and near islands

❏ *†Leporillus apicalis* **Lesser Stick-nest Rat**

Fomerly W, C & SE Australia; Extinct (last collected in 1933)

❏ *Leporillus conditor* **Greater Stick-nest Rat**

Formerly widespread in W & S Australia, now confined to a very few sites; Endangered

❏ *Mesembriomys gouldii* **Black-footed Tree Rat**

N Australia, with Melville and Bathurst Is.

❏ *Mesembriomys macrurus*
Golden-backed Tree Rat

N Western Australia and N Northern Territory (Australia)

❏ *Zyzomys argurus* **Silver-tailed Rock Rat**

N Australia and near islands

❏ *Zyzomys maini* **Arnhem Land Rock Rat**

Arnhem Land (Northern Territory, Australia)

❏ *Zyzomys palatalis* **Carpentarian Rock Rat**

Echo Gorge (Gulf Country, Northern Territory, Australia);
Critically Endangered

❏ *Zyzomys pedunculatus* **Central Rock Rat**

S Northern Territory (Australia); Critically Endangered

❏ *Zyzomys woodwardi* **Kimberley Rock Rat**

NE Western Australia and N Northern Territory (Australia)

❏ *Leggadina forresti* **Forrest's Mouse**

Interior Australia

❏ *Leggadina lakedownensis*
Lakeland Downs Mouse

NE Western Australia and N Queensland (Australia)

❏ *Xeromys myoides* **False Water Rat**

Northern Territory and Queensland (Australia)

❏ *Hydromys chrysogaster* **Australasian Water Rat**

Obi (S Moluccas), Kai and Aru Is. (SE Moluccas), New
Guinea, D'Entrecasteaux Is. and Australia incl. Tasmania

❏ *Hydromys habbema* **Mountain Water Rat**

Between Lake Habbema and Mt Wilhelmina (W New Guinea)

❏ *Hydromys hussoni* **Western Water Rat**

Wissel Lakes and Maprik region (W & NC New Guinea)

❏ *Hydromys neobritannicus* **New Britain Water Rat**

New Britain (Bismarck Archipelago)

❏ *Hydromys shawmayeri* **Shaw Mayer's Water Rat**

Mountains of C New Guinea

❏ *Crossomys moncktoni* **Earless Water Rat**

Fast-flowing mountain streams in E New Guinea

❏ *Microhydromys musseri* **Musser's Shrew-Mouse**

Undisturbed moss forest on Mt Somoro (Torricelli Mtns, NC
New Guinea)

❏ *Microhydromys richardsoni*
Groove-toothed Shrew-Mouse

Hill forests of New Guinea

❏ *Parahydromys asper* **Waterside Rat**

Highlands of New Guinea

❏ *Paraleptomys rufilatus* **Northern Hydromyine**

Upper slopes of Cyclops Mtns and North Coast Ranges (NC
New Guinea)

❏ *Paraleptomys wilhelmina* **Short-haired Hydromyine**

Medium- to high-elevation forests in W Central Cordillera (C
New Guinea)

❏ *Mayermys ellermani* **One-toothed Shrew-Mouse**

Scattered localities in montane moss forests of New Guinea

❏ *Neohydromys fuscus* **Mottled-tailed Shrew-Mouse**

Montane moss forests of C & E New Guinea

❏ *Pseudohydromys murinus* **Eastern Shrew-Mouse**

Montane forests from Mt Wilhelm to Wau region (E New
Guinea); Critically Endangered

❏ *Pseudohydromys occidentalis*
Western Shrew-Mouse

High-elevation moss forests of W & C New Guinea

❏ *Leptomys elegans* **Large Leptomys**

Mt Sisa, Owen Stanley Range and Mt Victory (New Guinea);
Critically Endangered

❏ *Leptomys ernstmayri* **Ernst Mayr's Leptomys**

Highlands of New Guinea

❏ *Leptomys signatus* **Fly River Leptomys**

Stuart Island Camp (Lower Fly R. region, S New Guinea);
Critically Endangered

❏ *Lorentzimys nouhuysi* **Long-footed Tree Mouse**

Forests of New Guinea

❏ *Melomys aerosus* **Dusky Melomys**

Seram (S Moluccas)

❏ *Melomys bannisteri* Bannister's Melomys

Kai Besar (Kai Is., S Moluccas)

❏ *Melomys bougainville* Bougainville Melomys

Buka and Bougainville (New Guinea) and Choiseul (Solomon Is.)

❏ *Melomys burtoni* Burton's Melomys

Coastal N & NE Australia

❏ *Melomys capensis* Cape York Melomys

Iron and McIlwraith Ranges (Cape York Peninsula, N Queensland)

❏ *Melomys caurinus* Short-tailed Talaud Melomys

Karakelong (Talaud Is., N Moluccas)

❏ *Melomys cervinipes* Fawn-footed Melomys

Coastal E Australia

❏ *Melomys cooperae* Cooper's Melomys

Yamdana I. (Tanimbar Is., Lesser Sundas)

❏ *Melomys fellowsi* Red-bellied Melomys

High-altitude moss forest in highlands of E New Guinea

❏ *Melomys fraterculus* Manusela Melomys

Mt Manusela (Seram, S Moluccas)

❏ *Melomys fulgens* Orange Melomys

Lowlands of Seram (S Moluccas)

❏ *Melomys gracilis* Long-tailed Melomys

Highlands of E New Guinea

❏ *Melomys gressitti* Gressitt's Melomys

New Guinea

❏ *Melomys howi* How's Melomys

Riama I. (Tanimbar Is., Lesser Sundas)

❏ *Melomys lanosus* Large-scaled Melomys

Montane forests of W & C New Guinea

❏ *Melomys leucogaster* White-bellied Melomys

Forests on S side of Central Cordillera (New Guinea) and Rossel (E Louisiade Archipelago)

❏ *Melomys levipes* Long-nosed Melomys

Sogeri Plateau and Astrolabe Range (SE New Guinea), ?New Britain (Bismarck Archipelago)

❏ *Melomys lorentzii* Lorentz's Melomys

Lowlands to mid-elevations of Arus Is. (SE Moluccas) and New Guinea

❏ *Melomys lutillus* Grassland Melomys

Lowland to montane grasslands of New Guinea, and Misima, Sudest and Woodlark (D'Entrecasteaux Is.)

❏ *Melomys matambuai* Manus Melomys

Manus (Bismarck Archipelago)

❏ *Melomys mollis* Thomas' Melomys

Scattered localities in montane forests of New Guinea

❏ *Melomys moncktoni* Monckton's Melomys

Lowland forests of Japen, Sideia and New Guinea

❏ *Melomys obiensis* Obi Melomys

Bisa and Obi (C Moluccas)

❏ *Melomys platyops* Lowland Melomys

Lowland and hill forests of Geelvinck Is., New Guinea, New Britain (Bismarck Archipelago) and D'Entrecasteaux Is.

❏ *Melomys rattoides* Large Melomys

Forests of Japen and N New Guinea

❏ *Melomys rubex* Mountain Melomys

Highlands of New Guinea

❏ *Melomys rubicola* Bramble Cay Melomys

Bramble Cay (Torres Strait, N Queensland, Australia); Critically Endangered

❏ *Melomys rufescens* Black-tailed Melomys

Salawati, Waigeo, Japen, New Guinea and Bismarck Archipelago

❏ *Melomys spechti* Specht's Melomys

Buka I. (Solomon Is.)

❏ *Melomys talaudium*
Long-tailed Talaud Melomys

Talaud Is. (N Moluccas)

❏ *Pogonomelomys bruijni* **Large Pogonomelomys**

Few localities in lowland rainforests of Salawati and New Guinea; Critically Endangered

❏ *Pogonomelomys mayeri*
Shaw Mayer's Pogonomelomys

Hill forests of N New Guinea

❏ *Pogonomelomys sevia*
Highland Pogonomelomys (Menzies' Mouse)

Forests to alpine zone in highlands of New Guinea

❏ *Coccymys albidens* **White-toothed Mouse**

Lake Habbema region (W New Guinea); Endangered

❏ *Coccymys ruemmleri* **Rümmler's Mouse**

Highlands of Central Cordillera (New Guinea)

Uromys **(Cyromys)**

❏ †*Uromys imperator* **Emperor Rat**

Formerly Guadalcanal (Solomon Is.), Extinct (not seen since c.1960s)

❏ †*Uromys porculus* **Guadalcanal Rat**

Formerly Guadalcanal (Solomon Is.), Extinct (not seen since c.1960s)

❏ *Uromys rex* **King Rat**

Primary forests of Guadalcanal (Solomon Is.); Critically Endangered

Uromys **(Uromys)**

❏ *Uromys anak* **Black-tailed Giant Rat**

Montane forests of New Guinea

❏ *Uromys boeadii* **Biak Giant Rat**

Biak (Geelvinck Is., New Guinea)

❏ *Uromys caudimaculatus* **Mottled-tailed Giant Rat**

Kai and Aru Is. (SE Moluccas), Waigeo, New Guinea, D'Entrecasteaux Is. and coastal NE Queensland (Australia)

❏ *Uromys emmae* **Emma's Giant Rat**

Owi (Geelvinck Is.) and W New Guinea

❏ *Uromys hadrourus* **Masked White-tailed Rat**

Thornton Peak (NE Queensland, Australia)

❏ *Uromys neobritannicus* **Bismarck Giant Rat**

New Britain (Bismarck Archipelago)

❏ *Solomys ponceleti* **Poncelet's Naked-tailed Rat**

Buka and Bougainville (New Guinea) and Choiseul (Solomon Is.); Endangered

❏ *Solomys salamonis* **Florida Naked-tailed Rat**

Florida I. (Solomon Is.); possibly extinct, only known from holotype collected in 19th Century

❏ *Solomys salebrosus* **Bougainville Naked-tailed Rat**

Buka and Bougainville (New Guinea) and Choiseul (Solomon Is.)

❏ *Solomys sapientis* **Isabel Naked-tailed Rat**

Santa Isabel (Solomon Is.)

❏ *Xenuromys barbatus* **Rock-dwelling Giant Rat**

Rocky areas in scattered localities in C & E New Guinea

❏ *Hyomys dammermani*
Western White-eared Giant Rat

Highlands of Arfak Mtns and W Central Cordillera (W & C New Guinea)

❏ *Hyomys goliath* **Eastern White-eared Giant Rat**

Highlands of E New Guinea

❏ *Anisomys imitator* **Uneven-toothed Rat**

Rainforests in Central Cordillera (New Guinea)

❏ *Mallomys aroaensis* **De Vis' Woolly Rat**

Highlands of New Guinea

❏ *Mallomys gunung* **Alpine Woolly Rat**

High-elevation alpine zone from Mt Wilhelmina to Mt Carstenz (W Central Cordillera, New Guinea); Critically Endangered

❏ *Mallomys istapantap* **Subalpine Woolly Rat**

Upper montane forests and subalpine grasslands of Central Cordillera (New Guinea)

❏ *Mallomys rothschildi* **Rothschild's Woolly Rat**

Montane forests of Central Cordillera (New Guinea)

❏ *Macruromys elegans* **Lesser Small-toothed Rat**

Mt Kunupi (Weyland Range, New Guinea); Critically Endangered

❏ *Macruromys major* Greater Small-toothed Rat

Mid-elevation forests in highlands of New Guinea; Endangered

❏ *Chiruromys forbesi* Forbes' Tree Mouse

Hill forests of SE New Guinea and forests of D'Entrecasteaux Is.

❏ *Chiruromys lamia* Broad-headed Tree Mouse

Mid-elevation forests of Owen Stanley Range (SE New Guinea)

❏ *Chiruromys vates* Lesser Tree Mouse

Lowland and foothill forests on S side of Central Cordillera (C & E New Guinea)

❏ *Pogonomys championi* Champion's Tree Mouse

Telefomin and Tifalmin valleys (C New Guinea)

❏ *Pogonomys loriae*
Large Tree Mouse (Prehensile-tailed Rat)

Forests and regrowth areas of New Guinea, D'Entrecasteaux Is. and coastal NE Queensland (Australia)

❏ *Pogonomys macrourus* Chestnut Tree Mouse

Japen, New Guinea and New Britain (Bismarck Archipelago)

❏ *Pogonomys sylvestris* Grey-bellied Tree Mouse

Montane forests of New Guinea

Subfamily: PLATACANTHOMYINAE (Pygmy Dormice—3)

❏ *Typhlomys chapensis* Chapa Pygmy Dormouse

Chapa region (N Vietnam); Critically Endangered

❏ *Typhlomys cinereus* Chinese Pygmy Dormouse

Montane forests of EC & S China

❏ *Platacanthomys lasiurus*
Malabar Spiny Dormouse

Forests of S India

Subfamily: MYOSPALACINAE (Zokors—7)

❏ *Myospalax aspalax* False Zokor

Amur basin (E Siberia) and Nei Mongol (N China)

❏ *Myospalax epsilanus* Manchurian Zokor

E Siberia and NE China

❏ *Myospalax fontanierii* Chinese Zokor

Grasslands of C China

❏ *Myospalax myospalax* Siberian Zokor

Kazakhstan and C Siberia

❏ *Myospalax psilurus* Transbaikal Zokor

E Siberia, E Mongolia, C & NE China

❏ *Myospalax rothschildi* Rothschild's Zokor

Gansu and Hubei (China)

❏ *Myospalax smithii* Smith's Zokor

Gansu and Ningxia (China)

Subfamily: SPALACINAE (Eurasian Mole-Rats—8)

❏ *Spalax arenarius* Sandy Mole-Rat

S Ukraine

❏ *Spalax giganteus* Giant Mole-Rat

Steppes of S Russia and W Kazakhstan

❏ *Spalax graecus* Balkan Mole-Rat

Steppes and fields of Romania and S Ukraine

❏ *Spalax microphthalmus* Greater Mole-Rat

Steppes of Ukraine and S Russia

❏ *Spalax zemni* Podolsk Mole-Rat

SE Poland and Ukraine

❏ *Nannospalax ehrenbergi* Palestine Mole-Rat

Arid soils of N Libya, N Egypt and Middle East

❏ *Nannospalax leucodon* Lesser Mole-Rat

Steppes, meadows and pastures of SE Europe, NW Turkey and SW Ukraine

❏ *Nannospalax nehringi*
Nehring's Blind Mole-Rat

Turkey, Georgia and Armenia

Subfamily: RHIZOMYINAE (African and Oriental Mole-Rats—15)

❏ *Tachyoryctes ankoliae* Ankole Mole-Rat

S Uganda

❏ *Tachyoryctes annectens* Mianzini Mole-Rat

E side of Lake Naivasha (Kenya); Endangered

❏ *Tachyoryctes audax* Audacious Mole-Rat

Aberdare Mtns (Kenya)

❏ *Tachyoryctes daemon* Demon Mole-Rat

N Tanzania

❏ *Tachyoryctes macrocephalus*
Big-headed Mole-Rat

Highlands of S Ethiopia

❏ *Tachyoryctes naivashae* Naivasha Mole-Rat

W & S sides of Lake Naivasha (Kenya); Endangered

❏ *Tachyoryctes rex* King Mole-Rat

Mid altitude slopes of Mt Kenya (Kenya); Endangered

❏ *Tachyoryctes ruandae* Rwanda Mole-Rat

E D.R. Congo, Rwanda and Burundi

❏ *Tachyoryctes ruddi* Rudd's Mole-Rat

SE Uganda and SW Kenya

❏ *Tachyoryctes spalacinus* Embi Mole-Rat

Plains near Mt Kenya (Kenya)

❏ *Tachyoryctes splendens*
East African Mole-Rat

Ethiopia, Somalia and NW Kenya

❏ *Rhizomys pruinosus* Hoary Bamboo Rat

Assam (NE India) to S China, S to Malay Peninsula

❏ *Rhizomys sinensis* Chinese Bamboo Rat

C & S China, N Myanmar and Vietnam

❏ *Rhizomys sumatrensis* Large Bamboo Rat

Yunnan (China), Indochina and Malay Peninsula, Sumatra

❏ *Cannomys badius* Lesser Bamboo Rat

E Nepal and N India to Yunnan (China) and Cambodia

Family: GLIRIDAE (Dormice—27)

❏ *Graphiurus christyi* Christy's Dormouse

S Cameroon and N D.R. Congo

❏ *Graphiurus crassicaudatus* Jentink's Dormouse

Liberia to Cameroon

❏ *Graphiurus hueti* Huet's Dormouse

Senegal to Cameroon, Central African Republic and Gabon

❏ *Graphiurus kelleni* Kellen's Dormouse

Angola, Zambia, Zimbabwe and Malawi

❏ *Graphiurus lorraineus* Lorrain Dormouse

Sierra Leone to N Angola, D.R. Congo, SW Tanzania and Uganda

❏ *Graphiurus microtis* Small-eared Dormouse

Zambia, Malawi and Tanzania

❏ *Graphiurus monardi* Monard's Dormouse

S D.R. Congo, E Angola and NW Zambia

❏ *Graphiurus murinus* Woodland Dormouse

Savannas of S Angola, D.R. Congo, Sudan and Ethiopia, S to South Africa

❏ *Graphiurus ocularis* Spectacled Dormouse

Rocky habitats and woods of W South Africa

❏ *Graphiurus olga* Olga's Dormouse

Scattered localities in N Niger, N Nigeria and NE Cameroon

❏ *Graphiurus parvus* Savanna Dormouse

Savannas of Gambia, Ivory Coast and Mali to Somalia, S in E Africa to Tanzania

❏ *Graphiurus platyops* Rock Dormouse

Rocky habitats of Angola, S D.R. Congo, Malawi and Mozambique, S to N South Africa

❏ *Graphiurus rupicola* Stone Dormouse

Namibia and NW South Africa

❏ *Graphiurus surdus* Silent Dormouse

S Cameroon and Equatorial Guinea

❏ *Eliomys melanurus* Asian Garden Dormouse

Woods, oases and rocky habitats of Morocco to Egypt; Turkey, Middle East, Iraq and Saudi Arabia

❏ *Eliomys quercinus* Garden Dormouse

Woods, plantations, scrub and gardens of Europe (except NW) to Ural Mtns and Ukraine

❏ *Dryomys laniger* Woolly Dormouse

W & C Taurus Mtns (S Turkey)

❏ *Dryomys niethammeri* Niethammer's Dormouse

Baluchistan (W Pakistan)

❏ *Dryomys nitedula* Forest Dormouse

Forests, scrub and meadows of EC & SE Europe to C Kazakhstan, NW China, N Pakistan and Saudi Arabia

❏ *Dryomys sichuanensis* Chinese Dormouse

Wang-lang region (N Sichuan, C China); Endangered

❏ *Selevinia betpakdalaensis* Desert Dormouse

Lake Balkhash region (E Kazakhstan); Endangered

❏ *Myomimus personatus*
Masked Mouse-tailed Dormouse

NE Iran, Turkmenistan and Uzbekistan

❏ *Myomimus roachi*
Roach's Mouse-tailed Dormouse

Dry woods and scrub of SE Bulgaria, E Greece and Turkey

❏ *Myomimus setzeri*
Setzer's Mouse-tailed Dormouse

W Iran; Endangered

❏ *Glirulus japonicus* Japanese Dormouse

Honshu to Kyushu (C & S Japan); Endangered

❏ *Glis glis* Fat Dormouse (Edible Dormouse)

Deciduous forests and rocky scrub of C & S Europe to N Turkey, N Iran and SW Turkmenistan; feral S England

❏ *Muscardinus avellanarius* Hazel Dormouse

Deciduous forests and scrub of WC Europe to Russia, Ukraine and N Turkey

Family: GEOMYIDAE
(Pocket Gophers—99)

❏ *Geomys arenarius* Desert Pocket Gopher

Deep sandy soils in river bottoms of S New Mexico and W Texas (USA) to N Chihuahua (Mexico)

❏ *Geomys attwateri* Attwater's Pocket Gopher

Coastal plain grasslands of SE Texas (USA)

❏ *Geomys breviceps* Baird's Pocket Gopher

Sandy prairie grasslands of E Oklahoma, S Arkansas, E Texas and W Louisiana (USA)

❏ *Geomys bursarius* Plains Pocket Gopher

Deep sandy or loam soils of Great Plains from SC Manitoba (Canada) to N Texas (USA)

❏ *Geomys knoxjonesi* Knox Jones' Pocket Gopher

Grasslands of SE New Mexico and WC Texas (USA)

❏ *Geomys personatus* Texas Pocket Gopher

Coastal plain grasslands of S Texas (USA) and NE Tamaulipas (Mexico)

❏ *Geomys pinetis* Southeastern Pocket Gopher

Sandy pine-forest soils in S Alabama and S Georgia to C Florida (USA)

❏ *Geomys texensis* Central Texas Pocket Gopher

Deep brown sandy or gravelly loam soils of C Texas (USA)

❏ *Geomys tropicalis* Tropical Pocket Gopher

SE Tamaulipas (Mexico)

Pappogeomys (Cratogeomys)

❏ *Pappogeomys castanops*
Yellow-faced Pocket Gopher

Deep sandy soils of SE Colorado and SW Kansas (USA) to C San Luis Potosi (Mexico)

❏ *Pappogeomys fumosus* Smoky Pocket Gopher

Plains of E Colima (Mexico)

❏ *Pappogeomys gymnurus* Llano Pocket Gopher

C Jalisco to NE Michoacán (Mexico)

❏ *Pappogeomys merriami* Merriam's Pocket Gopher

EC Mexico

❑ *Pappogeomys neglectus*
Querétaro Pocket Gopher

Cerro de la Calentura (Querétaro, Mexico); Critically Endangered

❑ *Pappogeomys tylorhinus*
Naked-nosed Pocket Gopher

WC & C Mexico

❑ *Pappogeomys zinseri* Zinser's Pocket Gopher

NE Jalisco (Mexico)

Pappogeomys (Pappogeomys)

❑ *Pappogeomys alcorni* Alcorn's Pocket Gopher

Sierra del Tigre (S Jalisco, Mexico)

❑ *Pappogeomys bulleri* Buller's Pocket Gopher

Nayarit, Jalisco and Colima (Mexico)

Orthogeomys (Heterogeomys)

❑ *Orthogeomys hispidus* Hispid Pocket Gopher

Sandy forest clearing and farmland soils of E Mexico to Guatemala, Belize and NW Honduras

❑ *Orthogeomys lanius* Big Pocket Gopher

Mt Orizaba (Veracruz, Mexico)

Orthogeomys (Macrogeomys)

❑ *Orthogeomys cavator* Chiriqui Pocket Gopher

Forest clearing and farmland soils of SE Costa Rica and NW Panama

❑ *Orthogeomys cherriei* Cherrie's Pocket Gopher

Forest, regrowth and farmland soils of N & NE Costa Rica

❑ *Orthogeomys dariensis* Darién Pocket Gopher

Forest clearing and farmland soils of E Panama

❑ *Orthogeomys heterodus* Variable Pocket Gopher

Montane forest clearing and farmland soils of Cordillera Central and Cordillera de Talamanca (C Costa Rica)

❑ *Orthogeomys matagalpae*
Nicaraguan Pocket Gopher

NC Honduras and NC Nicaragua

❑ *Orthogeomys thaeleri* Thaeler's Pocket Gopher

Serranía de Baudó (NW Colombia)

❑ *Orthogeomys underwoodi*
Underwood's Pocket Gopher

Forest clearing and farmland soils of Pacific lowlands of C Costa Rica

Orthogeomys (Orthogeomys)

❑ *Orthogeomys cuniculus* Oaxacan Pocket Gopher

Zanatepec region (Oaxaca, Mexico); Critically Endangered

❑ *Orthogeomys grandis* Giant Pocket Gopher

Forest and farmland soils of WC Mexico, S on Pacific slope to El Salvador and S Honduras

❑ *Zygogeomys trichopus* Michoacán Pocket Gopher

Lago Pátzcuaro region (Michoacán, Mexico); Endangered

Thomomys (Megascapheus)

❑ *Thomomys bottae* Botta's Pocket Gopher

W USA and N Mexico

❑ *Thomomys bulbivorus* Camas Pocket Gopher

Willamette Valley and tributaries (NW Oregon, USA)

❑ *Thomomys townsendii*
Townsend's Pocket Gopher

Deep soils in river valleys of SE Oregon, C Idaho, NE California and N & C Nevada (USA)

❑ *Thomomys umbrinus* Southern Pocket Gopher

Locally in SC Arizona and SW New Mexico (USA) to EC Mexico

Thomomys (Thomomys)

❑ *Thomomys clusius* Wyoming Pocket Gopher

SE Idaho and WC Wyoming (USA)

❑ *Thomomys idahoensis* Idaho Pocket Gopher

Locally in SW Montana, E Idaho and WC Wyoming (USA)

❑ *Thomomys mazama* Western Pocket Gopher

Deep soils in alpine meadows of NW Washington, W Oregon and N California (USA)

❑ *Thomomys monticola* Mountain Pocket Gopher

Montane meadows of Sierra Nevada of N & C California and extreme WC Nevada (USA)

❑ *Thomomys talpoides* Northern Pocket Gopher

SW & SC Canada, and W USA

❏ *Liomys adspersus*
Panamanian Spiny Pocket Mouse

Dry forests and scrub of Pacific slope of C Panama

❏ *Liomys irroratus* **Mexican Spiny Pocket Mouse**

Dense brush and forests of extreme S Texas (USA) to S Mexico

❏ *Liomys pictus* **Painted Spiny Pocket Mouse**

Dry forests and scrub of W & E Mexico to W Guatemala

❏ *Liomys salvini* **Salvin's Spiny Pocket Mouse**

Forests, scrub and regrowth areas of E Oaxaca (Mexico) to C Costa Rica

❏ *Liomys spectabilis* **Jaliscan Spiny Pocket Mouse**

SE Jalisco (Mexico)

Heteromys (Heteromys)

❏ *Heteromys anomalus*
Trinidad Spiny Pocket Mouse

Forests of N Colombia, N Venezuela; Margarita; Trinidad & Tobago

❏ *Heteromys australis*
Southern Spiny Pocket Mouse

Rainforests of E Panama, W Colombia and NW Ecuador

❏ *Heteromys desmarestianus*
Forest Spiny Pocket Mouse

Forests and regrowth areas of EC Mexico to NW Colombia

❏ *Heteromys gaumeri*
Gaumer's Spiny Pocket Mouse

Lowland semideciduous forests, scrub and regrowth areas of Yucatán Peninsula (SE Mexico, N Belize and N Guatemala)

❏ *Heteromys goldmani*
Goldman's Spiny Pocket Mouse

Forests and plantations of Pacific slope of SE Chiapas (SE Mexico) and SW Guatemala

❏ *Heteromys teleus*
Southern Chocó Spiny Pocket Mouse

Forests of WC Ecuador

Heteromys (Xylomys)

❏ *Heteromys nelsoni* **Nelson's Spiny Pocket Mouse**

Wet high-elevation montane forests of E Chiapas (SE Mexico) and W Guatemala; Critically Endangered

❏ *Heteromys oresterus*
Mountain Spiny Pocket Mouse

Wet montane oak forests of Talamanca Range (SE Costa Rica)

❏ *Perognathus alticola* **White-eared Pocket Mouse**

Locally in submontane grassland and scrub in W San Bernardino Mtns (SC California, USA)

❏ *Perognathus amplus* **Arizona Pocket Mouse**

Desert flats of Arizona (USA) and NW Sonora (Mexico)

❏ *Perognathus fasciatus*
Olive-backed Pocket Mouse

Grassland and desert scrub of SC Canada, S to NE Utah and SE Colorado (USA)

❏ *Perognathus flavescens* **Plains Pocket Mouse**

Sandy habitats of interior SW & C USA and extreme NC Mexico

❏ *Perognathus flavus* **Silky Pocket Mouse**

Grasslands and scrub of C & SW USA to C Mexico

❏ *Perognathus inornatus*
San Joaquin Pocket Mouse

Grasslands and desert scrub of Central and Salinas Valleys (C California, USA)

❏ *Perognathus longimembris* **Little Pocket Mouse**

Grassland, sagebrush and desert flats of SE Oregon and W Utah (USA) to N Sonora and Baja California Norte (Mexico)

❏ *Perognathus merriami* **Merriam's Pocket Mouse**

Shortgrass prairies and scrub of SE New Mexico and W Texas (USA) to NE Mexico

❏ *Perognathus parvus* **Great Basin Pocket Mouse**

Sandy habitats of SC British Columbia (Canada) and interior W USA

❏ *Chaetodipus arenarius*
Little Desert Pocket Mouse

Baja California (Mexico)

❏ *Chaetodipus artus* **Narrow-skulled Pocket Mouse**

S Sonora, Sinaloa, SW Chihuahua and W Durango (Mexico)

❏ *Chaetodipus baileyi* **Bailey's Pocket Mouse**

Gravel soils of extreme SE California to extreme SW New Mexico (USA), S to Baja California and N Sinaloa (Mexico)

❏ *Chaetodipus californicus*
California Pocket Mouse

Arid grassland and scrub of C California (USA) to N Baja California Norte (Mexico)

❏ *Chaetodipus eremicus*
Chihuahuan Desert Pocket Mouse

Chihuahuan Desert of S New Mexico and W Texas (USA) to NC Mexico

❏ *Chaetodipus fallax* **San Diego Pocket Mouse**

Scrub habitats of extreme S Calfornia (USA) to W Baja California Norte (Mexico)

❏ *Chaetodipus formosus* **Long-tailed Pocket Mouse**

Arid rocky habitats of SW USA and coastal E Baja California (Mexico)

❏ *Chaetodipus goldmani* **Goldman's Pocket Mouse**

Sonora, N Sinaloa and SW Chihuahua (Mexico)

❏ *Chaetodipus hispidus* **Hispid Pocket Mouse**

Grasslands and scrub of Great Plains region of C USA, S to EC Mexico

❏ *Chaetodipus intermedius* **Rock Pocket Mouse**

Arid rocky slopes of SW & SC USA, S to C Sonora and C Chihuahua (Mexico)

❏ *Chaetodipus lineatus* **Lined Pocket Mouse**

San Luis Potosi (Mexico)

❏ *Chaetodipus nelsoni* **Nelson's Pocket Mouse**

Rocky scrub of extreme SE New Mexico and SW Texas (USA) to Jalisco and San Luis Potosi (Mexico)

❏ *Chaetodipus penicillatus*
Sonoran Desert Pocket Mouse

Sandy soils of Sonoran Desert of SW USA and NW Mexico

❏ *Chaetodipus pernix* **Sinaloan Pocket Mouse**

S Sonora to N Nayarit (Mexico)

❏ *Chaetodipus spinatus*
Spiny-rumped Pocket Mouse

Rocky deserts of extreme S Nevada, SE California (USA) and Baja California (Mexico)

❏ *Dipodomys agilis* **Agile Kangaroo-Rat**

Chaparral and coastal scrub of SW California (USA) and Baja California (Mexico)

❏ *Dipodomys californicus* **California Kangaroo-Rat**

Scrub habitats of SC Oregon and N California (USA)

❏ *Dipodomys compactus* **Gulf Coast Kangaroo-Rat**

Coastal sandy plains of S Texas (USA) and barrier islands of N Tamaulipas (Mexico)

❏ *Dipodomys deserti* **Desert Kangaroo-Rat**

Arid sandy deserts of interior SW USA and NW Mexico

❏ *Dipodomys elator* **Texas Kangaroo-Rat**

Very locally in clay grasslands of extreme SW Oklahoma and extreme NC Texas (USA)

❏ *Dipodomys gravipes* **San Quintin Kangaroo-Rat**

NW Baja California Norte (Mexico); Endangered

❏ *Dipodomys heermanni* **Heermann's Kangaroo-Rat**

C California (USA)

❏ *Dipodomys ingens* **Giant Kangaroo-Rat**

Dry sandy-loam plains of SW San Joaquin Valley and Inner Coastal range (WC California, USA); Critically Endangered

❏ *Dipodomys insularis*
San José Island Kangaroo-Rat

San José I. (Baja California Sur, Mexico); Critically Endangered

❏ *Dipodomys margaritae*
Santa Margarita Island Kangaroo-Rat

Santa Margarita I. (Baja California Sur, Mexico); Critically Endangered

❏ *Dipodomys merriami* **Merriam's Kangaroo-Rat**

SW & SC USA to C Mexico

❏ *Dipodomys microps* **Chisel-toothed Kangaroo-Rat**

Saltbush flats of Great Basin region (interior W USA)

❏ *Dipodomys nelsoni* **Nelson's Kangaroo-Rat**

S Chihuahua and N Coahuila to N San Luis Potosi and S Nuevo Leon (Mexico)

❏ *Dipodomys nitratoides* **San Joaquin Kangaroo-Rat**

Alkaline flats of S San Joaquin Valley (S California, USA)

❏ *Dipodomys ordii* **Ord's Kangaroo-Rat**

Fine-sand semiarid grasslands and scrub of SC Canada and interior W & C USA to C Mexico

❏ *Dipodomys panamintinus* Panamint Kangaroo-Rat

Sandy and gravelly desert flats of E California and W & S Nevada (USA)

❏ *Dipodomys phillipsii* Phillips' Kangaroo-Rat

C Durango to N Oaxaca (Mexico)

❏ *Dipodomys simulans* Dulzura Kangaroo-Rat

Sandy or gravelly scrub of coastal S California and Baja Califormia (Mexico)

❏ *Dipodomys spectabilis*
Banner-tailed Kangaroo-Rat

Desert grasslands of SC USA and N Mexico

❏ *Dipodomys stephensi* Stephens' Kangaroo-Rat

Sandy grasslands and scrub of Riverside, San Bernardino and San Diego Cos. (S California, USA)

❏ *Dipodomys venustus* Narrow-faced Kangaroo-Rat

Sandy scrub-covered slopes of coastal WC California (USA)

❏ *Microdipodops megacephalus*
Dark Kangaroo-Mouse

Gravelly soils in sagebrush of SE Oregon to EC California, C Nevada and WC Utah (USA)

❏ *Microdipodops pallidus* Pale Kangaroo-Mouse

Stabilized sand dunes of extreme EC California, W & SC Nevada (USA)

Family: PEDETIDAE (Spring-Hare—1)

❏ *Pedetes capensis* Spring-Hare

Sparsely vegetated hard-sand habitats of Angola, S D.R. Congo and Kenya, S to South Africa

Family: ANOMALURIDAE (Scaly-tailed Squirrels—7)

❏ *Anomalurus beecrofti*
Beecroft's Scaly-tailed Squirrel

Rainforests and forest edges of Senegal to Equatorial Guinea, D.R. Congo and Uganda

❏ *Anomalurus derbianus*
Lord Derby's Scaly-tailed Squirrel

Forests and woods of Sierra Leone to Kenya, S to Angola, Zambia and Mozambique

❏ *Anomalurus pelii* Pel's Scaly-tailed Squirrel

Lowland primary rainforests of Sierra Leone to Ghana

❏ *Anomalurus pusillus* Dwarf Scaly-tailed Squirrel

Very locally in lowland rainforests of S Cameroon, Gabon and D.R. Congo

❏ *Zenkerella insignis* Cameroon Scaly-tail

Very locally in lowland rainforests of S Cameroon, SW Central African Republic, Equatorial Guinea and Gabon

❏ *Idiurus macrotis*
Long-eared Scaly-tailed Flying Squirrel

Lowlan rainforests of Sierra Leone to E D.R. Congo

❏ *Idiurus zenkeri* Pygmy Scaly-tailed Flying Squirrel

Lowland rainforests of S Cameroon to Uganda

Family: CTENODACTYLIDAE (Gundis—5)

❏ *Pectinator spekei* Speke's Pectinator

Rocky habitats of Ethiopia, Djibouti and Somalia

❏ *Massoutiera mzabi* Mzab Gundi

Locally in rocky habitats in very arid areas of Hoggar and Tibesti regions of C Sahara Desert, from NE Mali to N Chad

❏ *Felovia vae* Felou Gundi

Rocky habitats in arid areas of Senegal, Mauritania and Mali

❏ *Ctenodactylus gundi* Common Gundi

Rocky habitats of N Morocco to NW Libya

❏ *Ctenodactylus vali* Val's Gundi

Rocky habitats of S Morocco, W Algeria and NW Libya

Family: HYSTRICIDAE (Old World Porcupines—11)

❏ *Hystrix africaeaustralis* Cape Porcupine

Widely in damp areas of Rep. Congo to Kenya, S to South Africa

❏ *Hystrix brachyura* Malayan Porcupine

Forests and grasslands of Nepal and NE India to C China, S to Indochina, Sumatra and Borneo

❏ *Hystrix crassispinis* Thick-spined Porcupine

N Borneo

❏ *Hystrix cristata* Crested Porcupine

Woods, savannas, scrub and farmland of Italy incl. Sicily; Morocco to Egypt; Senegal to Ethiopia and Tanzania

❏ *Hystrix indica* Indian Crested Porcupine

Forests, scrub and grasslands of Transcaucasia and Middle East to S Kazakhstan, Tibet, India and Sri Lanka

❏ *Hystrix javanica* Sunda Porcupine

Java, Lesser Sundas and S Sulawesi

❏ *Hystrix pumila* Philippine Porcupine

Busuanga and Palawan (WC & SW Philippines)

❏ *Hystrix sumatrae* Sumatran Porcupine

Sumatra

❏ *Atherurus africanus*
African Brush-tailed Porcupine

Rainforests of Gambia and Sierra Leone to D.R. Congo, S Sudan, Uganda and Kenya

❏ *Atherurus macrourus*
Asiatic Brush-tailed Porcupine

NE India to C China, S to Indochina, Malay Peninsula and Sumatra

❏ *Trichys fasciculata* Long-tailed Porcupine

Malaya, Sumatra and Borneo

Family: ERETHIZONTIDAE (New World Porcupines—17)

❏ *Erethizon dorsatum* North American Porcupine

Forests and grasslands of Alaska, Canada, W & NE USA, and extreme N Mexico

Coendou (Coendou)

❏ *Coendou bicolor* Bicolour-spined Porcupine

Rainforests of Andes from N Colombia to Bolivia

❏ *Coendou nycthemera* Koopman's Porcupine

Rainforests of Amazon basin S of lower R. Amazonas, E to Marajó I. (EC Amazonian Brazil)

❏ *Coendou prehensilis* Brazilian Porcupine

Forests, plantations and gardens of E Venezuela to French Guiana, and E of Andes S to Bolivia, Paraguay, N Argentina and SE Brazil; Trinidad

❏ *Coendou rothschildi* Rothschild's Porcupine

Lowland forests and gardens of Panama

Coendou (Echinoprocta)

❏ *Coendou rufescens* Stump-tailed Porcupine

Mid-elevations of Cordillera Oriental (NC Colombia)

Coendou (Sphiggurus)

❏ *Coendou ichillus*
Long-tailed Hairy Dwarf Porcupine

Amazon basin of E Ecuador

❏ *Coendou insidiosus* Bahia Hairy Dwarf Porcupine

Forests and regrowth areas of Bahia (EC Brazil)

❏ *Coendou melanurus*
Black-tailed Hairy Dwarf Porcupine

Lowland rainforests of Amazon basin of Colombia to Suriname and N Brazil

❏ *Coendou mexicanus*
Mexican Hairy Dwarf Porcupine

Forests of NE Mexico to W Panama

❏ *Coendou paragayensis*
Paraguay Hairy Dwarf Porcupine

E Paraguay

❏ *Coendou pruinosus*
Frosted Hairy Dwarf Porcupine

Very few records from montane forests of Cordillera Central of Colombia (and W Venezuela?)

❏ *Coendou quichua* Quichua Hairy Dwarf Porcupine

Andes of Ecuador

❏ *Coendou roosmalenorum*
Van Roosmalens' Hairy Dwarf Porcupine

R. Madeira drainage (C Amazonian Brazil)

❏ *Coendou spinosus*
Orange-spined Hairy Dwarf Porcupine

Atlantic coastal forests of Minas Gerais to Rio Grande do Sul (SE Brazil); ?NE Argentina and Uruguay

❏ *Coendou vestitus* Brown Hairy Dwarf Porcupine

Very few records from the upper R. Magdalena valley (C Colombia)

❏ *Chaetomys subspinosus* Bristle-spined Porcupine

Atlantic coastal forests of S Bahía and N Espírito Santo (EC Brazil)

Family: PETROMURIDAE (Dassie-Rat—1)

❏ *Petromus typicus* Dassie-Rat

Rocky habitats of SW Angola, Namibia and W Northern Cape Prov. (South Africa)

Family: THRYONOMYIDAE (Cane Rats—2)

❏ *Thryonomys gregorianus* Lesser Cane Rat

Dense grasslands of Cameroon to Ethiopia, S to Zimbabwe and W Mozambique

❏ *Thryonomys swinderianus* Greater Cane Rat

Reedbeds and dense seasonally wet grasslands of subsaharan Africa, S to E South Africa

Family: BATHYERGIDAE (African Mole-Rats—14)

❏ *Cryptomys anselli* Ansell's Mole-Rat

Zambia

❏ *Cryptomys bocagei* Bocage's Mole-Rat

C Angola, S D.R. Congo and NW Zambia

❏ *Cryptomys damarensis* Damara Mole-Rat

Sandy soils of S Angola and S Zambia to E Namibia, Botswana and W Zimbabwe

❏ *Cryptomys foxi* Nigerian Mole-Rat

C Nigeria

❏ *Cryptomys hottentotus* African Mole-Rat

Sandy and light soils of S D.R. Congo and Tanzania to Namibia and South Africa

❏ *Cryptomys kafuensis* Kafue Mole-Rat

Zambia

❏ *Cryptomys mechowi* Mechow's Mole-Rat

S D.R. Congo and Tanzania to Angola, Zambia and Malawi

❏ *Cryptomys ochraceocinereus* Ochre Mole-Rat

E Nigeria to S Sudan, N D.R. Congo and NW Uganda

❏ *Cryptomys zechi* Togo Mole-Rat

EC Ghana and WC Togo

❏ *Georychus capensis* Cape Mole-Rat

Soils of Western and Eastern Cape Provs. (S South Africa)

❏ *Bathyergus janetta* Namaqua Dune Mole-Rat

Sandy soils of S Namibia and W Northern Cape Prov. (South Africa)

❏ *Bathyergus suillus* Cape Dune Mole-Rat

Sandy soils of Western Cape Prov. (S South Africa)

❏ *Heliophobius argenteocinereus* Silvery Mole-Rat

Sandy soils in Miombo (*Brachystegia*) woods of D.R. Congo and Kenya to Zimbabwe and N Mozambique

❏ *Heterocephalus glaber* Naked Mole-Rat

Dry hard-sandy soils in open woods, savannas and grasslands of C Ethiopia to C Somalia and S Kenya

Family: AGOUTIDAE (Agoutis and Pacas—16)

❏ *Myoprocta acouchy* Red Acouchi

Lowland rainforests of SE Colombia; also Guianas and E of Rs. Negro and Tapajós (E Amazonian Brazil)

❏ *Myoprocta exilis* Green Acouchi

Lowland rainforests of SE Colombia to E Peru, E to Rs. Negro and Madeira (C Amazonian Brazil)

❏ *Dasyprocta azarai* Azara's Agouti

Rainforests, cerrado and chaco of E Bolivia, S Brazil, Paraguay and NE Argentina

❏ *Dasyprocta coibae* Coiba Island Agouti

Forests of Coiba I. (Panama); Endangered

❏ *Dasyprocta cristata* Crested Agouti

Guyana, Suriname and French Guiana

❏ *Dasyprocta fuliginosa* Black Agouti

Forests of Colombia, Venezuela, Ecuador and NE Peru, E to Rs. Negro and Madeira (C Amazonian Brazil)

❏ *Dasyprocta guamara* Orinoco Agouti

Delta of Orinoco R. (NC Venezuela)

❏ *Dasyprocta kalinowskii* Kalinowski's Agouti

SE Peru

❏ *Dasyprocta leporina* Red-rumped Agouti

Forests, plantations and gardens of Venezuela to French Guiana, S to Amazonian & EC Brazil; Lesser Antilles

❏ *Dasyprocta mexicana* Mexican Black Agouti

Lowland forests and regrowth areas of C Veracruz to NW Chiapas (Mexico); feral Cuba

❏ *Dasyprocta prymnolopha* Black-rumped Agouti

Forests, scrub, cerrado and caatinga of E Brazil

❏ *Dasyprocta punctata* Central American Agouti

Forests, plantations and gardens of SE Mexico to W Venezuela, and W of Andes to C Ecuador; feral Cuba and Cayman Is.

❏ *Dasyprocta ruatanica* Roatán Island Agouti

Forests of Roatán I. (Honduras); Endangered

❏ *Dasyprocta variegata* Brown Agouti

Forests, plantations and gardens of C Peru to Bolivia and NW Argentina

❏ *Cuniculus paca* Paca

Forests, plantations and gardens of C Mexico to Suriname, S to Paraguay, Misiones Prov. (NE Argentina) and S Brazil; feral Cuba

❏ *Cuniculus taczanowskii* Mountain Paca

Montane cloud forests of NW Venezuela, Colombia, Ecuador and Peru

Family: DINOMYIDAE (Pacarana—1)

❏ *Dinomys branickii* Pacarana

Rainforests of E Andean foothills of Colombia and Venezuela to Bolivia, and Amazon basin of Peru and W Brazil; Endangered

Family: CAVIIDAE (Cavies and Guinea-Pigs—16)

❏ *Galea flavidens* Brandt's Yellow-toothed Cavy

Brazil? (type locality unknown); status uncertain

❏ *Galea musteloides* Common Yellow-toothed Cavy

Riparian habitats and farmland of S Peru, N Chile, Bolivia, W Paraguay and NW & EC Argentina

❏ *Galea spixii* Spix's Yellow-toothed Cavy

E Bolivia to caatinga of E Brazil

❏ *Microcavia australis* Southern Mountain Cavy

Sandy flats and farmland of extreme S Bolivia, S Chile and W Argentina

❏ *Microcavia niata* Andean Mountain Cavy

Andes of SW Bolivia

❏ *Microcavia shiptoni* Shipton's Mountain Cavy

Locally in montane brush and grasslands of NW Argentina

❏ *Cavia aperea* Brazilian Guinea-Pig

Grasslands of Colombia to French Guiana, S to Paraguay, NE & EC Argentina and Uruguay

❏ *Cavia fulgida* Shiny Guinea-Pig

Coastal EC Brazil

❏ *Cavia intermedia* Intermediate Guinea-Pig

Moleques do Sul Is. (Santa Catarina, SE Brazil)

❏ *Cavia magna* Greater Guinea-Pig

Uruguay and SE Brazil

❏ *Cavia porcellus* Domestic Guinea-Pig

Domesticated worldwide; wild ancestor unknown

❏ *Cavia tschudii* Montane Guinea-Pig

Moist rocky montane grasslands and riparian habitats of Peru, S Bolivia, N Chile and NW Argentina

❏ *Kerodon acrobata* Acrobat Cavy

Rio Sao Mateus (Goias, Brazil)

❏ *Kerodon rupestris* Rock Cavy

Dry rocky areas with low scrub of E Brazil

❏ *Dolichotis patagonum* Patagonian Mara

Open brush of sandy washes and flats of C Argentina

❏ *Dolichotis salinicola* Chacoan Mara

Arid lowland thorn scrub of Gran Chaco region of extreme S Bolivia, Paraguay and NW & NC Argentina

Family: HYDROCHOERIDAE (Capybara—1)

❑ *Hydrochoerus hydrochaeris* Capybara

Lowland riparian forests and seasonally flooded grasslands of Panama to Suriname, S to Peru, SE Paraguay, NE & EC Argentina, Uruguay and S Brazil

Family: OCTODONTIDAE (Degus and Tuco-tucos—53)

❑ *Spalacopus cyanus* Coruro

Sandy soils of lowlands of C Chile

❑ *Pithanotomys fuscus* Chilean Rock Rat

Montane forests and bunchgrass of Andes of C Chile and WC Argentina

❑ *Pithanotomys sagei* Sage's Rock Rat

Montane forests and bunchgrass of Andes of Neuquén Prov. (WC Argentina) and Malleco Prov. (C Chile)

❑ *Octodon bridgesi* Bridges' Degu

Forests and bamboo thickets of Andes of C Chile and WC Argentina

❑ *Octodon degus* Common Degu

Semiarid scrub of W slope of Andes of NC Chile

❑ *Octodon lunatus* Moon-toothed Degu

Dense thorn scrub of coastal mountains of NC Chile

❑ *Octodon pacificus* Isla Mocha Degu

Isla Mocha (Chile)

❑ *Octodontomys gliroides* Mountain Degu

Open dry rocky habitats of Andes of SW Bolivia, N Chile and NW Argentina

❑ *Octomys mimax* Mountain Viscacha-Rat

Arid rocky slopes of Andean foothills of NW Argentina

❑ *Pipanacoctomys aureus* Golden Viscacha-Rat

Salar de Pipanaco (Catamarca Prov., NW Argentina)

❑ *Tympanoctomys barrerae* Plains Viscacha-Rat

Arid halogenous plains of Mendoza Prov. (Argentina)

❑ *Salinoctomys loschalchalerosorum* Chalchalero Viscacha-Rat

Salinas Grandes (La Rioja Prov., NW Argentina)

❑ *Ctenomys argentinus* Argentine Tuco-tuco

NC Chaco Prov. (NE Argentina)

❑ *Ctenomys australis* Southern Tuco-tuco

Vegetated sand dunes of Buenos Aires Prov. (E Argentina)

❑ *Ctenomys azarai* Azara's Tuco-tuco

La Pampa Prov. (E Argentina)

❑ *Ctenomys boliviensis* Bolivian Tuco-tuco

C Bolivia, W Paraguay and Formosa Prov. (Argentina)

❑ *Ctenomys bonettoi* Bonetto's Tuco-tuco

Chaco Prov. (NE Argentina)

❑ *Ctenomys brasiliensis* Brazilian Tuco-tuco

Minas Gerais (E Brazil)

❑ *Ctenomys colburni* Colburn's Tuco-tuco

Extreme W Santa Cruz Prov. (Argentina)

❑ *Ctenomys conoveri* Conover's Tuco-tuco

Gran Chaco region of Paraguay and N Argentina

❑ *Ctenomys coyhaiquensis* Coyhaique Tuco-tuco

Chile

❑ *Ctenomys dorsalis* Chacoan Tuco-tuco

Gran Chaco region of W Paraguay

❑ *Ctenomys emilianus* Emily's Tuco-tuco

Sand dunes of Neuquén Prov. (WC Argentina)

❑ *Ctenomys frater* Forest Tuco-tuco

Forests, meadows and riparian flats of S Bolivia and NW Argentina

❑ *Ctenomys fulvus* Tawny Tuco-tuco

Sandy plains in mountains of N Chile and NW Argentina

❑ *Ctenomys haigi* Haig's Tuco-tuco

Arid grasslands of Chubut and Río Negro Provs. (WC Argentina)

❑ *Ctenomys knighti* Catamarca Tuco-tuco

Mountains of NW & WC Argentina

❏ *Ctenomys latro* Mottled Tuco-tuco

Dry sandy areas of NW Argentina

❏ *Ctenomys leucodon* White-toothed Tuco-tuco

Lake Titicaca region (SE Peru and WC Bolivia)

❏ *Ctenomys lewisi* Lewis' Tuco-tuco

S Bolivia

❏ *Ctenomys magellanicus* Magellanic Tuco-tuco

Grasslands of extreme S Chile and S Argentina

❏ *Ctenomys maulinus* Maule Tuco-tuco

Forests and open sandy areas of C Chile and WC Argentina

❏ *Ctenomys mendocinus* Mendoza Tuco-tuco

C Argentina

❏ *Ctenomys minutus* Tiny Tuco-tuco

S Brazil, NE Argentina and W Uruguay

❏ *Ctenomys nattereri* Natterer's Tuco-tuco

Mato Grosso (SW Brazil)

❏ *Ctenomys occultus* Furtive Tuco-tuco

Arid habitats of NW Argentina

❏ *Ctenomys opimus* Highland Tuco-tuco

Dry slopes of altiplano of Andes of S Peru, SW Bolivia, N Chile and NW Argentina

❏ *Ctenomys osvaldoreigi*
Osvaldo Reig's Tuco-tuco

Argentina

❏ *Ctenomys pearsoni* Pearson's Tuco-tuco

S Uruguay

❏ *Ctenomys perrensis* Goya Tuco-tuco

NE Argentina

❏ *Ctenomys peruanus* Peruvian Tuco-tuco

Altiplano of extreme S Peru

❏ *Ctenomys pontifex* San Luis Tuco-tuco

Andean foothills of WC Argentina

❏ *Ctenomys porteousi* Porteous' Tuco-tuco

EC Argentina

❏ *Ctenomys roigi* Roig's Tuco-tuco

Corrientes Prov. (NE Argentina)

❏ *Ctenomys saltarius* Salta Tuco-tuco

Dry slopes and plains of NW Argentina

❏ *Ctenomys sericeus* Silky Tuco-tuco

S Argentina

❏ *Ctenomys sociabilis* Social Tuco-tuco

Neuquén Prov. (WC Argentina)

❏ *Ctenomys steinbachi* Steinbach's Tuco-tuco

E Andes of W Santa Cruz Dept. (Bolivia)

❏ *Ctenomys talarum* Talas Tuco-tuco

Grasslands and pastures of EC Argentina

❏ *Ctenomys torquatus* Collared Tuco-tuco

Sandy savannas of NE Argentina, Uruguay and extreme S Brazil

❏ *Ctenomys tuconax* Robust Tuco-tuco

Moist plains of Tucumán Prov. (NW Argentina)

❏ *Ctenomys tucumanus* Tucumán Tuco-tuco

Moist plains of Tucumán Prov. (NW Argentina)

❏ *Ctenomys validus* Strong Tuco-tuco

Mendoza Prov. (WC Argentina)

Family: ECHIMYIDAE (Spiny Rats—81)

❏ *Thrichomys apereoides* Punar

Rocky outcrops, cerrado and caatinga of Paraguay and E Brazil

Proechimys (Proechimys)

❏ *Proechimys amphichoricus*
Venezuelan Spiny Rat

S Venezuela (and NW Brazil?)

❏ *Proechimys barinas* Barinas Spiny Rat

WC Venezuela

❏ *Proechimys bolivianus* Bolivian Spiny Rat

Upper R. Beni drainage (NW Bolivia)

❏ *Proechimys brevicauda* Huallaga Spiny Rat

NW Brazil and E Peru

❏ *Proechimys canicollis* Colombian Spiny Rat

NC Colombia and Venezuela

❏ *Proechimys chrysaeolus* Boyaca Spiny Rat

E Colombia

❏ *Proechimys cuvieri* Cuvier's Spiny Rat

E Guyana, Suriname, French Guiana and E Amazonian Brazil

❏ *Proechimys decumanus* Pacific Spiny Rat

Tumbes region of SW Ecuador and NW Peru

❏ *Proechimys echinothrix* Hedgehog Spiny Rat

Lower R. Juruá drainage (W Amazonian Brazil)

❏ *Proechimys gardneri* Gardner's Spiny Rat

W Amazonian Brazil and N Bolivia

❏ *Proechimys goeldii* Goeldi's Spiny Rat

SW & C Amazonian Brazil

❏ *Proechimys gorgonae* Gorgona Spiny Rat

Gorgona I. (Colombia)

❏ *Proechimys guairae* Guaira Spiny Rat

Mountains of NC Venezuela

❏ *Proechimys gularis* Ecuadorian Spiny Rat

Amazon basin of E Ecuador

❏ *Proechimys guyannensis* Cayenne Spiny Rat

E Colombia to French Guiana, S to C Brazil

❏ *Proechimys hendeei* Hendee's Spiny Rat

Amazon basin of S Colombia to NE Peru

❏ *Proechimys hoplomyoides* Guyanan Spiny Rat

SE Venezuela, W Guyana and NW Brazil

❏ *Proechimys kulinae* Kulinas' Spiny Rat

Between R. Solimões and upper R. Juruá (W Amazonian Brazil)

❏ *Proechimys longicaudatus* Long-tailed Spiny Rat

E Peru, Bolivia, N Paraguay and S Brazil

❏ *Proechimys magdalenae* Magdalena Spiny Rat

W Colombia

❏ *Proechimys mincae* Minca Spiny Rat

N Colombia

❏ *Proechimys oconnelli* O'Connell's Spiny Rat

R. Meta and tributaries (EC Colombia)

❏ *Proechimys pattoni* Patton's Spiny Rat

Headwaters of R. Juruá (W Amazonian Brazil) and Amazonian SE Peru

❏ *Proechimys poliopus* Grey-footed Spiny Rat

NE Colombia and extreme NW Venezuela

❏ *Proechimys quadruplicatus* Napo Spiny Rat

Amazon basin of E Ecuador and NE Peru

❏ *Proechimys roberti* Pará Spiny Rat

Rainforests of Pará to Minas Gerais (EC Brazil)

❏ *Proechimys semispinosus* Tomes' Spiny Rat

Lowland forests and regrowth areas of E Honduras to W Colombia, S to NE Peru and C Brazil

❏ *Proechimys simonsi* Simons' Spiny Rat

S Colombia, E Ecuador, NE Peru and W Amazonian Brazil

❏ *Proechimys steerei* Steere's Spiny Rat

Amazon basin of E Peru and SW & C Amazonian Brazil

❏ *Proechimys trinitatis* Trinidad Spiny Rat

Trinidad

❏ *Proechimys urichi* Sucre Spiny Rat

Sucre region (N Venezuela)

❏ *Proechimys warreni* Warren's Spiny Rat

Guyana and Suriname

Proechimys (Trinomys)

❏ *Proechimys albispinus* White-spined Spiny Rat

Coast and near islands of Bahia (EC Brazil)

❏ *Proechimys dimidiatus* **Atlantic Spiny Rat**

Atlantic coastal forests of E Rio de Janeiro (E Brazil)

❏ *Proechimys iheringi* **Ihering's Spiny Rat**

Atlantic coastal forests of SE Brazil

❏ *Proechimys mirapatanga* **Mirapatanga Spiny Rat**

SE Brazil

❏ *Proechimys myosuros* **Mouse-tailed Spiny Rat**

Bahia (EC Brazil)

❏ *Proechimys setosus* **Hairy Spiny Rat**

Forests of Minas Gerais (EC Brazil)

❏ *Proechimys yonenagae*
Yonenaga-Yassuda's Spiny Rat

Sand dunes of middle R. São Francisco region (Bahia, Brazil)

❏ *Euryzygomatomys spinosus* **Guiara**

Wet grasslands of Paraguay, S & E Brazil and NE Argentina

❏ *Carterodon sulcidens* **Owl's Spiny Rat**

Cerrado of E Brazil

❏ *Clyomys bishopi* **Bishop's Fossorial Spiny Rat**

Itapetininga region (São Paulo, Brazil)

❏ *Clyomys laticeps* **Broad-headed Spiny Rat**

Savannas and gardens of E Paraguay and S Brazil

❏ *Lonchothrix emiliae* **Tuft-tailed Spiny Tree Rat**

Lowland rainforests of lower Rs. Madeira and Tapajós (C Amazonian Brazil)

❏ *Hoplomys gymnurus* **Armoured Rat**

Lowland riparian rainforests of E Honduras to W Colombia and NW Ecuador

❏ *Mesomys hispidus* **Common Spiny Tree Rat**

Lowland and foothill rainforests of Amazon basin of S Venezuela, Colombia, E Ecuador, N Brazil and N & E Peru

❏ *Mesomys leniceps* **Woolly-headed Spiny Tree Rat**

Andes of Peru

❏ *Mesomys obscurus* **Dusky Spiny Tree Rat**

"Brazil" (exact type locality unknown); status unknown

❏ *Mesomys occultus* **Furtive Spiny Tree Rat**

Primary rainforests of Amazonian Brazil

❏ *Mesomys stimulax* **Suriname Spiny Tree Rat**

Rainforests of Suriname and N Brazil

❏ *Myocastor coypus* **Coypu (Nutria)**

Wetlands of SC Bolivia, SE Paraguay, S Chile, Argentina, Uruguay and SE Brazil; feral NW & S USA, C & SE Europe, E Africa, C & N Asia and Japan

❏ *Kannabateomys amblyonyx* **Southern Bamboo Rat**

Forests and bamboo thickets of SE Paraguay, Misiones Prov. (NE Argentina) and SE Brazil

❏ *Dactylomys boliviensis* **Bolivian Bamboo Rat**

SE Peru and C Bolivia

❏ *Dactylomys dactylinus* **Amazon Bamboo Rat**

Lowland rainforests and bamboo thickets of upper Amazon basin of SE Colombia and S Venezuela to N Bolivia, E to mouth of R. Amazonas (Brazil)

❏ *Dactylomys peruanus* **Peruvian Bamboo Rat**

Cloud forests of SE Peru and Bolivia

❏ *Olallamys albicauda* **White-tailed Olalla Rat**

Dense montane bamboo thickets of NW & C Colombia

❏ *Olallamys edax* **Greedy Olalla Rat**

High elevations of NC Colombia and W Venezuela

❏ *Echimys blainvillei* **Golden Atlantic Tree Rat**

Caatinga of Ceará and Bahia and near islands (E Brazil)

❏ *Echimys braziliensis* **Red-nosed Tree Rat**

Minas Gerais and Rio de Janeiro (EC Brazil)

❏ *Echimys chrysurus* **White-faced Tree Rat**

Primary rainforests of Guyana and Suriname, S to E of Rs. Negro and Xingu (E Amazonian Brazil)

❏ *Echimys dasythrix* **Drab Atlantic Tree Rat**

Atlantic coastal forests of São Paulo to Rio Grande do Sul (SE Brazil)

❏ *Echimys grandis* **Giant Tree Rat**

Riparian forests of lower R. Negro and both banks of R. Amazonas from R. Negro to I. Caviana (E Amazonian Brazil)

❑ *Echimys lamarum* Pallid Atlantic Tree Rat

Lowland Atlantic coastal forests of Bahia (EC Brazil)

❑ *Echimys macrurus* Long-tailed Tree Rat

Upper Amazon basin of Peru (and Brazil S of R. Amazonas?)

❑ *Echimys nigrispinus*
Black-spined Atlantic Tree Rat

Atlantic coastal rainforests of Rio de Janeiro to Santa Catarina (SE Brazil)

❑ *Echimys pattoni* Rusty-sided Atlantic Tree Rat

Atlantic coastal rainforests of SE Paraíba to São Paulo (E Brazil)

❑ *Echimys pictus* Painted Tree Rat

Atlantic coastal rainforests and plantations of S Bahia (EC Brazil)

❑ *Echimys rhipidurus* Peruvian Tree Rat

Lowland rainforests of Amazon basin of N & C Peru

❑ *Echimys saturnus* Dark Tree Rat

Few records from rainforests of Andean foothills and Amazon basin of E Ecuador and NE Peru

❑ *Echimys semivillosus* Speckled Tree Rat

Lowland dry deciduous forests and savannas of NE Colombia and NW Venezuela

❑ *Echimys thomasi* Giant Atlantic Tree Rat

Forests and plantations of I. de São Sabastião (Bahia, EC Brazil)

❑ *Echimys unicolor* Unicoloured Tree Rat

"Brazil" (exact type locality unknown); taxonomic status uncertain, possibly synonymous with *E. braziliensis*

❑ *Isothrix bistriata* Yellow-crowned Brush-tailed Rat

Lowland rainforests of Amazon basin of SE Colombia, S Venezuela, E Ecuador, E Peru and N Bolivia, E to Rs. Negro and Madeira (C Amazonian Brazil)

❑ *Isothrix pagurus* Plain Brush-tailed Rat

Lowland rainforests E of Rs. Negro and Madeira (E Amazonian Brazil)

❑ *Isothrix sinnamarensis*
Sinnamary Brush-tailed Rat

Lowland rainforests of R. Sinnamary region (French Guiana)

❑ *Diplomys caniceps* Arboreal Soft-furred Spiny Rat

Between Cordilleras Occidental and Central (W Colombia) (and S to N Ecuador?)

❑ *Diplomys labilis* Rufous Tree Rat

Forests, plantations and regrowth areas of Panama incl. I. del Rey, and W Colombia

❑ *Diplomys rufodorsalis* Red-crested Tree Rat

Very few records from montane forests of Sierra Nevada de Santa Marta (NE Colombia)

❑ *Makalata didelphoides* Armored Tree Rat

Lowland rainforests and gardens of Colombia to French Guiana, S to Peru, Bolivia and C Brazil; Trinidad & Tobago

❑ *Makalata occasius* Bare-tailed Tree Rat

Few records from lowland rainforests of E Ecuador and E Peru; Critically Endangered

Family: CAPROMYIDAE (Hutias—15)

Capromys (Capromys)

❑ *Capromys pilorides* Desmarest's Hutia

Cuba and near islands

Capromys (Mesocapromys)

❑ *Capromys angelcabrerai* Cabrera's Hutia

Cayos de Ana Maria (Cuba); Critically Endangered

❑ *Capromys auritus* Eared Hutia

Mangroves of Cayo Fragoso (Las Villas Prov., Cuba); Critically Endangered

❑ *Capromys nanus* Dwarf Hutia

Cienaga de Zapata (Matanzas Prov., Cuba); Critically Endangered

❑ *Capromys sanfelipensis* San Felipe Hutia

Grasslands of Cayo Juan Garcia (Pinar del Rio Prov., Cuba); Critically Endangered

Capromys (Mysateles)

❑ *Capromys garridoi* Garrido's Hutia

"Cayo Majá" (type locality uncertain) (Cuba); Critically Endangered

❑ *Capromys gundlachi* Gundlach's Hutia

N Isla de la Juventud (Cuba)

❏ *Capromys melanurus* Black-tailed Hutia

E Cuba

❏ *Capromys meridionalis* Southern Hutia

SW Isla de la Juventud (Cuba)

❏ *Capromys prehensilis* Prehensile-tailed Hutia

W & C Cuba

❏ *Isolobodon portoricensis* Puerto Rican Hutia

Hispaniola; feral Puerto Rico and near islands; Critically
Endangered

❏ *Geocapromys brownii* Brown's Hutia

Jamaica

❏ *Geocapromys ingrahami* Bahaman Hutia

East Plana Key (Bahamas); feral Little Wax and Warderick
Wells Cays (Bahamas)

❏ †*Geocapromys thoracatus* Swan Island Hutia

Formerly Little Swan I. (Honduras); Extinct (since 1950s)

❏ *Plagiodontia aedium* Hispaniolan Hutia

Hispaniola, incl. La Gonave I. (Haiti)

Family: CHINCHILLIDAE
(Viscachas and Chinchillas—6)

❏ *Lagostomus maximus* Plains Viscacha

Open grasslands and brush of SE Bolivia, W Paraguay, N & C
Argentina and S Uruguay

❏ *Lagidium peruanum* Northern Viscacha

Rocky montane areas in Andes of C & S Peru and extreme N Chile

❏ *Lagidium viscacia* Southern Viscacha

Rocky outcrops in Andes of S Peru, SW Bolivia, N & C Chile
and W Argentina

❏ *Lagidium wolffsohni* Wolffsohn's Viscacha

Extreme S Chile and SW Argentina

❏ *Chinchilla brevicaudata* Short-tailed Chinchilla

Montane grasslands and scrub of Andes of S Peru (extinct), S
Bolivia, N Chile and NW Argentina; Critically Endangered

❏ *Chinchilla lanigera* Chinchilla

Rocky areas in foothills of the Andes of NC Chile

Family: ABROCOMIDAE
(Chinchilla-Rats—9)

❏ *Cuscomys ashaninka*
White-fronted Cusco Rat

Peru

❏ *Abrocoma bennettii* Bennett's Chinchilla-Rat

Rocky slopes and brush of C Chile

❏ *Abrocoma boliviensis* Bolivian Chinchilla-Rat

Comarapa region (Santa Cruz Dept., Bolivia); only known
from type locality

❏ *Abrocoma budini* Budin's Chinchilla-Rat

Otro Cerro (Catamarca Prov., NW Argentina)

❏ *Abrocoma cinerea* Ashy Chinchilla-Rat

Arid rocky slopes of SE Peru, W Bolivia, N Chile and NW
Argentina

❏ *Abrocoma famatina* Famatina Chinchilla-Rat

Sierra de Famatina (LaRioja Prov., NW Argentina)

❏ *Abrocoma schistacea* Tontal Chinchilla-Rat

Sierra del Tontal (San Juan Prov., W Argentina)

❏ *Abrocoma uspallata*
Uspallata Chinchilla-Rat

Quebrada de la Vena (Mendoza Prov., W Argentina)

❏ *Abrocoma vaccarum* Vacas Chinchilla-Rat

Punta de Vacas (Mendoza Prov., W Argentina)

Grandorder: FERAE
Order: CIMOLESTA
Family: MANIDAE (Pangolins—7)

❏ *Manis crassicaudata* Indian Pangolin

Forests and arid scrub of Pakistan, Nepal and India to Yunnan
(SW China) and Sri Lanka

❏ *Manis javanica* Malayan Pangolin

Myanmar, Thailand, Indochina, Sumatra, Java, Borneo and SW
Philippines

❏ *Manis pentadactyla* Chinese Pangolin

Nepal to S China, N Indochina, Hainan and Taiwan

❏ *Smutsia gigantea* Giant Pangolin

Forests, savannas and grasslands of Senegal to W Kenya, S to
SW Angola, C D.R. Congo and Rwanda

❏ *Smutsia temminckii* Ground Pangolin

Woods, savannas and grasslands of Chad and S Sudan, S to N
South Africa

❏ *Phataginus tricuspis* Tree Pangolin

Lowland rainforests, regrowth areas and farmland of Senegal
to W Kenya, S to SW Angola and NE Zambia

❏ *Uromanis tetradactyla* Long-tailed Pangolin

Riparian habitats and swamps in forests of Gambia and
Senegal to W Uganda, S to SW Angola

Order: CARNIVORA
Family: NANDINIIDAE
(African Palm Civet—1)

❏ *Nandinia binotata* African Palm Civet

Rainforests and regrowth areas of Senegal to Sudan and
Kenya, S to Angola, Zambia, Zimbabwe and Mozambique

Family: VIVERRIDAE
(Genets and Civets—37)

❏ *Viverra civettina* Malabar Civet

S India; Critically Endangered

❏ *Viverra megaspila* Large-spotted Civet

Myanmar, Thailand, Vietnam and Malay Peninsula

❏ *Viverra tangalunga* Malayan Civet

Myanmar and China to Sumatra, Borneo, Philippines and
Sulawesi; feral Halmahera, Buru and Bacan (Moluccas)

❏ *Viverra tainguensis* Tainguen Civet

Tainguen Plateau (Vietnam)

❏ *Viverra zibetha* Large Indian Civet

Forests and scrub of India and Nepal to EC China, Indochina
and Indonesia

❏ *Genetta abyssinica* Ethiopian Genet

Dry woods of Ethiopia, Eritrea and Somalia (also Egypt and Sudan?)

❏ *Genetta angolensis* Angolan Genet

Miombo (*Brachystegia*) woods and riparian thickets of Angola
and D.R. Congo to Tanzania and Mozambique

❏ *Genetta bourloni* Bourlon's Genet

Rainforests of Sierra Leone, Guinea, Liberia and Ivory Coast

❏ *Genetta genetta* Common Genet

Widespread in SW Europe (?feral), Arabian Peninsula and N
and subsaharan Africa

❏ *Genetta johnstoni* Johnston's Genet

Rainforests of Guinea and Liberia (also Ivory Coast and
Ghana?)

❏ *Genetta maculata* Rusty-spotted Genet

Woods and savannas of subsaharan Africa

❏ *Genetta pardina* Forest Genet

Forests, woods and savannas of Gambia to Ghana

❏ *Genetta poensis* King Genet

Locally in rainforests of Liberia to Equatorial Guinea and Rep.
Congo

❏ *Genetta servalina* Servaline Genet

Rainforests of Niger; Cameroon and Rep. Congo to Kenya and
Tanzania

❏ *Genetta thierryi* Haussa Genet

Dry savannas of Gambia and Senegal to Niger and Cameroon

❏ *Genetta tigrina* Cape Genet

South Africa and Lesotho

❏ *Genetta victoriae* Giant Genet

Very locally in rainforests of NE D.R. Congo and extreme W
Uganda

❏ *Civettictis civetta* African Civet

Forests, riparian thickets and farmland in dry open country of
subsaharan Africa, S to NE South Africa

❏ *Viverricula indica* Small Indian Civet

Open woods, scrub, grasslands and built areas of India and SE
Asia; feral Yemen, Madagascar, Indian Ocean and SE Asian
islands

❏ *Poiana richardsonii* African Linsang

Rainforests of Liberia and Ivory Coast; Cameroon to Rep.
Congo and D.R. Congo

❏ *Osbornictis piscivora* Aquatic Genet

Clear shallow headwaters of forest sreams of NE D.R. Congo

❏ *Prionodon linsang* **Banded Linsang**

Malay Peninsula, Sumatra, Java, Borneo and near islands

❏ *Prionodon pardicolor* **Spotted Linsang**

Hill forests of N India, E Nepal and Bhutan to Yunnan (SW China), Indochina and Indonesia

❏ *Paguma larvata* **Masked Palm Civet**

Mid-elevation and montane forests of Pakistan and India to China, Taiwan, Indochina, Sumatra and Borneo; feral Japan

❏ *Macrogalidia musschenbroekii*
Sulawesi Palm Civet

Sulawesi

❏ *Paradoxurus hermaphroditus* **Asian Palm Civet**

Forests, plantations, scrub and built areas of India to China, Japan, Indochina, Philippines and New Guinea; feral Sula Is., Seram and Aru Is. (Moluccas)

❏ *Paradoxurus jerdoni* **Jerdon's Palm Civet**

S India

❏ *Paradoxurus zeylonensis* **Golden Palm Civet**

Sri Lanka

❏ *Arctogalidia trivirgata* **Small-toothed Palm Civet**

Forests of NE India to Yunnan (SW China), Indochina and Indonesia

❏ *Arctictis binturong* **Binturong**

Hill forests of India and Nepal to SW China, Indochina, Sumatra, Java, Borneo and Philippines

❏ *Hemigalus derbyanus* **Banded Palm Civet**

Malay Peninsula, Sumatra, Mentawai Is. and Borneo

❏ *Diplogale hosei* **Hose's Palm Civet**

N & W Borneo

❏ *Chrotogale owstoni* **Owston's Palm Civet**

SW China, Laos and Vietnam

❏ *Cynogale bennettii* **Otter-Civet**

Vietnam, Malay Peninsula, Sumatra and Borneo; Endangered

❏ *Fossa fossana* **Fanaloka (Malagasy Civet)**

Madagascar

❏ *Eupleres goudotii* **Falanouc**

Madagascar; Endangered

❏ *Cryptoprocta ferox* **Fosa**

Madagascar; Endangered

Family: FELIDAE (Cats—39)

Felis (Felis)

❏ *Felis bieti* **Chinese Desert Cat**

E Tibetan Plateau in C China

❏ *Felis catus* **Domestic Cat**

Domesticated worldwide, feral Madagascar, Solomon Is., Australia, New Zealand, New Caledonia, Canada, Brazil etc.; wild ancestor is *Felis silvestris* Wild Cat

❏ *Felis chaus* **Jungle Cat**

Forests, scrub, wet grasslands, deserts and farmland of N Egypt and Middle East to China and Vietnam

❏ *Felis margarita* **Sand Cat**

Sandy and rocky deserts of North Africa and Middle East to Uzbekistan and Pakistan

❏ *Felis nigripes* **Black-footed Cat**

Near cover in open sandy areas of SE Namibia, Botswana and South Africa

❏ *Felis silvestris* **Wild Cat**

Forests, scrub, and grasslands of Europe and Africa to China

Felis (Otocolobus)

❏ *Felis manul* **Pallas' Cat**

High-altitude stony deserts of Caspian Sea to E Siberia, Mongolia, C China and NW India

Felis (Prionailurus)

❏ *Felis bengalensis* **Leopard Cat**

Forests and forest edges of Afghanistan to Korea, Japan, Taiwan, Vietnam, Malaysia and the Philippines

❏ *Felis planiceps* **Flat-headed Cat**

Malay Peninsula, Sumatra and Borneo

❏ *Felis rubiginosa* **Rusty-spotted Cat**

Locally in forests, scrub, grassland and built areas of India and Sri Lanka

❏ *Felis viverrina* Fishing Cat

Forests, scrub, riparian habitats and grasslands of Pakistan, Nepal, India and Sri Lanka to S China, Taiwan, Indochina and Indonesia

Felis (Leptailurus)

❏ *Felis serval* Serval

Savannas, riparian grasslands and reedbeds of Morocco to Tunisia and subsaharan Africa S to E South Africa

Felis (Caracal)

❏ *Felis caracal* Caracal

Savannas, scrub and semi-deserts of Africa (except Sahara Desert); Turkey to Turkmenistan and India

Felis (Profelis)

❏ *Felis aurata* African Golden Cat

Forests and thickets of Gambia and Sierra Leone to N Angola, S D.R. Congo and Kenya

Felis (Catopuma)

❏ *Felis badia* Bay Cat

Borneo; Endangered

❏ *Felis temminckii* Asiatic Golden Cat

Hill forests and rocky habitats of NE India, Nepal, Bhutan and Bangladesh to C China, Indochina and Sumatra

Felis (Pardofelis)

❏ *Felis marmorata* Marbled Cat

Forests of NE India, Nepal, Bhutan and Bangaldesh to S China, Indochina, Sumatra and Borneo

Felis (Lynx)

❏ *Felis canadensis* Canadian Lynx

Boreal forests of Alaska, Canada, and locally in W & NE USA

❏ *Felis lynx* Eurasian Lynx

Montane forests of Europe to N Pakistan, N India, N Nepal, E Siberia and NC China

❏ *Felis pardina* Spanish Lynx

Woods and maquis of Portugal and SW Spain; Critically Endangered

❏ *Felis rufa* Bobcat

S Canada, USA and Mexico

Felis (Puma)

❏ *Felis concolor* Puma

Forests, deserts and mountains of SW Canada and W USA to French Guiana, S to S Chile and S Argentina

Felis (Herpailurus)

❏ *Felis yaguarondi* Jaguarundi

Forests, savannas, scrub and farmland of extreme SC Arizona and S Texas (USA) to French Guiana, S to Bolivia, Paraguay, N & C Argentina and S Brazil

Felis (Leopardus)

❏ *Felis braccata* Pantanal Cat

Paraguay, Uruguay and SW & S Brazil

❏ *Felis colocolo* Colocolo

Highlands of N Chile and forests of C Chile on W slope of Andes

❏ *Felis geoffroyi* Geoffroy's Cat

Savannas, brush and marshes of Bolivia, Paraguay, S Chile, Argentina, Uruguay and S Brazil

❏ *Felis guigna* Kodkod

Moist coniferous forests and clearings of SC Chile and SW Argentina

❏ *Felis pajeros* Pampas Cat

Highlands on E slope of Andes from Ecuador to NW Argentina, and lowlands of Argentina and S Chile

❏ *Felis pardalis* Ocelot

Forests, regrowth areas and scrub of extreme S Texas (USA) and Mexico to French Guiana, S to N Argentina, Paraguay and SE Brazil (and Uruguay?); Trinidad

❏ *Felis tigrina* Oncilla (Little Spotted Cat)

Rainforests and scrub of Costa Rica to French Guiana, S to Paraguay, N Argentina and SE Brazil

❏ *Felis wiedii* Margay

Locally in dense undisturbed forests of Mexico to Suriname, S to N Argentina, SE Paraguay, Uruguay and SE Brazil

Felis (Oreailurus)

❏ *Felis jacobita* Andean Cat

High altitude barren rocky regions of Andes of S Peru, SW Bolivia, N Chile and NW Argentina; Endangered

❏ *Acinonyx jubatus* Cheetah

Savannas and grasslands of N Iran and subsaharan Africa S to N South Africa (formerly to N Africa, C Asia and India)

❏ *Neofelis nebulosa* Clouded Leopard

Forests of India and Nepal to China, Indochina, Malay Peninsula and Indonesia

Panthera (Uncia)

❏ *Panthera uncia* Snow Leopard

Montane forests, scrub and grasslands of Afghanistan and Pakistan to Mongolia, China and Bhutan; Endangered

Panthera (Panthera)

❏ *Panthera pardus* Leopard

Forests, scrub and open areas of Africa and Middle East to India, Sri Lanka, China, Korea, Indochina and Java

Panthera (Leo)

❏ *Panthera leo* Lion

Woods, savannas and semi-deserts of subsaharan Africa; India; formerly N Africa

Panthera (Jaguarius)

❏ *Panthera onca* Jaguar

Undisturbed forests, scrub and grassland of S Arizona (USA) and Mexico to French Guiana, S to N Argentina, Paraguay and S Brazil

Panthera (Tigris)

❏ *Panthera tigris* Tiger

Forests, mangrove swamps and grasslands of Pakistan to China, Korea, Indochina and Sumatra; Endangered

Family: HERPESTIDAE (Mongooses—38)

Herpestes (Galerella)

❏ *Herpestes flavescens* Black Slender Mongoose

S Angola, N & C Namibia

❏ *Herpestes pulverulentus* Cape Grey Mongoose

Woods, scrub and rocky slopes of S Namibia and W South Africa

❏ *Herpestes sanguineus*
Common Slender Mongoose

Forests, savannas and marshes of subsaharan Africa; Cape Verde Is.

❏ *Herpestes swalius* Namaqua Slender Mongoose

C & S Namibia

Herpestes (Herpestes)

❏ *Herpestes auropunctatus* Small Indian Mongoose

Scrub and farmland of Iraq and Afghanistan to China, S to Malay Peninsula; feral SE Europe, Tanzania, Japan, Pacific Ocean islands, West Indies and Suriname

❏ *Herpestes brachyurus* Short-tailed Mongoose

S India, Sri Lanka, Vietnam, Malay Peninsula, Sumatra, Borneo and Philippines

❏ *Herpestes edwardsii* Indian Grey Mongoose

Scrub, semi-deserts, farmland and built areas of Arabian Peninsula to India and Indonesia; feral Italy (now extinct), Japan and Indian Ocean islands

❏ *Herpestes ichneumon* Egyptian Mongoose

Riparian habitats, moist savannas and forest clearings of SW Europe (?feral), Africa and Turkey to S Israel and Jordan

❏ *Herpestes javanicus* Javan Mongoose

Java

❏ *Herpestes naso* Long-nosed Mongoose

Clear forest streams of Niger; Cameroon and Rep. Congo to Kenya and Tanzania

❏ *Herpestes palustris* Bengal Mongoose

West Bengal (NE India); Endangered

❏ *Herpestes semitorquatus* Collared Mongoose

Sumatra and Borneo

❏ *Herpestes smithii* Ruddy Mongoose

Forests and scrub of India and Sri Lanka

❏ *Herpestes urva* Crab-eating Mongoose

Riparian habitats in hill forests of India and Nepal to China, Taiwan, Indochina and Malay Peninsula

❏ *Herpestes vitticollis* Stripe-necked Mongoose

Forests, scrub, farmland and built areas of SW India and Sri Lanka

❏ *Atilax paludinosus* Marsh Mongoose

Wetlands of Algeria and subsaharan Africa

❑ *Helogale dybowskii* Pousargues' Mongoose

Forest edge and moist savannas of Central African Republic, S Sudan, NE D.R. Congo and W Uganda

❑ *Helogale hirtula* Desert Dwarf Mongoose

Deciduous woods and scrub of S Ethiopia, C & S Somalia and N & C Kenya

❑ *Helogale parvula* Dwarf Mongoose

Woods and savannas of Sudan and Ethiopia to NE South Africa

❑ *Mungos gambianus* Gambian Mongoose

Woods, savannas and grasslands of Gambia, Senegal and Sierra Leone to Niger and Nigeria

❑ *Mungos mungo* Banded Mongoose

Woods, savannas and grasslands of subsaharan Africa, S to E South Africa

❑ *Cynictis penicillata* Yellow Mongoose

Open grasslands and semi-desert scrub of S Angola, Namibia, Botswana, SW Zimbabwe and South Africa

❑ *Cynictis selousi* Selous' Mongoose

Shortgrass areas, open woods, savannas and farmlands of Angola to Mozambique, S to NE Namibia, Botswana and NE South Africa

❑ *Ichneumia albicauda* White-tailed Mongoose

Woods, regrowth areas, savannas, farmland and built areas of S Arabian Peninsula and subsaharan Africa, S to E South Africa

❑ *Suricata suricatta* Suricate (Meerkat)

Arid grasslands of Angola, Namibia, S Botswana and South Africa

❑ *Crossarchus alexandri* Alexander's Cusimanse

Swamp forests of Central African Republic, Rep. Congo and D.R. Congo; Mt Elgon (Uganda)

❑ *Crossarchus ansorgei* Ansorge's Cusimanse

Primary rainforests of N Angola and SE D.R. Congo

❑ *Crossarchus obscurus* Long-nosed Cusimanse

Rainforests of Sierra Leone and Liberia to Central African Republic

❑ *Liberiictis kuhni* Liberian Mongoose

Sandy stream banks in forests of Liberia and Ivory Coast; Endangered

❑ *Rhynchogale melleri* Meller's Mongoose

Miombo (*Brachystegia*) woods and savannas of D.R. Congo and Tanzania to Zimbabwe, Mozambique and NE South Africa

❑ *Bdeogale crassicauda* Bushy-tailed Mongoose

Savannas, coastal thickets and rocky habitats of Kenya to NE Zimbabwe and C Mozambique

❑ *Bdeogale jacksoni* Jackson's Mongoose

Forests of E Uganda (S of Mt Elgon) and montane forests and bamboo thickets of C Kenya

❑ *Bdeogale nigripes* Black-footed Mongoose

Lowland rainforests from Nigeria to N Angola

❑ *Galidictis fasciata* Broad-striped Mongoose

Madagascar

❑ *Galidictis grandidieri* Giant-striped Mongoose

Spiny desert of SW Madagascar; Endangered

❑ *Mungotictis decemlineata* Narrow-striped Mongoose

Madagascar; Endangered

❑ *Salanoia concolor* Brown-tailed Mongoose

Madagascar

❑ *Galidia elegans* Ring-tailed Mongoose

Madagascar

Family: HYAENIDAE (Hyenas—4)

❑ *Hyaena hyaena* Striped Hyena

Arid savannas, grasslands, semi-deserts and rocky slopes of N & NE Africa S to Tanzania, Middle East, Arabian Peninsula, Iran to India (formerly E to China)

❑ *Pachycrocuta brunnea* Brown Hyena

Locally in arid open areas of Namibia, Botswana, S Zimbabwe, SW Mozambique and N South Africa

❑ *Crocuta crocuta* Spotted Hyena

Open savannas and grasslands of subsaharan Africa, S to N South Africa

❑ *Proteles cristatus* Aardwolf

Open savannas and grasslands of S Egypt, Sudan and Central African Republic, S to South Africa

Family: CANIDAE (Dogs and Foxes—35)

Vulpes (Alopex)

❏ *Vulpes lagopus* Arctic Fox

Tundra zone of North America, N Europe and N Siberia, incl. Arctic islands

Vulpes (Vulpes)

❏ *Vulpes bengalensis* Bengal Fox

Open habitats and scrub of Pakistan, India, S Nepal, Bhutan and Bangladesh

❏ *Vulpes cana* Blanford's Fox

Rocky montane habitats of NE Egypt, NE Iran, Turkmenistan, Afghanistan and Pakistan

❏ *Vulpes chama* Cape Fox

Open savannas, grasslands, scrub and farmland of S Angola and S Zambia, S to South Africa

❏ *Vulpes corsac* Corsac Fox

Kazakhstan and Afghanistan to E Siberia, Mongolia and N China

❏ *Vulpes ferrilata* Tibetan Fox

Nepal, Tibet and C China

❏ *Vulpes macrotis* Kit Fox

Deserts and grasslands of SW USA and N Mexico

❏ *Vulpes pallida* Pale Fox

Sahel zone of subsaharan Africa from Senegal to Sudan and Somalia

❏ *Vulpes rueppellii* Rüppell's Fox

Sandy and stony deserts of N Africa to Somalia, Arabian Peninsula, Iran, Afghanistan and Pakistan

❏ *Vulpes velox* Swift Fox

Prairies of C USA

❏ *Vulpes vulpes* Red Fox

Woods, scrub, open habitats, farmland and built areas of Europe, N Africa, Asia, N India, Indochina, Japan and N America; feral Australia and Pacific Ocean islands

❏ *Vulpes zerda* Fennec Fox

Sandy deserts of Morocco to Arabian Peninsula

❏ *Urocyon cinereoargenteus* Grey Fox

Forests and brush of SE Canada and USA to montane forests of N Colombia and Venezuela

❏ *Urocyon littoralis* Island Grey Fox

Channel Is. (S California, USA)

❏ *Otocyon megalotis* Bat-eared Fox

Dry open savannas, grasslands and scrub of Ethiopia to Tanzania; Namibia, Botswana, S Zimbabwe, SW Mozambique and South Africa (except SE)

❏ *Pseudalopex culpaeus* Culpeo

Arid and semiarid regions, farmland and subantarctic forests of Andes from Colombia to S Chile and S Argentina

❏ *Pseudalopex gymnocercus* Pampas Fox (Argentine Grey Fox)

Arid regions, grassland, pampas and forests of E Bolivia, Paraguay, S Brazil, Uruguay, Chile and Argentina

❏ *Pseudalopex sechurae* Sechura Fox

Arid lowlands of Tumbes region of SW Ecuador and NW Peru

❏ *†Dusicyon australis* Falkland Island Wolf

Formerly Falkland Is.; Extinct (last recorded 1876)

❏ *Lycalopex vetulus* Hoary Fox

Cerrado and caatinga of highlands of interior EC Brazil

❏ *Chrysocyon brachyurus* Maned Wolf

Grassland, cerrado and scrub of Bolivia, Paraguay, S Brazil and extreme NE Argentina

❏ *Cerdocyon thous* Crab-eating Fox

Forests, forest edge, savannas and llanos of Colombia to Suriname, S to N Argentina and Uruguay

❏ *Nyctereutes procyonoides* Raccoon Dog

Damp woods and riparian habitats of E Siberia, Korea, China, Japan and N Indochina; feral C & E Europe, Belarus, Russia and Ukraine

❏ *Atelocynus microtis* Short-eared Dog

Forests of Amazon basin of E Colombia to E Peru and S of R. Amazonas (C Brazil); Paraguay, S Brazil and N Argentina

❏ *Speothos venaticus* Bush Dog

Forests and savannas of E Panama to French Guiana, S to N Bolivia, Paraguay, NE Argentina and S Brazil

❑ *Canis adustus* Side-striped Jackal

Woods, savannas, scrub and farmland of subsaharan Africa, S to E South Africa

❑ *Canis aureus* Golden Jackal

Woods, scrub, maquis, semi-deserts and built areas of SE Europe, N & E Africa, Middle East, C & S Asia to Sri Lanka and Thailand

❑ *Canis familiaris* Domestic Dog (Dingo)

Domesticated worldwide, feral Australia, New Zealand, N & S America, etc.; wild ancestor is *Canis lupus* Grey Wolf

❑ *Canis latrans* Coyote

Widespread in Alaska, Canada and USA, S to W Panama

❑ *Canis lupus* Grey Wolf

Open forests, tundra and mountains of S & E Europe, Asia and N North America

❑ *Canis mesomelas* Black-backed Jackal

Dry *Acacia* savannas of Sudan and Ethiopia to Tanzania; open arid areas of Angola and Mozambique, S to South Africa

❑ *Canis rufus* Red Wolf

SC & SE USA; Critically Endangered

❑ *Canis simensis* Ethiopian Wolf (Simien Fox)

High-altitude montane meadows and moorlands of C Ethiopia; Critically Endangered

❑ *Cuon alpinus* Dhole

Forests and open woods of Pakistan and India to E Siberia, Korea, China, Indochina, Sumatra and Java

❑ *Lycaon pictus* African Wild Dog

Locally in woods, savannas and grasslands of subsaharan Africa, S to NE South Africa; Endangered

Family: URSIDAE (Bears—8)

❑ *Ailuropoda melanoleuca* Giant Panda

Bamboo thickets of C China; Endangered

❑ *Ursus americanus* American Black Bear

Forests and swamps of Alaska, Canada, locally in USA and N Mexico

❑ *Ursus arctos* Brown Bear (Grizzly Bear)

Forests, tundra and montane meadows of S & E Europe, NW Africa and Asia, E to Hokkaido (N Japan); NW North America, S to NW Wyoming (USA)

❑ *Ursus thibetanus* Asiatic Black Bear

Forests and open rocky areas of Afghanistan and Pakistan to Korea, China, Taiwan, Japan and Indochina

❑ *Melursus ursinus* Sloth Bear

Forests and rocky areas of India, Nepal, Bhutan, Bangladesh and Sri Lanka

❑ *Helarctos malayanus* Sun Bear

India to C China, Taiwan, Indochina, Sumatra and Borneo

❑ *Thalarctos maritimus* Polar Bear

Permanent ice and pack ice of Arctic Ocean

❑ *Tremarctos ornatus*
Spectacled Bear (Andean Bear)

Montane forests of NW Venezuela, Colombia, Ecuador, Peru and W Bolivia; formerly W Argentina

Family: OTARIIDAE (Eared Seals—14)

❑ *Callorhinus ursinus* Northern Fur Seal

N Pacific Ocean, Bering Sea and Sea of Okhotsk, S to NE China, Japan and S California (USA)

❑ *Arctocephalus australis* South American Fur Seal

Coasts of S South America, incl. Falkland Is., N to C Peru and Uruguay; non-breeders N to SE Brazil

❑ *Arctocephalus forsteri* New Zealand Fur Seal

Coasts of SW, S & SE Australia, New Zealand and subantarctic islands

❑ *Arctocephalus galapagoensis* Galápagos Fur Seal

Galápagos Is.

❑ *Arctocephalus gazella* Antarctic Fur Seal

S Southern Ocean islands (S Shetland Is. to Kerguelen and Heard I.); vagrant Isla Hoste (extreme S Chile)

❑ *Arctocephalus philippii* Juan Fernández Fur Seal

San Félix Is. and Juan Fernández Is. (Chile); vagrant to S Peru and S Chile

❑ *Arctocephalus pusillus*
South African Fur Seal (Australian Fur Seal)

Coasts of Angola to SC South Africa; SE Australia and Tasmania

❑ *Arctocephalus townsendi* Guadalupe Fur Seal

Channel Is. (S California, USA) and Guadalupe I. (Baja California Norte, Mexico); vagrant to C California

❏ *Arctocephalus tropicalis* Subantarctic Fur Seal

N Southern Ocean islands (Tristan da Cunha to Macquarie Is.); vagrant to South Africa

❏ *Eumetopias jubatus*
Northern Sea-Lion (Steller's Sea-Lion)

Coasts of N Pacific Ocean and S Bering Sea, S to Hokkaido (N Japan) and S California (USA); Endangered

❏ *Otaria flavescens* South American Sea-Lion

Coasts of S South America, incl. Falkland Is., N to N Peru and S Brazil

❏ *Zalophus californianus*
California Sea-Lion (Galápagos Sea-Lion)

Coasts of W USA and S Mexico; Galápagos Is.; formerly NW Pacific Ocean S to Japan (exterminated early 1950s)

❏ *Neophoca cinerea* Australian Sea-Lion

Coasts of W & S Australia

❏ *Phocarctos hookeri* New Zealand Sea-Lion

Subantarctic islands of New Zealand; non-breeders N to S South Island (New Zealand)

Family: PHOCIDAE (Earless Seals—20)

❏ *Odobenus rosmarus* Walrus

Drifting pack ice of Arctic Ocean, Bering Sea, NE Canada and N Greenland; vagrants S to NW Europe, Japan and NE USA

❏ *Mirounga angustirostris* Northern Elephant Seal

Coasts of Califrornia (USA) and Baja California (Mexico); non-breeders N to Alaska; vagrants Japan and Hawaiian Is.

❏ *Mirounga leonina* Southern Elephant Seal

Coasts of extreme S Chile, S Argentina, Falkland Is. and subantarctic islands; non-breeders widely in Southern Ocean; vagrant to South Africa

❏ *Monachus monachus* Mediterranean Monk Seal

Coasts of Madeira, NW Africa and Mediterranean Sea; formerly Black Sea (extinct); Critically Endangered

❏ *Monachus schauinslandi* Hawaiian Monk Seal

NW Hawaiian Is.; Endangered

❏ *†Monachus tropicalis* West Indian Monk Seal

Formerly coasts of Yucatán (SE Mexico) and Caribbean Sea; Extinct (last confirmed sighting in 1952)

❏ *Ommatophoca rossii* Ross Seal

Circumpolar on Antarctic pack ice

❏ *Lobodon carcinophagus* Crabeater Seal

Circumpolar on Antarctic pack ice; vagrants N to S Australia, New Zealand, S Brazil and South Africa

❏ *Hydrurga leptonyx* Leopard Seal

Circumpolar in Southern Ocean; vagrants N to S Africa, Australia, New Zealand and S South America

❏ *Leptonychotes weddellii* Weddell Seal

Circumpolar on Antarctic pack ice, South Georgia and South Orkney Is.; vagrants N to S Australia and South America

❏ *Histriophoca fasciata* Ribbon Seal

Chukchi and Bering Seas, Sea of Okhotsk (E Siberia) and Sea of Japan (Korea/Japan)

❏ *Pagophilus groenlandicus* Harp Seal

Drifting pack ice of Arctic and N Atlantic Oceans from E Canada to N Siberia; vagrants S to NW Europe

❏ *Phoca largha* Largha Seal (Spotted Seal)

Pack ice of N Pacific Ocean, Bering Sea and Sea of Okhotsk, S to NE China and Japan

❏ *Phoca vitulina* Harbour Seal

Coasts of N Atlantic and N Pacific Oceans, S to Portugal, NE China, Mexico and NE USA

❏ *Pusa caspica* Caspian Seal

Caspian Sea (S Russia and W Kazakhstan to N Iran)

❏ *Pusa hispida* Ringed Seal

Coasts and pack ice of Arctic, N Atlantic and N Pacific Oceans, Baltic and Bering Seas, and N Canada; lakes in Finland and NW Russia

❏ *Pusa sibirica* Baikal Seal

Lake Baikal (E Siberia)

❏ *Erignathus barbatus* Bearded Seal

Circumpolar on drifting pack ice of Arctic, N Atlantic and N Pacific Oceans; vagrants S to NW Europe, Japan and E Canada

❏ *Halichoerus grypus* Grey Seal

Coasts of N Atlantic Ocean and Baltic Sea, S to E Canada and British Isles; vagrants S to Portugal and Germany

❑ *Cystophora cristata* **Hooded Seal**

Drifting pack ice of NW Atlantic Ocean from E Canada to Jan Mayen I. (Norway); vagrants S to Portugal and Florida (USA)

Family: MUSTELIDAE
(Otters, Weasels and Badgers—68)

Lutra (Lontra)

❑ *Lutra canadensis* **Northern River Otter**

Wetlands and coasts of Alaska, Canada and NW & E USA

❑ *Lutra felina* **Marine Otter**

Rocky coasts from N Peru to S Chile and extreme S Argentina; Endangered

❑ *Lutra longicaudis* **Neotropical River Otter**

Locally in clear, fast-flowing rivers and streams, and coastal lagoons of N Mexico to Suriname, S to Peru, N Argentina and Uruguay

❑ *Lutra provocax* **Southern River Otter**

Rocky coasts, lakes and streams of S Chile and S Argentina; Endangered

Lutra (Lutra)

❑ *Lutra lutra* **Eurasian Otter**

Rivers and lakes of Europe and Morocco to Korea, China, Taiwan, Japan, Indochina and Indonesia

❑ *Lutra maculicollis* **Spotted-necked Otter**

Clear rivers, lakes and marshes of subsaharan Africa (except SW deserts)

❑ *Lutra sumatrana* **Hairy-nosed Otter**

Indochina, Sumatra and Borneo

❑ *Pteronura brasiliensis* **Giant Otter**

Locally in lowland-forest rivers and lakes of SE Colombia to French Guiana, S to Bolivia, N Argentina, Uruguay and S Brazil; Endangered

❑ *Lutrogale perspicillata* **Smooth-coated Otter**

Rivers, lakes, coastal lagoons and canals of Iraq to China, Indochina, Malay Peninsula, Sumatra, Java and Borneo

❑ *Aonyx capensis* **African Clawless Otter**

Rivers, lakes and marshes of subsaharan Africa incl. Bioko (Equatorial Guinea)

❑ *Aonyx congicus* **Congo Clawless Otter**

Riparian habitats and swamps in rainforests of Nigeria to Kenya, S to Angola, D.R. Congo and Uganda

❑ *Amblonyx cinereus* **Oriental Small-clawed Otter**

Rivers, lakes, estuaries and flooded rice fields of India to S China, Hainan, Taiwan, Indochina, Sumatra, Java, Borneo and Philippines

❑ *Enhydra lutris* **Sea Otter**

Locally in shallow coastal waters of E Siberia, Japan, Alaska, W USA and NW Mexico; Endangered

❑ *Mydaus javanensis* **Sunda Stink Badger**

Sumatra, Java, Borneo and Natuna Is. (between Malaya and Borneo)

❑ *Mydaus marchei* **Palawan Stink Badger**

Calamian Is. and Palawan (SW Philippines)

❑ *Spilogale gracilis* **Western Spotted Skunk**

Rocky areas and brush of SW British Columbia (Canada) and W USA to C Mexico

❑ *Spilogale putorius* **Eastern Spotted Skunk**

Woodlands, brush and farmland of C & SE USA and E Mexico to Costa Rica

❑ *Spilogale pygmaea* **Pygmy Spotted Skunk**

Sinaloa to Oaxaca (Mexico)

❑ *Mephitis macroura* **Hooded Skunk**

Deciduous forests, scrub, rocky canyons and riparian habitats of SW & SC USA and Mexico to NW Costa Rica

❑ *Mephitis mephitis* **Striped Skunk**

Woods, fields and gardens of Canada, USA and N Mexico

❑ *Conepatus chinga* **Molina's Hog-nosed Skunk**

Open grasslands of Peru, SW Bolivia, Paraguay, N & C Chile, N & WC Argentina, Uruguay (and S Brazil?)

❑ *Conepatus humboldtii*
Humboldt's Hog-nosed Skunk

Grasslands and scrub of S Chile and S Argentina (and Paraguay?)

❑ *Conepatus leuconotus* **Eastern Hog-nosed Skunk**

Grasslands and brush of SE Texas to Veracruz (EC Mexico)

❑ *Conepatus mesoleucus*
Western Hog-nosed Skunk

Rocky semiarid grasslands and brush of SC USA and Mexico to Nicaragua

❑ *Conepatus semistriatus*
Striped Hog-nosed Skunk

Forest clearings, forest edge, scrub and pastures of EC Mexico to W Panama; NE Colombia and N Venezuela; NW Peru; caatinga and cerrado of EC Brazil

❑ *Martes americana* **American Marten**

Coniferous forests of Alaska, Canada and mountains of W & extreme NE USA

❑ *Martes flavigula* **Yellow-throated Marten**

Forests of Russia and Pakistan to China, Korea, Taiwan, Indochina, Sumatra, Java and Borneo

❑ *Martes foina* **Beech Marten**

Woods, montane meadows, farmland and built areas of C & S Europe to Kazakhstan, Mongolia and N China

❑ *Martes gwatkinsii* **Nilgiri Marten**

Forests of S India

❑ *Martes martes* **Pine Marten**

Forests and scrub of Europe to W Siberia, Caucasus Mtns and Elburz Mtns (NC Iran)

❑ *Martes melampus* **Japanese Marten**

Korea and Japan; feral Sado Is. (Japan)

❑ *Martes pennanti* **Fisher**

Forests of Canada and mountains of W & NE USA

❑ *Martes zibellina* **Sable**

E Europe, Siberia, Mongolia, NE China, N Korea and Hokkaido (N Japan)

❑ *Mustela africana* **Amazon Weasel**

Forests, plantations and gardens of Amazon basin of E Ecuador, E Peru and C Brazil

❑ *Mustela altaica* **Mountain Weasel**

Mountains of E Kazakhstan, S Siberia, Mongolia, Tibet, N & W China and Korea

❑ *Mustela erminea* **Stoat (Ermine)**

Widespread in C & N Europe to N China and Japan; Alaska, Canada, and N USA; feral New Zealand

❑ *Mustela eversmannii* **Steppe Polecat**

Steppes, grasslands and fields of EC Europe to Mongolia and NE China

❑ *Mustela felipei* **Colombian Weasel**

Very few records from mountains of Cordillera Central (C Colombia); Endangered

❑ *Mustela formosana*
Taiwan Mountain Weasel

Mountains of Taiwan

❑ *Mustela frenata* **Long-tailed Weasel**

Forests, regrowth areas and farmland of S Canada and USA to Venezuela, and W of the Andes S to W Bolivia

❑ *Mustela furo* **Ferret**

Domesticated worldwide, feral New Zealand; wild ancestor is *Mustela putorius* European Polecat

❑ *Mustela kathiah* **Yellow-bellied Weasel**

Mountains of N Pakistan and Nepal to E China and Indochina

❑ *Mustela lutreola* **European Mink**

Formerly widely in wetlands of Europe, now confined to N Spain, W France, and locally in E Europe, E to R. Ob (Russia and Kazakhstan); Endangered

❑ *Mustela lutreolina* **Indonesian Mountain Weasel**

Highlands of Sumatra and Java; Endangered

❑ *Mustela nigripes* **Black-footed Ferret**

Formerly SC Canada and WC USA; reintrod. to a few localities in WC USA; Extinct in the Wild

❑ *Mustela nivalis* **Least Weasel**

Grasslands, farmland and semi-deserts of Europe to Japan; Alaska, Canada and NC & NE USA; feral New Zealand

❑ *Mustela nudipes* **Malayan Weasel**

Thailand, Malay Peninsula, Sumatra, Java and Borneo

❑ *Mustela putorius*
European Polecat

Woods, forest edge, marshes and farmland of Europe and Russia, E to Ural Mtns

❑ *Mustela sibirica* **Siberian Weasel**

Forests and montane meadows of Russia and Siberia, S to Pakistan, Myanmar, Thailand, Taiwan and Japan

❏ *Mustela strigidorsa* Black-striped Weasel

Nepal to Yunnan (SW China) and Laos

❏ *Mustela vison* American Mink

Wetlands of Alaska, Canada and NW, NC & E USA; feral Europe and Siberia

❏ *Vormela peregusna* Marbled Polecat

Steppes and deserts of SE Europe and Caucasus Mtns to Mongolia and N China

❏ *Taxidea taxus* American Badger

Deserts, grasslands and damp meadows of S Canada and USA (except SE) to C Mexico

❏ *Lyncodon patagonicus* Patagonian Weasel

Few records from Andean foothills of S Chile and W Argentina

❏ *Poecilictis libyca* Saharan Striped Polecat

Dry stony and sandy deserts of S Morocco and Mauritania to Egypt and N Sudan

❏ *Galictis cuja* Lesser Grison

Scrub and dry chaco of Peru, Bolivia, Paraguay, Chile, Argentina, Uruguay and S Brazil

❏ *Galictis vittata* Greater Grison

Locally in rainforests, savannas and llanos of EC Mexico to Suriname, S to Peru, Bolivia, S Brazil and Argentina

❏ *Ictonyx striatus* Zorilla (Striped Polecat)

Locally in grasslands of subsaharan Africa

❏ *Poecilogale albinucha* African Striped Weasel

Grasslands of D.R. Congo, Uganda and Kenya, S to South Africa

❏ *Eira barbara* Tayra

Forests, plantations and gardens of Mexico to French Guiana, S to Bolivia, Paraguay, N Argentina and SE Brazil; Trinidad

❏ *Mellivora capensis* Honey Badger

Forests, open woods, scrub and stony deserts of Africa and Middle East to Turkmenistan, Nepal and India

❏ *Gulo gulo* Wolverine

Boreal forests and tundra of Scandinavia, Siberia and North America

❏ *Meles meles* Eurasian Badger

Woods, parks and farmland of Europe and Middle East to S Siberia, Korea, China and Japan

❏ *Arctonyx collaris* Hog Badger

NE India and Myanmar to China, Indochina and Sumatra

❏ *Melogale everetti* Everett's Ferret-Badger

Borneo

❏ *Melogale moschata* Chinese Ferret-Badger

NE India, Bangladesh and Myanmar to SE China, Taiwan, Hainan, N Thailand, C Laos and C Vietnam

❏ *Melogale orientalis* Javan Ferret-Badger

Java

❏ *Melogale personata* Burmese Ferret-Badger

Nepal and NE India to Indochina and Malay Peninsula

Family: PROCYONIDAE (Raccoons—19)

❏ *Bassariscus astutus* Ringtail

Woods, arid scrub and riversides of S USA to S Mexico, incl. Gulf of California islands

❏ *Bassariscus sumichrasti* Cacomistle

Forests and regrowth areas of C Mexico to W Panama

❏ *Bassaricyon alleni* Allen's Olingo

Forests of Amazon basin of E Ecuador, E Peru and N Bolivia

❏ *Bassaricyon beddardi* Beddard's Olingo

Guyana; status uncertain

❏ *Bassaricyon gabbii* Bushy-tailed Olingo

Forests and forest edge of C Nicaragua to W Colombia and W Ecuador

❏ *Bassaricyon lasius* Harris' Olingo

Estrella de Cartago region (Costa Rica); Endangered

❏ *Bassaricyon pauli* Chiriqui Olingo

Cerro Pando region (Chiriqui Prov., Panama); Endangered

❏ *Potos flavus* Kinkajou

Forests, plantations, farmland and gardens of NE Mexico to French Guiana, S to Peru, Bolivia and SC Brazil

❏ *Nasua narica* White-nosed Coati

Forests, regrowth areas and scrub of S Arizona to S Texas (USA), S to N Colombia

❏ *Nasua nasua* South American Coati

Forests and scrub of Colombia to French Guiana, S to Bolivia, N Argentina, Paraguay, Uruguay and Brazil

❏ *Procyon cancrivorus* Crab-eating Raccoon

Lowland undisturbed riparian and coastal habitats of Costa Rica to French Guiana, S to Bolivia, N Argentina, Uruguay and S Brazil; Trinidad & Tobago

❏ *†Procyon gloveralleni* Barbados Raccoon

Formerly Barbados (Lesser Antilles); Extinct (last seen in 1964)

❏ *Procyon insularis* Tres Marías Raccoon

Tres Marías Is. (Nayarit, Mexico); Endangered

❏ *Procyon lotor* Northern Raccoon

Damp woods, parks and gardens of S Canada and USA to Panama; feral C Europe and W Asia

❏ *Procyon maynardi* Bahaman Raccoon

Nassau I. (Bahamas); Endangered

❏ *Procyon minor* Guadeloupe Raccoon

Guadeloupe I. (Lesser Antilles); Endangered

❏ *Procyon pygmaeus* Cozumel Raccoon

Mangroves and forests of Cozumel I. and Isla la Pasión (Quintana Roo, Mexico); Endangered

❏ *Nasuella olivacea* Mountain Coati

High montane forests of Colombia, NW Venezuela and Ecuador

❏ *Ailurus fulgens* Red Panda

Forests and bamboo thickets of Nepal, Sikkim (NE India), N Myanmar and C China; Endangered

Grandorder: LIPOTYPHLA (Insectivores)
Order: CHRYSOCHLORIDEA
Family: CHRYSOCHLORIDAE
(Golden-Moles—21)

❏ *Chrysochloris asiatica* Cape Golden-Mole

Cape Prov. (South Africa)

❏ *Chrysochloris stuhlmanni*
Stuhlmann's Golden-Mole

Cameroon, N D.R. Congo, Uganda, Kenya and Tanzania

❏ *Chrysochloris visagiei* Visagie's Golden-Mole

Gouna, near Calvinia (SC Northern Cape Prov., South Africa); only known from holotype specimen, Critically Endangered

❏ *Chrysospalax trevelyani* Giant Golden-Mole

Deep forest soils of Eastern Cape Prov. (South Africa); Endangered

❏ *Chrysospalax villosus* Rough-haired Golden-Mole

Damp marshy soils in arid areas of Mpumalanga to Eastern Cape Prov. (E South Africa)

❏ *Amblysomus corriae* Western Cape Golden-Mole

Western Cape region (South Africa)

❏ *Amblysomus gunningi* Gunning's Golden-Mole

Woodbush and New Agatha Forests (Northern Prov., South Africa)

❏ *Amblysomus hottentotus* Hottentot Golden-Mole

Coastal Western Cape to Northern Provs. (South Africa)

❏ *Amblysomus julianae* Juliana's Golden-Mole

Locally in Mpumalanga Prov. (NE South Africa); Critically Endangered

❏ *Amblysomus marleyi* Marley's Golden-Mole

Extreme NE KwaZulu-Natal (E South Africa)

❏ *Amblysomus obtusirostris* Yellow Golden-Mole

Sandy soils of SE Zimbabwe, S Mozambique and extreme E South Africa

❏ *Amblysomus robustus* Robust Golden-Mole

Mpumalanga Highveld (South Africa)

❏ *Amblysomus septentrionalis*
Highveld Golden-Mole

South Africa

❏ *Chlorotalpa arendsi* Arend's Golden-Mole

Extreme E Zimbabwe and adjacent WC Mozambique

❏ *Chlorotalpa duthieae* Duthie's Golden-Mole

Knysna region (SE Western Cape Prov., South Africa)

❏ *Chlorotalpa leucorhina* Congo Golden-Mole

Cameroon, Central African Republic, D.R. Congo and N Angola

❏ *Chlorotalpa sclateri* Sclater's Golden-Mole

Interior South Africa and Lesotho

❏ *Chlorotalpa tytonis* Somali Golden-Mole

Giohar (Somalia); only known from holotype specimen, Critically Endangered

❏ *Cryptochloris wintoni* De Winton's Golden-Mole

Sand dunes of Port Nolloth region (NW Northern Cape Prov., South Africa)

❏ *Cryptochloris zyli* Van Zyl's Golden-Mole

Sand dunes of Compagnies Drift, near Lambert's Bay (W Northern Cape Prov., South Africa); only known from holotype specimen, Critically Endangered

❏ *Eremitalpa granti* Grant's Golden-Mole

Coastal sand dunes of W Namibia to W Western Cape Prov. (South Africa)

Order: ERINACEOMORPHA
Family: ERINACEIDAE
(Moonrats and Hedgehogs—21)

❏ *Echinosorex gymnura* Moonrat

Malay Peninsula, Sumatra and Borneo

❏ *Hylomys hainanensis* Hainan Gymnure

Hainan (S China); Endangered

❏ *Hylomys sinensis* Shrew Gymnure

N Myanmar, SW China and N Vietnam

❏ *Hylomys suillus* Lesser Gymnure

Yunnan (SW China), Malay Peninsula, Indochina, Sumatra, Java and Borneo

❏ *Podogymnura aureospinula* Dinagat Gymnure

Dinagat (EC Philippines); Endangered

❏ *Podogymnura truei* Mindanao Gymnure

Mindanao (SE Philippines); Endangered

Erinaceus (Atelerix)

❏ *Erinaceus albiventris* Four-toed Hedgehog

Subsaharan Africa, S to Angola, Zambia and N Mozambique

❏ *Erinaceus algirus* Algerian Hedgehog

Dry grasslands, scrub and farmland of Western Sahara to N Libya; probably feral SW Europe, Mediterranean islands and Canary Is.

❏ *Erinaceus frontalis* Southern African Hedgehog

SW Angola, Namibia, Botswana, W Zimbabwe and South Africa

❏ *Erinaceus sclateri* Somali Hedgehog

N Somalia

Erinaceus (Erinaceus)

❏ *Erinaceus amurensis* Amur Hedgehog

Russia and S Siberia to Korea and E China

❏ *Erinaceus concolor* Eastern Hedgehog

Forest edge, scrub and built areas of E & SE Europe, Middle East, S Russia and W Siberia

❏ *Erinaceus europaeus* Western Hedgehog

Forest edge, scrub, grasslands and built areas of W & N Europe incl. Mediterranean islands, and NW Russia; feral New Zealand

Erinaceus (Mesechinus)

❏ *Erinaceus dauuricus* Daurian Hedgehog

E Siberia, N Mongolia and NE China

❏ *Erinaceus hughi* Hugh's Hedgehog

Shaanxi and Shanxi Provs. (C China)

❏ *Hemiechinus aethiopicus* Desert Hedgehog

Hot dry deserts of Sahara Desert from Mauritania to Egypt and S Ethiopia; Arabian Peninsula

❏ *Hemiechinus auritus* Long-eared Hedgehog

Cool deserts of E Ukraine to Mongolia, S to E Libya and Pakistan

❏ *Hemiechinus collaris*
Indian Long-eared Hedgehog

Pakistan and NW India

❏ *Hemiechinus hypomelas* Brandt's Hedgehog

Uzbekistan and Iran to Pakistan; S Yemen, Oman and Persian Gulf islands

❏ *Hemiechinus micropus* Indian Hedgehog

Pakistan and NW India

❏ *Hemiechinus nudiventris* Bare-bellied Hedgehog

Kerala and Tamil Nadu Provs. (S India)

Family: TALPIDAE
(Shrew-Moles, Moles and Desmans—42)

❏ *Uropsilus andersoni* Anderson's Shrew-Mole

C Sichuan (C China)

❏ *Uropsilus gracilis* Gracile Shrew-Mole

Sichuan and Yunnan (C & SW China) and N Myanmar

❏ *Uropsilus investigator* Inquisitive Shrew-Mole

Yunnan (SW China); Endangered

❏ *Uropsilus soricipes* Chinese Shrew-Mole

C Sichuan (C China); Endangered

❏ *Scaptonyx fusicaudus* Long-tailed Mole

N Myanmar, C & S China

Talpa (Mogera)

❏ *Talpa etigo* Echigo Mole

Echigo Plain (Honshu, C Japan); Endangered

❏ *Talpa insularis* Insular Mole

SE China, Hainan and Taiwan

❏ *Talpa kobeae* Kobe Mole

S Honshu, Shikoku and Kyushu (C & S Japan)

❏ *Talpa minor* Small Japanese Mole

Honshu (C Japan)

❏ *Talpa robusta* Large Mole

E Siberia, Korea and NE China

❏ *Talpa tokudae* Tokuda's Mole

Sado Is. (Japan); Endangered

❏ *Talpa wogura* Japanese Mole

Honshu, Kyushu and near islands (C & S Japan)

Talpa (Parascaptor)

❏ *Talpa leucura* White-tailed Mole

Assam (NE India), Myanmar and Yunnan (SW China)

Talpa (Talpa)

❏ *Talpa altaica* Siberian Mole

Taiga zone of C Siberia, S to N Mongolia

❏ *Talpa caeca* Blind Mole

Locally in soils of woods and pastures of SC & SE Europe

❏ *Talpa caucasica* Caucasian Mole

NW Caucasus Mtns (Russia)

❏ *Talpa europaea* European Mole

Moist soils of woods and pastures of C & NE Europe, E to C Siberia

❏ *Talpa levantis* Levantine Mole

SE Europe, Turkey and SW Caucasus Mtns (Georgia)

❏ *Talpa occidentalis* Iberian Mole

Soils in meadows of Portugal and Spain (except NE)

❏ *Talpa romana* Roman Mole

Soils in forests and fields of S Italy; formerly Sicily (now extinct)

❏ *Talpa stankovici* Balkan Mole

Soils in grasslands and pastures of extreme S Montenegro (Yugoslavia), W Macedonia and NW Greece incl. Corfu

❏ *Talpa streeti* Persian Mole

N Iran; Critically Endangered

❏ *Scaptochirus moschatus* Short-faced Mole

Shanxi, Shaanxi, Hebei and Shandong (EC & NE China)

❏ *Euroscaptor grandis* Greater Chinese Mole

C China and Vietnam

❏ *Euroscaptor klossi* Kloss' Mole

Highlands of Thailand, Laos and Malay Peninsula

❏ *Euroscaptor longirostris* Long-nosed Mole

C & S China

❏ *Euroscaptor micrura* Himalayan Mole

Nepal and NE India

❏ *Euroscaptor mizura* Japanese Mountain Mole

Mountains of Honshu (C Japan)

❏ *Euroscaptor parvidens* Small-toothed Mole

Di Linh and Rakho regions (Vietnam); Critically
Endangered

❏ *Nesoscaptor uchidai* Ryukyu Mole

Uotsuri-jima island (Ryukyu Is., S Japan); Endangered

❏ *Scapanulus oweni* Gansu Mole

Montane forests of C China

❏ *Parascalops breweri* Hairy-tailed Mole

Moist sandy soils of SE Canada and NE USA

❏ *Scapanus latimanus* Broad-footed Mole

Moist soils of W USA and extreme N Baja California Norte
(Mexico)

❏ *Scapanus orarius* Coast Mole

Soils in grasslands and woods of SW British Columbia
(Canada) to NW California and WC Idaho (USA)

❏ *Scapanus townsendii* Townsend's Mole

SW British Columbia (Canada) to NW California (USA)

❏ *Scalopus aquaticus* Eastern Mole

Moist loam soils of C & E USA and N Tamaulipas (NE
Mexico)

❏ *Urotrichus pilirostris* True's Shrew-Mole

Montane forests of Honshu, Shikoku and Kyushu (C & S
Japan)

❏ *Urotrichus talpoides* Japanese Shrew-Mole

Honshu, Shikoku, Kyushu and Dogo I. (C & S Japan)

❏ *Neurotrichus gibbsii* American Shrew-Mole

Leaf or dense grass litter habitats from SW British Columbia
(Canada) to WC California (USA)

❏ *Condylura cristata* Star-nosed Mole

Wetlands of E Canada and NE USA

❏ *Desmana moschata* Russian Desman

River basins of S Russia and W Kazakhstan; feral Ukraine and
W Siberia

❏ *Galemys pyrenaicus* Pyrenean Desman

Montane streams and lakes of N Portugal, N & C Spain and
SW France

Order: SORICOMORPHA
Family: SOLENODONTIDAE
(Solenodons—2)

❏ *Solenodon cubanus* Cuban Solenodon

Oriente Prov. (E Cuba); Endangered

❏ *Solenodon paradoxus* Hispaniolan Solenodon

Haiti and Domincan Republic; Endangered

Family: SORICIDAE (Shrews—335)

❏ *Blarinella quadraticauda*
Sichuan Short-tailed Shrew

Montane forests of C & SW China

❏ *Blarinella wardi* Ward's Short-tailed Shrew

N Myanmar and Yunnan (SW China)

❏ *Sorex alaskanus* Glacier Bay Water Shrew

Streams of Glacier Bay region (S Alaska, USA); status
uncertain, not seen since 1899

❏ *Sorex alpinus* Alpine Shrew

Rocky areas and streamsides in mountains of C Europe

❏ *Sorex araneus* Common Shrew

Dense forests and grasslands of Europe (except Ireland,
Iberian Peninsula and Mediterranean region) to C Siberia

❏ *Sorex arcticus* Arctic Shrew

Swamps and clearings in boreal coniferous forests of Canada
and NC & NE USA

❏ *Sorex arizonae* Arizona Shrew

Locally in submontane forests of SE Arizona and SW New
Mexico (USA), and Chihuahua (NC Mexico)

❏ *Sorex arunchi* Italian Shrew

NE Italy

❏ *Sorex asper* Tien Shan Shrew

Mountains of Tien Shan (Kazakhstan and W China)

❏ *Sorex bairdii* Baird's Shrew

Moist coniferous forests of SW Washington (USA)

❏ *Sorex bedfordiae* Lesser Striped Shrew

Montane forests of Nepal, N Myanmar, C & SW China

❑ *Sorex bendirii* **Marsh Shrew**

Marshes and streams in coastal forests of SE British Columbia (Canada) to NW California (USA)

❑ *Sorex buchariensis* **Pamir Shrew**

Mountains of Pamirs (Tajikistan)

❑ *Sorex caecutiens* **Masked Shrew**

Forests, taiga and tundra zones of NE Europe to E Siberia, S to Ukraine, Kazakhstan, C China and Korea

❑ *Sorex camtschatica* **Kamchatka Shrew**

S Kamchatka Peninsula (E Siberia)

❑ *Sorex cansulus* **Gansu Shrew**

Lintan region (Gansu, NC China); Critically Endangered

❑ *Sorex cinereus* **Cinereous Shrew**

Alaska, Canada and N USA

❑ *Sorex coronatus* **Millet's Shrew**

Woods, marshes and meadows of C Germany, S to Alps and N Spain

❑ *Sorex cylindricauda* **Stripe-backed Shrew**

Montane forests of N Sichuan (C China); Endangered

❑ *Sorex daphaenodon*
Siberian Large-toothed Shrew

Siberia, incl. Kamchatka Peninsula and Kuril Is., and NE China

❑ *Sorex dispar* **American Long-tailed Shrew**

Rocky areas in forests of Appalachian Mtns from E Canada to North Carolina (USA)

❑ *Sorex emarginatus* **Zacatecas Shrew**

Durango, Zacatecas and Jalisco (WC Mexico)

❑ *Sorex excelsus* **Lofty Shrew**

Sichuan and Yunnan (C & SW China)

❑ *Sorex fumeus* **Smoky Shrew**

Wet forests of SE Canada and NE & EC USA

❑ *Sorex gaspensis* **Gaspé Shrew**

Rocky areas of Gaspé Peninsula to Nova Scotia and Cape Breton I. (E Canada)

❑ *Sorex gracillimus* **Slender Shrew**

E Siberia, incl. Sakhalin I. (Russia), N Korea and Hokkaido (N Japan)

❑ *Sorex granarius* **Spanish Shrew**

Woods and scrub of N Portugal and NW & WC Spain

❑ *Sorex haydeni* **Prairie Shrew**

Prairie grasslands of SC Canada and NC USA

❑ *Sorex hosonoi* **Azumi Shrew**

Montane forests of C Honshu (C Japan)

❑ *Sorex isodon* **Taiga Shrew**

Dense wet coniferous and mixed forests of NE Europe and Siberia, E to Kamchatka Peninsula, Sakhalin and Kuril Is. (Russia)

❑ *Sorex kozlovi* **Kozlov's Shrew**

Zi Qu R. region (Tibet); Critically Endangered

❑ *Sorex leucogaster* **Paramushir Shrew**

Paramushir Is. (S of Kamchatka Peninsula, E Siberia)

❑ *Sorex longirostris* **Southeastern Shrew**

SE USA

❑ *Sorex lyelli* **Mount Lyell Shrew**

Mountains of C Sierra Nevada range (California, USA)

❑ *Sorex macrodon* **Mexican Large-toothed Shrew**

Veracruz and Puebla (EC Mexico)

❑ *Sorex maritimensis* **Maritime Shrew**

New Brunswick and Nova Scotia (E Canada)

❑ *Sorex merriami* **Merriam's Shrew**

Sagebrush, prairies and woods of interior W USA

❑ *Sorex milleri* **Carmen Mountain Shrew**

N Sierra Madre Oriental (NE Mexico)

❑ *Sorex minutissimus* **Eurasian Least Shrew**

Wet coniferous forests, clearings and peat bogs of N Europe to E Siberia, Korea, Mongolia, China and Hokkaido (N Japan)

❑ *Sorex minutus* **Eurasian Pygmy Shrew**

Marshes, meadows, dunes and montane forests of Europe to C Asia

❏ *Sorex mirabilis* Ussuri Shrew

Ussuri region (E Siberia), N Korea and NE China

❏ *Sorex monticolus* Montane Shrew

Wet montane habitats of Alaska, SW Canada, W USA and NW Mexico

❏ *Sorex nanus* American Dwarf Shrew

Arid prairies and rocky tundra from Montana and South Dakota to Arizona and New Mexico (USA)

❏ *Sorex oreopolus* Mexican Long-tailed Shrew

Jalisco (WC Mexico)

❏ *Sorex ornatus* Ornate Shrew

Streams and montane forests of coastal C & S California (USA) and Baja California (Mexico)

❏ *Sorex pacificus* Pacific Shrew

Marshes and streamsides in coastal forests of W Oregon and extreme N California (USA)

❏ *Sorex palustris* American Water Shrew

Streams in boreal forest zone of SE Alaska, Canada, W, NC & NE USA

❏ *Sorex planiceps* Kashmir Shrew

N Pakistan and Kashmir

❏ *Sorex portenkoi* Portenko's Shrew

NE Siberia

❏ *Sorex preblei* Preble's Shrew

Marshes and grasslands of NW USA

❏ *Sorex pribilofensis* Pribilof Island Shrew

St Paul (Pribilof Is., Alaska, USA); Endangered

❏ *Sorex raddei* Radde's Shrew

N Turkey and Caucasus Mtns (Georgia)

❏ *Sorex roboratus* Flat-skulled Shrew

C & E Siberia and Mongolia

❏ *Sorex sadonis* Sado Shrew

Sado Is. (Japan); Endangered

❏ *Sorex samniticus* Apennine Shrew

Riparian habitats in Appenine Mtns (Italy)

❏ *Sorex satunini* Caucasian Shrew

N Turkey and Caucasus Mtns (Georgia)

❏ *Sorex saussurei* Saussure's Shrew

Montane forests of C & S Mexico and W Guatemala

❏ *Sorex sclateri* Sclater's Shrew

Tumbalá region (N Chiapas, S Mexico); Endangered

❏ *Sorex shinto* Shinto Shrew

Hokkaido, Honshu and Shikoku (Japan)

❏ *Sorex sinalis* Chinese Shrew

W & C China

❏ *Sorex sonomae* Fog Shrew

Forests of fog zone of coastal Oregon and N California (USA)

❏ *Sorex stizodon* San Cristobal Shrew

Wet montane forests of San Cristobal region (N Chiapas, Mexico); Endangered

❏ *Sorex tenellus* Inyo Shrew

Semiarid areas of E Sierra Nevada and Great Basin of EC California and WC Nevada (USA)

❏ *Sorex thibetanus* Tibetan Shrew

NE Tibet

❏ *Sorex trowbridgii* Trowbridge's Shrew

Coastal range forests from SW British Columbia (Canada) to C California (USA)

❏ *Sorex tundrensis* Tundra Shrew

Tundra zone of Siberia incl. Sakhalin I. (Russia), S to Mongolia and NE China; Alaska (USA) and extreme NW Canada

❏ *Sorex ugyunak* Barren Ground Shrew

Tundra zone of N Alaska (USA) and extreme N Canada

❏ *Sorex unguiculatus* Large-clawed Shrew

Coastal SE Siberia incl. Sakhalin I. (Russia) and Hokkaido (N Japan)

❏ *Sorex vagrans* Vagrant Shrew

Marshes of S British Columbia (Canada) and NW USA

❏ *Sorex ventralis* Chestnut-bellied Shrew

NW Puebla to Oaxaca (SC Mexico)

❏ *Sorex veraepacis* Verapaz Shrew

Wet montane forests of EC & S Mexico to C Guatemala

❏ *Sorex volnuchini* Caucasian Pygmy Shrew

S Ukraine and Caucasus Mtns (Russia)

❏ *Sorex yukonicus* Alaska Tiny Shrew

W Alaska (USA)

❏ *Microsorex hoyi* American Pygmy Shrew

Taiga zone and high mountains of Alaska, Canada and USA

❏ *Notiosorex cockrumi* Cockrum's Desert Shrew

Riparian scrub of SC & SE Arizona (USA) to C Sonora (N Mexico)

❏ *Notiosorex crawfordi* Desert Shrew

Arid habitats of SW & SC USA to C Mexico

❏ *Notiosorex evotis* Large-eared Grey Shrew

Cactus and dense thorn scrub deserts of WC Mexico

❏ *Notiosorex villai* Villa's Grey Shrew

Mountain valleys of Tamaulipas (NE Mexico)

❏ *Megasorex gigas* Mexican Shrew

Nayarit to Oaxaca (WC & S Mexico)

❏ *Anourosorex squamipes* Mole Shrew

Assam (NE India), Bhutan and Myanmar to C China, Taiwan and N Vietnam

Soriculus (Chodsigoa)

❏ *Soriculus hypsibius* De Winton's Shrew

C & SW China

❏ *Soriculus lamula* Lamulate Shrew

C China

❏ *Soriculus parca* Lowe's Shrew

N Myanmar, SW China, Thailand and N Vietnam

❏ *Soriculus salenskii* Salenski's Shrew

Liangfu region (Sichuan, C China); Critically Endangered

❏ *Soriculus smithii* Smith's Shrew

C China

Soriculus (Episoriculus)

❏ *Soriculus caudatus*
Hodgson's Brown-toothed Shrew

Kashmir to N Myanmar and SW China

❏ *Soriculus fumidus* Taiwan Brown-toothed Shrew

Highland forests of Taiwan

❏ *Soriculus leucops*
Long-tailed Brown-toothed Shrew

C Nepal and NE India to N Myanmar, S China and Vietnam

❏ *Soriculus macrurus* Long-tailed Mountain Shrew

C Nepal and W China to N Myanmar and Vietnam

Soriculus (Soriculus)

❏ *Soriculus nigrescens* Himalayan Shrew

Nepal, Tibet, Assam (NE India) and SW China

❏ *Neomys anomalus* Miller's Water Shrew

Riparian habitats of C & S Europe and Russia

❏ *Neomys fodiens* Eurasian Water Shrew

Riparian habitats, wet woods and fields of C & N Europe to E Siberia incl. Sakhalin I. (Russia), N Korea, NW Mongolia and NE China

❏ *Neomys schelkovnikovi*
Transcaucasian Water Shrew

Caucasus Mtns (Georgia, Armenia and Azerbaijan)

❏ *Nectogale elegans* Elegant Water Shrew

Mountain streams from Nepal and Tibet to N Myanmar and C China

❏ *Chimarrogale hantu* Malayan Water Shrew

Forest streams of Malaya; Critically Endangered

❏ *Chimarrogale himalayica* Himalayan Water Shrew

Kashmir to C China, Taiwan and Indochina

❏ *Chimarrogale phaeura* Bornean Water Shrew

Forest streams of Borneo; Endangered

❏ *Chimarrogale platycephala*
Flat-headed Water Shrew

Japan

❑ *Chimarrogale styani* Styan's Water Shrew

N Myanmar and C China

❑ *Chimarrogale sumatrana* Sumatran Water Shrew

Forest streams of Sumatra; Critically Endangered

❑ *Cryptotis colombiana*
Colombian Small-eared Shrew

Colombia

❑ *Cryptotis endersi* Enders' Small-eared Shrew

Dense wet forests of Atlantic slope of Talamanca Mtns (W Panama); Endangered

❑ *Cryptotis goldmani*
Goldman's Small-eared Shrew

Forests of C & S Mexico and WC Guatemala

❑ *Cryptotis goodwini*
Goodwin's Small-eared Shrew

Wet montane forests of Atlantic slope of SE Chiapas (SE Mexico) to S Guatemala and W El Salvador

❑ *Cryptotis gracilis*
Talamancan Small-eared Shrew

Montane oak forests and páramo of Central and Talamanca Mtns (Costa Rica and W Panama)

❑ *Cryptotis hondurensis*
Honduran Small-eared Shrew

Montane forests of Francisco Morazán Dept. (S Honduras)

❑ *Cryptotis magna* Big Small-eared Shrew

Highlands of NC Oaxaca (Mexico)

❑ *Cryptotis mayensis* Maya Small-eared Shrew

Lowland dry forests and scrub of Yucatán Peninsula (SE Mexico, NE Guatemala and N Belize)

❑ *Cryptotis mera* Darién Small-eared Shrew

Dense fern-covered streambanks in wet evergreen forests of Darién Highlands (E Panama)

❑ *Cryptotis meridensis* Merida Small-eared Shrew

Highlands of NW Venezuela

❑ *Cryptotis merriami* Merriam's Small-eared Shrew

Montane forests and fields of E Chiapas (SE Mexico) to N Nicaragua; NW Costa Rica

❑ *Cryptotis mexicana* Mexican Small-eared Shrew

Damp grasslands and forest edge habitats of Tamaulipas to NE Chiapas (E & SE Mexico)

❑ *Cryptotis montivaga*
Ecuadorian Small-eared Shrew

Dense montane forests of Andes of S Ecuador

❑ *Cryptotis nigrescens* Blackish Small-eared Shrew

Montane forests and meadows of Costa Rica and W Panama (and NW Colombia?)

❑ *Cryptotis parva* American Least Shrew

Forests and grasslands of extreme SE Canada, C & E USA and Mexico to Costa Rica

❑ *Cryptotis peruviensis* Peruvian Small-eared Shrew

Cloud forests of Andes of N Peru

❑ *Cryptotis squamipes*
Scaly-footed Small-eared Shrew

Montane forests of Cordillera Occidental of W Colombia and NW Ecuador

❑ *Cryptotis tamensis*
Páramo de Tamá Small-eared Shrew

Páramo de Tamá (Colombia and Venezuela)

❑ *Cryptotis thomasi* Thomas' Small-eared Shrew

Páramo of Cordillera Oriental of C Colombia and NW Venezuela (and S to Ecuador?)

❑ *Blarina brevicauda* Northern Short-tailed Shrew

SC & SE Canada, NC & NE USA

❑ *Blarina carolinensis* Southern Short-tailed Shrew

SC & SE USA

❑ *Blarina hylophaga* Elliot's Short-tailed Shrew

SC USA

❑ *Crocidura aleksandrisi* Alexandrian Shrew

Cyrenaica region (Libya)

❑ *Crocidura allex* Highland Shrew

Highlands of SW Kenya and N Tanzania

❑ *Crocidura andamanensis* Andaman Shrew

South Andaman I. (Andaman Is., India)

❏ *Crocidura ansellorum* Ansell's Shrew

Isombu R. region (Mwinilunga Dist., N Zambia); Critically Endangered

❏ *Crocidura arabica* Arabian Shrew

Coastal plains of Yemen and Oman

❏ *Crocidura armenica* Armenian Shrew

Garni region (Armenia)

❏ *Crocidura attenuata* Indochinese Shrew

India and Nepal to S China, Taiwan, Indochina and Philippines

❏ *Crocidura attila* Hun Shrew

Highlands of Cameroon to E D.R. Congo

❏ *Crocidura baileyi* Bailey's Shrew

Highlands of W Ethiopia

❏ *Crocidura batesi* Bates' Shrew

Lowland rainforests of S Cameroon and Gabon

❏ *Crocidura beatus* Mindanao Shrew

Primary rainforests and regrowth areas of Leyte, Maripipi and Mindanao (EC & SE Philippines)

❏ *Crocidura beccarii* Beccari's Shrew

Mt Singgalang (W Sumatra); Endangered

❏ *Crocidura bottegi* Bottego's Shrew

Scattered localities from Guinea to Ethiopia and N Kenya

❏ *Crocidura bottegoides* Bale Shrew

Bale Mtns and Mt Albasso (Ethiopia)

❏ *Crocidura brunnea* Thick-tailed Shrew

Java

❏ *Crocidura buettikoferi* Buettikofer's Shrew

Forests of Guinea-Bissau to Liberia; Nigeria

❏ *Crocidura caliginea* African Foggy Shrew

Medje region (NE D.R. Congo); Critically Endangered

❏ *Crocidura canariensis* Canary Shrew

Lava fields and rocky areas of Fuerteventura, Lanzarote, Lobos and Mt Clara (E Canary Is.)

❏ *Crocidura cinderella* Cinderella Shrew

Senegal, Gambia, Mali and Niger

❏ *Crocidura congobelgica* Congo Shrew

NE D.R. Congo

❏ *Crocidura cossyrensis* Pantellerian Shrew

Pantelleria I. (Italy)

❏ *Crocidura crenata* Long-footed Shrew

Rainforests of S Cameroon, N Gabon and W Cameroon

❏ *Crocidura crossei* Crosse's Shrew

Lowland forests from Sierra Leone to W Cameroon

❏ *Crocidura cyanea* Reddish-grey Musk Shrew

S Angola, Namibia, E Botswana, Zimbabwe, Mozambique and South Africa

❏ *Crocidura denti* Dent's Shrew

Cameroon, Gabon and NE D.R. Congo

❏ *Crocidura desperata* Desperate Shrew

Relict forests in Rungwe and Udzungwa Mtns (S Tanzania); Critically Endangered

❏ *Crocidura dhofarensis* Dhofarian Shrew

Dhofar region (Oman); Critically Endangered

❏ *Crocidura dolichura* Long-tailed Musk Shrew

Forests from Nigeria to Uganda, S to Gabon, Rep. Congo, D.R. Congo and Burundi

❏ *Crocidura douceti* Doucet's Musk Shrew

Guinea, Ivory Coast and Nigeria

❏ *Crocidura dsinezumi* Dsinezumi Shrew

Densely vegetated foothills and riparian habitats of Cheju do (Korea) and Japan

❏ *Crocidura eisentrauti* Eisentraut's Shrew

Mt Cameroun (Cameroon); Critically Endangered

❏ *Crocidura elgonius* Elgon Shrew

Mt Elgon (W Kenya); NE Tanzania

❏ *Crocidura elongata* Elongated Shrew

N & C Sulawesi

❏ *Crocidura erica* Heather Shrew

W Angola

❏ *Crocidura fischeri* Fischer's Shrew

Nguruman region (S Kenya) and Himo region (Tanzania)

❏ *Crocidura flavescens* Greater Musk Shrew

Coastal South Africa and extreme S Mozambique

❏ *Crocidura floweri* Flower's Shrew

Upper Nile Valley and Wadi el Natrun (Egypt); Endangered

❏ *Crocidura foetida* Lowland Bornean Shrew

Borneo

❏ *Crocidura foxi* Fox's Shrew

Jos Plateau (Nigeria)

❏ *Crocidura fuliginosa* Southeast Asian Shrew

N India, Myanmar, SW & E China and Malay Peninsula

❏ *Crocidura fulvastra* Savanna Shrew

Savanna zone from Mali to Kenya

❏ *Crocidura fumosa* Smoky White-toothed Shrew

Mt Kenya and Aberdare Mtns (Kenya)

❏ *Crocidura fuscomurina* Tiny Musk Shrew

Locally in savannas of subsaharan Africa, S to E South Africa

❏ *Crocidura glassi* Glass' Shrew

Highlands of E Ethiopia

❏ *Crocidura gmelini* Gmelin's Shrew

Steppes and semi-desert areas of C Iran to C China

❏ *Crocidura goliath* Goliath Shrew

Rainforests of S Cameroon, Gabon and D.R. Congo

❏ *Crocidura gracilipes* Peters' Musk Shrew

"Kilimanjaro, Tanzania" (type locality uncertain); Critically Endangered

❏ *Crocidura grandiceps* Large-headed Shrew

Forests of Guinea, Ivory Coast, Ghana and Nigeria

❏ *Crocidura grandis* Mount Malindang Shrew

Mt Malindang (Mindanao, SE Philippines); Endangered

❏ *Crocidura grassei* Grasse's Shrew

Scattered localities in forests of Cameroon, Central African Republic and Gabon

❏ *Crocidura grayi* Luzon Shrew

Luzon and Mindoro (N & WC Philippines)

❏ *Crocidura greenwoodi* Greenwood's Shrew

S Somalia

❏ *Crocidura gueldenstaedtii* Güldenstädt's Shrew

E Europe to China and Taiwan

❏ *Crocidura harenna* Harenna Shrew

Harenna Forest (Bale Mtns, Ethiopia); Critically Endangered

❏ *Crocidura hildegardeae* Hildegarde's Shrew

Nigeria and Cameroon to East Africa

❏ *Crocidura hilliana* Hill's Shrew

Thailand

❏ *Crocidura hirta* Lesser Red Musk Shrew

D.R. Congo, Kenya and Somalia, S to N & E South Africa

❏ *Crocidura hispida* Andaman Spiny Shrew

Middle Andaman I. (Andaman Is., India); Endangered

❏ *Crocidura horsfieldii* Horsfield's Shrew

Nepal, India and Sri Lanka to S China, Taiwan, Indochina and Ryukyu Is. (Japan)

❏ *Crocidura hutanis* Hutan Shrew

N & W Sumatra

❏ *Crocidura jacksoni* Jackson's Shrew

E D.R. Congo, Uganda, Kenya and N Tanzania

❏ *Crocidura jenkinsi* Jenkins' Shrew

Wright Myo region (South Andaman I., Andaman Is., India)

❏ *Crocidura kivuana* Kivu Shrew

Montane swamps in Kahuzi-Biega region (Kivu, D.R. Congo)

❏ *Crocidura lamottei* Lamotte's Shrew

Savanna zone from Senegal to W Cameroon

❏ *Crocidura lanosa* Lemera Shrew

Kivu, Lemera and Irangi regions (D.R. Congo) and Uinka region (Rwanda)

❏ *Crocidura lasiura* Ussuri White-toothed Shrew

E Siberia, Korea and NE China

❏ *Crocidura latona* Latona Shrew

Lowland rainforests of Medje region (NE D.R. Congo)

❏ *Crocidura lea* Sulawesi Shrew

Rainforests of N & C Sulawesi

❏ *Crocidura lepidura* Sumatran Giant Shrew

Sumatra

❏ *Crocidura leucodon*
Bicoloured White-toothed Shrew

Forest edge, farmland and gardens of C & SE Europe to S Russia, Caucasus Mtns, Elburz Mtns (NC Iran) and Middle East

❏ *Crocidura levicula* Brown Shrew

Rainforests of C & SE Sulawesi

❏ *Crocidura littoralis* Butiaba Naked-tailed Shrew

Rainforests of D.R. Congo, Uganda and Kenya

❏ *Crocidura longipes* Savanna Swamp Shrew

Swamps in savanna region of W Nigeria; Endangered

❏ *Crocidura lucina* Moorland Shrew

Montane moorlands of E Ethiopia

❏ *Crocidura ludia* Dramatic Shrew

Medje and Tandala regions (N D.R. Congo)

❏ *Crocidura luna* Greater Grey-brown Musk Shrew

D.R. Congo, Uganda and Kenya, S to E Angola, Zambia, E Zimbabwe and WC Mozambique

❏ *Crocidura lusitania* Mauritanian Shrew

S Morocco, Mauritania and Senegal to Sudan and Ethiopia

❏ *Crocidura macarthuri* MacArthur's Shrew

Savannas of Kenya and Somalia

❏ *Crocidura macmillani* Macmillan's Shrew

Kotelee region (Ethiopia); Critically Endangered

❏ *Crocidura macowi* Macow's Shrew

Mt Nyiro (Kenya); Critically Endangered

❏ *Crocidura malayana* Malayan Shrew

Malaya and near islands; Endangered

❏ *Crocidura manengubae* Manenguba Shrew

Bamenda, Adamaoua and Yaounde highlands of Cameroon

❏ *Crocidura maquassiensis* Maquassie Musk Shrew

Locally in Zimbabwe and Northern Prov. (South Africa)

❏ *Crocidura mariquensis* Swamp Musk Shrew

SC Angola, Zambia, Zimbabwe and Mozambique, S to E South Africa

❏ *Crocidura maurisca* Dark Shrew

Echuya Swamp (Uganda) and Kaimosi region (Kenya)

❏ *Crocidura mindorus* Mindoro Shrew

Mt Halcon (Mindoro, WC Philippines); Endangered

❏ *Crocidura miya* Sri Lankan Long-tailed Shrew

Highlands of C Sri Lanka; Endangered

❏ *Crocidura monax* Rombo Shrew

Montane forests in W Kenya and N Tanzania

❏ *Crocidura monticola* Sunda Shrew

Malay Peninsula, Java, Borneo and Lesser Sundas; feral Ambon, Obi, Aru and Kai Is. (S Moluccas)

❏ *Crocidura montis* Montane White-toothed Shrew

Imatong Mtns (Sudan), Mt Ruwenzori (Uganda) and Mt Meru (Tanzania)

❏ *Crocidura muricauda* Mouse-tailed Shrew

Forests from Guinea to Ghana

❏ *Crocidura musseri* Mossy Forest Shrew

Gunung Rorekatimbo (C Sulawesi)

❏ *Crocidura mutesae*
Uganda Large-toothed Shrew

Uganda

❏ *Crocidura nana* Dwarf White-toothed Shrew

Ethiopia and Somalia

❏ *Crocidura nanilla* Tiny White-toothed Shrew

Savannas from Mauritania to Uganda and Kenya

❏ *Crocidura neglecta* Neglected Shrew

Sumatra

❏ *Crocidura negrina* Negros Shrew

S Negros (C Philippines); Critically Endangered

❏ *Crocidura nicobarica* Nicobar Shrew

Great Nicobar I. (Nicobar Is., India)

❏ *Crocidura nigeriae* Nigerian Shrew

Rainforests of Nigeria, Cameroon and Bioko (Equatorial Guinea)

❏ *Crocidura nigricans* Black White-toothed Shrew

Angola

❏ *Crocidura nigripes* Black-footed Shrew

Rainforests of N & C Sulawesi

❏ *Crocidura nigrofusca* Tenebrous Shrew

Sudan and S Ethiopia, S to Angola and Zambia

❏ *Crocidura nimbae* Nimba Shrew

Guinea, Sierra Leone and Liberia

❏ *Crocidura niobe* Stony Shrew

Montane forests of D.R. Congo and Uganda

❏ *Crocidura obscurior*
Obscure White-toothed Shrew

Sierra Leone to Ivory Coast

❏ *Crocidura olivieri* Giant Musk Shrew

Egypt and subsaharan Africa, S to N Botswana and C Mozambique

❏ *Crocidura orientalis* Oriental Shrew

Java

❏ *Crocidura orii* Amami Shrew

Ryukyu Is. (Japan); Endangered

❏ *Crocidura osorio* Osorio Shrew

Damp evergreen forests and regrowth areas of N Gran Canaria I.; ?feral Tenerife (Canary Is.)

❏ *Crocidura palawensis* Palawan Shrew

Palawan (SW Philippines)

❏ *Crocidura paradoxura* Paradox Shrew

Sumatra; Endangered

❏ *Crocidura parvipes* Small-footed Shrew

Cameroon to S Sudan and Ethiopia, S to Angola, Zambia and Tanzania

❏ *Crocidura pasha* Pasha Shrew

Sudan and Ethiopia

❏ *Crocidura pergrisea* Pale Grey Shrew

Kashmir

❏ *Crocidura phaeura* Guramba Shrew

Mt Guramba region (Sidamo, Ethiopia); Critically Endangered

❏ *Crocidura picea* Pitch Shrew

Assumbo region (Mamfe Div., Cameroon); Critically Endangered

❏ *Crocidura pitmani* Pitman's Shrew

N & C Zambia

❏ *Crocidura planiceps* Flat-headed Shrew

Nigeria; Sudan, Ethiopia, D.R. Congo and Uganda

❏ *Crocidura poensis* Fraser's Musk Shrew

Liberia to Cameroon; Bioko and Principe (Equatorial Guinea)

❏ *Crocidura polia* Fuscous Shrew

Medje region (D.R. Congo); Critically Endangered

❏ *Crocidura pullata* Dusky White-toothed Shrew

Afghanistan, Pakistan and Kashmir (NW India) to Yunnan (SW China) and Thailand

❏ *Crocidura raineyi* Rainey Shrew

Mt Garguez (Kenya); Critically Endangered

❏ *Crocidura ramona* Negev Shrew

C Negev Desert (Israel)

❏ *Crocidura rapax* Rapacious Shrew

SW China

❏ *Crocidura religiosa* Egyptian Pygmy Shrew

Nile Valley (Egypt)

❏ *Crocidura rhoditis* Temboan Shrew

Rainforests of N, C & SW Sulawesi

❏ *Crocidura roosevelti* Roosevelt's Shrew

Cameroon, Central African Republic, Uganda, Rwanda, Angola, D.R. Congo and Tanzania

❏ *Crocidura russula*
Greater White-toothed Shrew

Open rocky habitats and built areas of W Europe incl. W Mediterranean islands, and Morocco to Tunisia

❏ *Crocidura selina* Moon Shrew

Lowland forests of Uganda; Endangered

❏ *Crocidura serezkyensis* Serezkaya Shrew

Azerbaijan, Iran, Turkmenistan, Tajikistan and Kazakhstan

❏ *Crocidura shantungensis* Shantung Shrew

E Asia and C China

❏ *Crocidura sibirica* Siberian Shrew

C Siberia

❏ *Crocidura sicula* Sicilian Shrew

Scrub, farmland and built areas of Sicily and Egadi Is. (Italy), and Gozo (Malta)

❏ *Crocidura silacea*
Lesser Grey-brown Musk Shrew

SE Botswana, Zimbabwe, extreme S Mozambique and NE South Africa

❏ *Crocidura smithii* Desert Musk Shrew

Arid regions of Senegal and Ethiopia

❏ *Crocidura somalica* Somali Shrew

Semi-desert regions of Mali, Sudan and Ethiopia (not certainly found in Somalia)

❏ *Crocidura stenocephala* Narrow-headed Shrew

Montane swamps of Mt Kahuzi (E D.R. Congo)

❏ *Crocidura suaveolens* Lesser White-toothed Shrew

Grasslands, farmland and built areas of C & S Europe incl. E Mediterranean islands, and Asia, E to South Korea, E China and Taiwan

❏ *Crocidura susiana* Iranian Shrew

Dezful region (SW Iran); Endangered

❏ *Crocidura tansaniana* Tanzanian Shrew

Usambara Mtns (Tanzania)

❏ *Crocidura tarella* Ugandan Shrew

Uganda

❏ *Crocidura tarfayensis* Tarfaya Shrew

Atlantic coast of S Morocco, Western Sahara and Mauritania

❏ *Crocidura telfordi* Telford's Shrew

Montane forests of Morningside region (Uluguru Mtns, Tanzania); Critically Endangered

❏ *Crocidura tenuis* Thin Shrew

Timor (Lesser Sundas)

❏ *Crocidura thalia* Thalia Shrew

Highlands of C Ethiopia

❏ *Crocidura theresae* Therese's Shrew

Savannas of Guinea to Ghana

❏ *Crocidura thomensis* São Tomé Shrew

São Tomé (Equatorial Guinea)

❏ *Crocidura turba* Tumultuous Shrew

Cameroon to Kenya, S to Angola, Zambia and Malawi

❏ *Crocidura ultima* Ultimate Shrew

Jombeni Range (Kenya); Critically Endangered

❏ *Crocidura usambarae* Usambara Shrew

W Usambara Mtns (Tanzania)

❏ *Crocidura viaria* Savanna Path Shrew

Savannas of S Morocco to Senegal, E to Sudan, Ethiopia and Kenya

❏ *Crocidura voi* Voi Shrew

Savannas of Mali, Nigeria, Sudan, Ethiopia, Somalia and Kenya

❏ *Crocidura vorax* Voracious Shrew

SW China

❏ *Crocidura vosmaeri* Banka Shrew

Sumatra

❏ *Crocidura whitakeri* Whitaker's Shrew

Coasts of Morocco to Tunisia and Egypt

❏ *Crocidura wimmeri* Wimmer's Shrew

S Ivory Coast; Endangered

❏ *Crocidura xantippe* Vermiculate Shrew

SE Kenya and Usambara Mtns (Tanzania)

❏ *Crocidura yankariensis* Yankari Shrew

Savannas of Nigeria, Cameroon, Sudan, Ethiopia, Somalia and Kenya

❏ *Crocidura zaphiri* Zaphir's Shrew

S Ethiopia and Kenya

❏ *Crocidura zarudnyi* Zarudny's Shrew

SE Iran, SE Afghanistan and SW Pakistan

❏ *Crocidura zimmeri* Zimmer's Shrew

Upemba region (D.R. Congo)

❏ *Crocidura zimmermanni* Zimmermann's Shrew

Very locally in Crete (Greece)

Myosorex (Congosorex)

❏ *Myosorex polli* Poll's Shrew

Miombo (*Brachystegia*) woods and gallery forest remnants of Lubondai region (S D.R. Congo); Critically Endangered

Myosorex (Myosorex)

❏ *Myosorex babaulti* Babault's Mouse-Shrew

EC D.R. Congo, Rwanda and Burundi

❏ *Myosorex blarina* Montane Mouse-Shrew

Mt Ruwenzori (D.R. Congo and Uganda)

❏ *Myosorex cafer* Dark-footed Forest Shrew

Highlands of E Zimbabwe, W Mozambique and E South Africa

❏ *Myosorex eisentrauti* Eisentraut's Mouse-Shrew

Montane forests of Bioko (Equatorial Guinea); Endangered

❏ *Myosorex geata* Geata Mouse-Shrew

Mountains of Tanzania; Endangered

❏ *Myosorex kihaulei* Udzungwa Mouse-Shrew

Forests of Udzungwa Mtns scarp (Tanzania)

❏ *Myosorex longicaudatus* Long-tailed Forest Shrew

Forests of SE Western Cape Prov. (South Africa)

❏ *Myosorex okuensis* Oku Mouse-Shrew

Montane forests of Bamenda plateau (Cameroon)

❏ *Myosorex rumpii* Rumpi Mouse-Shrew

Rumpi Hills (Cameroon); Critically Endangered

❏ *Myosorex schalleri* Schaller's Mouse-Shrew

Itombwe Mtns (E D.R. Congo); Critically Endangered

❏ *Myosorex sclateri* Sclater's Forest Shrew

KwaZulu-Natal region (E South Africa)

❏ *Myosorex tenuis* Zuurbron Forest Shrew

NE South Africa

❏ *Myosorex varius* Common Forest Shrew

South Africa, Lesotho and S Swaziland

Myosorex (Surdisorex)

❏ *Myosorex norae* Aberdare Shrew

Aberdare Mtns (Kenya)

❏ *Myosorex polulus* Mount Kenya Shrew

Mt Kenya (Kenya)

❏ *Suncus ater* Black Shrew

Gunung Kinabalu (N Borneo); Critically Endangered

❏ *Suncus dayi* Day's Shrew

S India

❏ *Suncus etruscus* Pygmy White-toothed Shrew

Scrub, farmland and built areas of S Europe, N Africa and Arabian Peninsula to Yunnan (SW China) and Thailand

❏ *Suncus fellowesgordoni* Sri Lanka Shrew

Highlands of C Sri Lanka; Endangered

❏ *Suncus hosei* Hose's Shrew

Lowland rainforests of Borneo

❏ *Suncus infinitesimus* Least Dwarf Shrew

Cameroon, Central African Republic and Kenya, S to South Africa

❏ *Suncus lixus* Greater Dwarf Shrew

Savannas of D.R. Congo and Kenya, S to NE South Africa

❏ *Suncus malayanus* Malayan Pygmy Shrew

Malay Peninsula (Thailand and Malaysia)

❏ *Suncus mertensi* Flores Shrew

Flores (Lesser Sundas); Critically Endangered

❏ *Suncus montanus* Sri Lanka Highland Shrew

Highlands of S India and Sri Lanka

❏ *Suncus murinus* Asian House Shrew

Afghanistan to China, Taiwan and Japan; introduced Indian and Pacific Ocean islands, Moluccas etc.

❏ *Suncus remyi* Remy's Shrew

Rainforests of NE Gabon; Critically Endangered

❏ *Suncus stoliczkanus* Anderson's Shrew

Arid regions of Pakistan, India, Nepal and Bangladesh

❏ *Suncus varilla* Lesser Dwarf Shrew

Savannas of Nigeria, D.R. Congo and Tanzania, S to South Africa

❏ *Suncus zeylanicus* Jungle Shrew

Highlands of Sri Lanka; Endangered

❏ *Sylvisorex granti* Grant's Shrew

Montane forests of Cameroon, D.R. Congo, Uganda, Rwanda, Kenya and Tanzania

❏ *Sylvisorex howelli* Howell's Shrew

Usambara and Uluguru Mtns (Tanzania)

❏ *Sylvisorex isabellae* Isabella Shrew

Bamenda plateau (Cameroon) and Bioko (Equatorial Guinea)

❏ *Sylvisorex johnstoni* Johnston's Shrew

Lowland forests from SW Cameroon to Uganda, S to Gabon and Tanzania

❏ *Sylvisorex konganensis* Central African Shrew

Central African Republic

❏ *Sylvisorex lunaris* Crescent Shrew

High mountains of E D.R. Congo, Uganda, Rwanda and Burundi

❏ *Sylvisorex megalura* Climbing Shrew

Dense moist forests from Guinea to Ethiopia, S to E Zimbabwe and WC Mozambique

❏ *Sylvisorex morio* Arrogant Shrew

Mt Cameroun (Cameroon); Endangered

❏ *Sylvisorex ollula* Forest Musk Shrew

Forests of S Nigeria, S Cameroon, Gabon and S D.R. Congo

❏ *Sylvisorex oriundus* Mountain Shrew

Medje region (NE D.R. Congo)

❏ *Sylvisorex pluvialis* Rainforest Shrew

Rainforests of SW Cameroon

❏ *Sylvisorex vulcanorum* Volcano Shrew

Montane forests of E D.R. Congo, Uganda, Rwanda and Burundi

❏ *Diplomesodon pulchellum* Piebald Shrew

Kazakhstan, Uzbekistan and Turkmenistan

❏ *Feroculus feroculus*
  Kelaart's Long-clawed Shrew

Highlands of C Sri Lanka; Endangered

❏ *Scutisorex somereni* Armoured Shrew

Seasonally flooded lowland and mid-elevation forests of D.R. Congo, Uganda, Rwanda and Burundi

❏ *Ruwenzorisorex suncoides* Ruwenzori Shrew

Montane forests of E D.R. Congo, Uganda, Rwanda and Burundi

❏ *Solisorex pearsoni*
  Pearson's Long-clawed Shrew

Highlands of C Sri Lanka; Endangered

❏ *Paracrocidura graueri* Grauer's Shrew

Montane forests of Itombwe Mtns (D.R. Congo); Critically Endangered

❏ *Paracrocidura maxima* Greater Shrew

Forests of D.R. Congo, Rwanda and Uganda

❏ *Paracrocidura schoutedeni* Schouteden's Shrew

Lowland forests of S Cameroon, Central African Republic, Gabon, Rep. Congo and D.R. Congo

Family: TENRECIDAE (Tenrecs—30)

❏ *Geogale aurita* Large-eared Tenrec

NE & SW Madagascar

❏ *Potamogale velox* Giant Otter-Shrew

Rivers, streams and swamps in forests of Nigeria, Gabon and Angola, E to Kenya; Endangered

❏ *Micropotamogale lamottei* Nimba Otter-Shrew

Streams, swamps and ditches in montane forests and forest edges of Mt Nimba region (Guinea, Liberia and Ivory Coast); Endangered

❏ *Micropotamogale ruwenzorii*
Ruwenzori Otter-Shrew

Streams in montane forests and forest edges of Ruwenzori Mtns (Uganda and D.R. Congo) and highlands of EC D.R. Congo; Endangered

❏ *Echinops telfairi* Lesser Hedgehog-Tenrec

S Madagascar

❏ *Tenrec ecaudatus* Tailless Tenrec

Madagascar and Comoro Is.; feral Réunion, Mauritius and Seychelles

❏ *Hemicentetes nigriceps* Highland Streaked Tenrec

Highlands of SC Madagascar

❏ *Hemicentetes semispinosus*
Lowland Streaked Tenrec

Lowland and mid-altitude forests and gardens of E Madagascar

❏ *Setifer setosus* Greater Hedgehog-Tenrec

C Madagascar

❏ *Limnogale mergulus* Aquatic Tenrec

Freshwater streams of E Madagascar; Endangered

❏ *Microgale brevicaudata* Short-tailed Shrew-Tenrec

Forests of Madagascar

❏ *Microgale cowani* Cowan's Shrew-Tenrec

N & E Madagascar

❏ *Microgale dobsoni* Dobson's Shrew-Tenrec

Forests of E Madagascar

❏ *Microgale drouhardi* Striped Shrew-Tenrec

Forests of E Madagascar

❏ *Microgale dryas* Tree Shrew-Tenrec

Ambatovaky Special Reserve (NE Madagascar); Critically Endangered

❏ *Microgale fotsifotsy* Pale-footed Shrew-Tenrec

Few scattered localities in forests of N, EC & SE Madagascar

❏ *Microgale gracilis* Gracile Shrew-Tenrec

Forests of E Madagascar

❏ *Microgale gymnorhyncha*
Naked-nosed Shrew-Tenrec

Madagascar

❏ *Microgale longicaudata*
Lesser Long-tailed Shrew-Tenrec

N & E Madagascar

❏ *Microgale monticola* Montane Shrew-Tenrec

Anjanaharibe-Sud Special Reserve (NE Madagascar)

❏ *Microgale nasoloi* Nasolo's Shrew-Tenrec

Forests of Vohibasia and Analavelona (SW Madagascar)

❏ *Microgale parvula* Pygmy Shrew-Tenrec

N & NE Madagascar; Endangered

❏ *Microgale principula*
Greater Long-tailed Shrew-Tenrec

E Madagascar; Endangered

❏ *Microgale pusilla* Least Shrew-Tenrec

E & S Madagascar

❏ *Microgale soricoides* Soricine Shrew-Tenrec

Few scattered localities in montane rainforests of E Madagascar

❏ *Microgale taiva* Taiva Shrew-Tenrec

Few localities in rainforests of EC? & SE Madagascar

❏ *Microgale talazaci* Talazac's Shrew-Tenrec

N & E Madagascar

❏ *Microgale thomasi* Thomas' Shrew-Tenrec

E Madagascar

❏ *Oryzorictes hova* Hova Rice Tenrec

Rainforests of N & EC Madagascar

❏ *Oryzorictes tetradactylus* Four-toed Rice Tenrec

Highlands of SC Madagascar

Grandorder: ARCHONTA
Order: CHIROPTERA
Family: PTEROPODIDAE
(Old World Fruit Bats—177)

❏ *Eidolon dupreanum*
Madagascar Straw-coloured Fruit Bat

Madagascar

❏ *Eidolon helvum* Straw-coloured Fruit Bat

Forests, riparian habitats and built areas of subsaharan Africa and Arabian Peninsula, incl. near islands; migrant to S Africa

Rousettus (Lissonycteris)

❏ *Rousettus angolensis* Angolan Rousette

Forests, savannas and riparian habitats of Senegal to Ethiopia, S to Angola, E Zimbabwe and WC Mozambique

Rousettus (Rousettus)

❏ *Rousettus amplexicaudatus* Geoffroy's Rousette

Indochina to Philippines, New Guinea, Bismarck Archipelago and Solomon Is.

❏ *Rousettus celebensis* Sulawesi Rousette

Sulawesi, Sangihe Is. and Sula Is. (NW & W Moluccas)

❏ *Rousettus egyptiacus* Egyptian Rousette

Forests, savannas and riparian habitats of subsaharan Africa, Cyprus, Turkey, Middle East and Arabian Peninsula to Pakistan

❏ *Rousettus leschenaulti* Leschenault's Rousette

Pakistan to Sri Lanka, S China, Indochina, Sumatra, Java and Bali

❏ *Rousettus madagascariensis*
Madagascar Rousette

Coastal WC, N & E Madagascar

❏ *Rousettus obliviosus* Comoro Rousette

Comoro Is.

❏ *Rousettus spinalatus* Bare-backed Rousette

Sumatra and Borneo

Rousettus (Stenonycteris)

❏ *Rousettus lanosus* Long-haired Rousette

Locally in montane forests of S Sudan and S Ethiopia, S to E D.R. Congo and Tanzania

❏ *Boneia bidens* Manado Fruit Bat

N Sulawesi

Myonycteris (Myonycteris)

❏ *Myonycteris relicta*
East African Collared Fruit Bat

Forests of Shimba Hills (Kenya), Nguru and Usambara Mtns (Tanzania)

❏ *Myonycteris torquata* Little Collared Fruit Bat

Forests of Sierra Leone to Uganda, S to Angola and Zambia

Myonycteris (Phrygetis)

❏ *Myonycteris brachycephala*
São Tomé Collared Fruit Bat

Forests of São Tomé (Equatorial Guinea); Endangered

❏ *Pteropus admiralitatum* Admiralty Flying Fox

Forests of Admiralty Is., Bismarck Archipelago and Solomon Is.

❏ *Pteropus aldabrensis* Aldabra Flying Fox

Aldabra Is.

❏ *Pteropus alecto* Black Flying Fox

Forests of Sulawesi, Lesser Sundas, lowlands of SC New Guinea and N & EC Australia

❏ *Pteropus anetianus* Vanuatu Flying Fox

Vanuatu

❏ *Pteropus argentatus* Ambon Flying Fox

Ambon (S Moluccas); status uncertain, only known from juvenile holotype collected in early 19th Century

❏ *Pteropus banakrisi* Torresian Flying Fox

Moa I. (Torres Strait, N Queensland, Australia)

❏ †*Pteropus brunneus* Dusky Flying Fox

Formerly Percy Is. (E Queensland, Australia); Extinct (only known from holotype specimen collected in 1859)

❏ *Pteropus caniceps* North Moluccan Flying Fox

Morotai, Halmahera, Ternate, Tidore, Bacan, Mangole etc. (N & W Moluccas)

❏ *Pteropus capistratus* Bismarck Flying Fox

Forests of Bismarck Archipelago

❏ *Pteropus chrysoproctus* Moluccan Flying Fox

Obi, Buru, Ambon, Seram and near islands (C & S Moluccas)

❏ *Pteropus cognatus* Makira Flying Fox

San Cristobal and near island (Solomon Is.)

❏ *Pteropus conspicillatus* Spectacled Flying Fox

Forests of Moluccas, Geelvinck Is., New Guinea, D'Entrecasteaux Is. and NE Queensland (Australia)

❏ *Pteropus dasymallus* Ryukyu Flying Fox

S Kyushu and Ryukyu Is. (S Japan) and Taiwan; Endangered

❏ *Pteropus faunulus* Nicobar Flying Fox

Nicobar Is. (India)

❏ *Pteropus fundatus* Banks Flying Fox

Vanua Lava and Mota (Banks Group, Vanuatu)

❏ *Pteropus giganteus* Indian Flying Fox

Pakistan, India incl. Maldives and Andaman Is., Sri Lanka and Myanmar

❏ *Pteropus gilliardorum* New Britain Flying Fox

Undisturbed montane moss forests of Whiteman Mtns (W New Britain, Bismarck Archipelago)

❏ *Pteropus griseus* Grey Flying Fox

Sulawesi and Lesser Sundas

❏ *Pteropus howensis* Ontong Java Flying Fox

Ontong Java Atoll (Solomon Is.)

❏ *Pteropus hypomelanus* Variable Flying Fox

Maldives (India); Indochina to Philippines, Bismarck Archipelago and Solomon Is.

❏ *Pteropus insularis* Chuuk Flying Fox

Atolls of Chuuk Lagoon (Fed. States of Micronesia); Critically Endangered

❏ *Pteropus leucopterus* White-winged Flying Fox

Luzon and Dinagat (N & EC Philippines); Endangered

❏ *Pteropus livingstonii* Comoro Black Flying Fox

Comoro Is.; Critically Endangered

❏ *Pteropus lombocensis* Lombok Flying Fox

Lombok, Flores and Alor (Lesser Sundas)

❏ *Pteropus lylei* Lyle's Flying Fox

Thailand and Vietnam

❏ *Pteropus macrotis* Big-eared Flying Fox

Lowland forests of Aru Is. (SE Moluccas), Salawati, Wokam and New Guinea

❏ *Pteropus mahaganus* Sanborn's Flying Fox

Bougainville (New Guinea) and Santa Isabel (Solomon Is.)

❏ *Pteropus mariannus* Marianas Flying Fox

Ryukyu Is. (S Japan), Mariana Is. and Caroline Is. (Palau and Fed. States of Micronesia); Endangered

❏ *Pteropus melanopogon*
Black-bearded Flying Fox

Timor (Lesser Sundas) and S & SE Moluccas incl. Kai and Aru Is.

❏ *Pteropus melanotus* Black-eared Flying Fox

Andaman and Nicobar Is. (India); Nias and Enggano (Sumatra)

❏ *Pteropus molossinus* Caroline Flying Fox

Ant and Pakin Atolls (Pohnpei, Fed. States of Micronesia) and ?Mortlock Is.; Critically Endangered

❏ *Pteropus neohibernicus* Great Flying Fox

Lowlands and mid-elevations of New Guinea, Admiralty Is. and Bismarck Archipelago

❏ *Pteropus niger* Greater Mascerene Flying Fox

Réunion and Mauritius

❏ *Pteropus nitendiensis* Temotu Flying Fox

Nendö and Tömotu Neo (Santa Cruz Is., Solomon Is.)

❏ *Pteropus ocularis* Seram Flying Fox

Buru and Seram (S Moluccas)

❏ *Pteropus ornatus* Ornate Flying Fox

Lifou and Maré (Loyalty Is.) and New Caledonia

❏ *Pteropus pelewensis* Palau Flying Fox

Palau

❏ *Pteropus personatus* Masked Flying Fox

Morotai, Halmahera, Ternate, Tidore, Bacan, Bisa and Obi (N Moluccas)

❏ *Pteropus phaeocephalus* Mortlock Flying Fox

Mortlock Is. (State of Chuuk, Fed. States of Micronesia); Critically Endangered

❏ †*Pteropus pilosus* Large Palau Flying Fox

Formerly Palau; Extinct (only known from two specimens collected prior to 1874)

❏ *Pteropus pohlei* Geelvink Bay Flying Fox

Rani, Numfoor, Biak and Japen (Geelvinck Is., New Guinea)

❏ *Pteropus poliocephalus* Grey-headed Flying Fox

E Australia

❏ *Pteropus pselaphon* Bonin Flying Fox

Bonin and Volcano Is. (Japan); Critically Endangered

❏ *Pteropus pumilus*
Little Golden-mantled Flying Fox

Philippines

❏ *Pteropus rayneri* Solomons Flying Fox

Buka and Bougainville (New Guinea) and Solomon Is.

❏ *Pteropus rennelli* Rennell Flying Fox

Rennell (Solomon Is.)

❏ *Pteropus rodricensis* Rodriguez Flying Fox

Rodriguez and Round I.; Critically Endangered

❏ *Pteropus rufus* Madagascar Flying Fox

Madagascar

❏ *Pteropus samoensis* Samoan Flying Fox

Fiji and Samoan Is. (Fiji, Samoa and American Samoa)

❏ *Pteropus sanctacrucis* Santa Cruz Flying Fox

Santa Cruz Is. (Solomon Is.)

❏ *Pteropus scapulatus* Little Red Flying Fox

S New Guinea and N & E Australia; vagrant New Zealand

❏ *Pteropus seychellensis* Seychelles Flying Fox

Mafia Is. (Tanzania); Seychelles, Aldabra Is. and Comoro Is.

❏ *Pteropus speciosus* Philippine Flying Fox

Basilan, Mindanao and Sulu Archipelago (S & SE Philippines), Talaud Is.; ?Java Sea islands

❏ †*Pteropus subniger*
Lesser Mascarene Flying Fox (Dark Flying Fox)

Formerly Mauritius and Réunion; Extinct (last seen 1860s)

❏ *Pteropus temmincki* Temminck's Flying Fox

Primary forests of Buru, Ambon and Seram (S Moluccas)

❏ †*Pteropus tokudae* Guam Flying Fox

Formerly Guam; Extinct (last specimen collected in 1967)

❏ *Pteropus tonganus* Pacific Flying Fox

Islands off NE New Guinea, Solomon Is. and New Caledonia, E to Cook Is.

❏ *Pteropus tuberculatus* Vanikolo Flying Fox

Vanikolo (Santa Cruz Is., Solomon Is.)

❏ *Pteropus ualanus* Kosrae Flying Fox

Kosrae (Fed. States of Micronesia)

❏ *Pteropus vampyrus* Large Flying Fox

Indochina, Malay Peninsula, Sumatra, Java, Borneo, Philippines and Lesser Sundas

❏ *Pteropus vetulus* New Caledonia Flying Fox

New Caledonia

❏ *Pteropus voeltzkowi* Pemba Flying Fox

Pemba I. (Tanzania); Critically Endangered

❏ *Pteropus woodfordi* Dwarf Flying Fox

W & S Solomon Is.

❏ *Pteropus yapensis* Yap Flying Fox

Yap (Fed. States of Micronesia)

❏ *Acerodon celebensis* Signal-winged Acerodon

Salayar, Sulawesi, Togian, Sangihe and Sula Is.

❏ *Acerodon humilis* Talaud Acerodon

Salebabu (Talaud Is., N Moluccas), status uncertain, only known from holotype specimen collected in 1897

❏ *Acerodon jubatus* Golden-capped Acerodon

Philippines; Endangered

❏ *Acerodon leucotis* Palawan Acerodon

Busuanga, Palawan and Balabac (WC & SW Philippines)

❏ †*Acerodon lucifer*
Panay Golden-capped Acerodon

Formerly Panay (C Philippines); Extinct (only known from a few specimens collected in 1888 and 1892)

❏ *Acerodon mackloti* Sunda Acerodon

Lombok, Sumbawa, Sumba, Flores, Alor and Timor (Lesser Sundas)

❏ *Pteralopex acrodonta* Fijian Monkey-faced Bat

Mossy forests on summit of Des Veaux Peak (Taveuni, Fiji Is.); ?Vanua Levu (Fiji Is.); Critically Endangered

❏ *Pteralopex anceps* Bougainville Monkey-faced Bat

Primary forests of Buka, Bougainville and Choiseul (Solomon Is.); Critically Endangered

❏ *Pteralopex atrata* Guadalcanal Monkey-faced Bat

Guadalcanal and ?Santa Isabel (Solomon Is.); Critically Endangered

❏ *Pteralopex pulchra* Montane Monkey-faced Bat

Mt Makarakomburu (Guadalcanal, Solomon Is.); Critically Endangered

❏ *Pteralopex taki* New Georgia Monkey-faced Bat

Lowland primary rainforests of New Georgia (Solomon Is.)

❏ *Styloctenium wallacei* Stripe-faced Fruit Bat

Sulawesi

❏ *Neopteryx frosti* Small-toothed Fruit Bat

N & W Sulawesi

❏ *Dobsonia anderseni*
Andersen's Bare-backed Fruit Bat

Admiralty Is. and Bismarck Archipelago

❏ *Dobsonia beauforti*
Beaufort's Bare-backed Fruit Bat

Salawati, Gebe, Waigeo, Batanta, Biak and Owi (New Guinea); Endangered

❏ †*Dobsonia chapmani*
Negros Bare-backed Fruit Bat

Formerly Negros (C Philippines); Extinct (since about 1970)

❏ *Dobsonia crenulata*
Halmahera Bare-backed Fruit Bat

Sulawesi and N & C Moluccas

❏ *Dobsonia emersa* Biak Bare-backed Fruit Bat

Biak, Numfoor and Owi (Geelvinck Is., New Guinea)

❏ *Dobsonia exoleta* Sulawesi Bare-backed Fruit Bat

Sulawesi, Togian Is. and Sanana (Sula Is.)

❏ *Dobsonia inermis* Solomons Bare-backed Fruit Bat

Bougainville (New Guinea) and Solomon Is.

❏ *Dobsonia magna* Great Bare-backed Fruit Bat

Aru Is. (SE Moluccas), islands NW of New Guinea incl. Salawati and Japen, New Guinea and N Queensland (Australia)

❏ *Dobsonia minor* Lesser Bare-backed Fruit Bat

Rainforests of Sulawesi, Japen, New Guinea and near islands

❏ *Dobsonia moluccensis*
Moluccan Bare-backed Fruit Bat

Buru, Ambon and Seram (S Moluccas)

❏ *Dobsonia pannietensis*
De Vis' Bare-backed Fruit Bat

D'Entrecasteaux Is. (New Guinea)

❏ *Dobsonia peroni* Western Bare-backed Fruit Bat

Penida to Babar (Lesser Sundas)

❏ *Dobsonia praedatrix*
Bismarck Bare-backed Fruit Bat

Bismarck Archipelago

❑ *Dobsonia viridis* Green Bare-backed Fruit Bat

Mangole (Sula Is.), S Moluccas and Kai Is,

❑ *Aproteles bulmerae* Bulmer's Fruit Bat

Hindenberg Wall area (Western Prov., Papua New Guinea); one small colony, Critically Endangered

❑ *Harpyionycteris celebensis*
Sulawesi Harpy Fruit Bat

Sulawesi

❑ *Harpyionycteris whiteheadi* Harpy Fruit Bat

Philippines

❑ *Plerotes anchietai* D'Anchieta's Fruit Bat

Very locally in Angola, S D.R. Congo and Zambia

❑ *Hypsignathus monstrosus*
Hammer-headed Fruit Bat

Shaded lowland forests and mangrove swamps of Sierra Leone to W Kenya, S to Angola and Zambia

❑ *Epomops buettikoferi*
Buettikofer's Epauletted Bat

Forests of Guinea to Nigeria

❑ *Epomops dobsoni* Dobson's Epauletted Bat

Forests of Angola to Tanzania, S to NE Botswana and Malawi

❑ *Epomops franqueti* Franquet's Epauletted Bat

Forests of Ivory Coast to Sudan, S to Angola and N Zambia

❑ *Epomophorus angolensis*
Angolan Epauletted Fruit Bat

Forests and riparian woods of W Angola and NW Namibia

❑ *Epomophorus gambianus*
Gambian Epauletted Fruit Bat

Forests and riparian woods of Senegal to W Ethiopia; Angola and S Tanzania to South Africa

❑ *Epomophorus grandis*
Lesser Angolan Epauletted Fruit Bat

Forests of S Rep. Congo and N Angola

❑ *Epomophorus labiatus*
Ethiopian Epauletted Fruit Bat

Forests of Nigeria to Ethiopia, S to Rep. Congo and Malawi

❑ *Epomophorus minimus*
East African Epauletted Fruit Bat

Forests of Ethiopia to Uganda and Tanzania

❑ *Epomophorus wahlbergi*
Wahlberg's Epauletted Fruit Bat

Forests and riparian woods of Cameroon to Somalia, S to Angola and E South Africa; Pemba and Zanzibar Is. (Tanzania)

❑ *Micropteropus intermedius*
Hayman's Dwarf Epauletted Fruit Bat

Woods and savannas of N Angola and SE D.R. Congo

❑ *Micropteropus pusillus*
Peters' Dwarf Epauletted Fruit Bat

Woods and savannas of Gambia to Ethiopia, S to Angola, Zambia and Tanzania

❑ *Nanonycteris veldkampi* Veldkamp's Bat

Guinea to Central African Republic

❑ *Scotonycteris ophiodon* Pohle's Fruit Bat

Lowland rainforests of Liberia to Rep. Congo

❑ *Scotonycteris zenkeri* Zenker's Fruit Bat

Lowland rainforests of Liberia to Rep. Congo and E D.R. Congo; Bioko (Equatorial Guinea)

❑ *Casinycteris argynnis* Short-palated Fruit Bat

Dense lowland forests of Cameroon to E D.R. Congo

❑ *Cynopterus brachyotis*
Lesser Short-nosed Fruit Bat

India, incl. Andaman and Nicobar Is., and Sri Lanka to Philippines and Moluccas

❑ *Cynopterus horsfieldi*
Horsfield's Short-nosed Fruit Bat

Thailand, Malaya, Sumatra, Java, Borneo and Lesser Sundas

❑ *Cynopterus luzoniensis*
Sulawesi Short-nosed Fruit Bat

Sulawesi and Philippines

❑ *Cynopterus minutus*
Small Short-nosed Fruit Bat

Borneo, Sumatra, Java and Sulawesi

❏ *Cynopterus nusatenggara*
Nusa Tenggara Short-nosed Fruit Bat

Lombok, Sumbawa, Sumba, Flores and near islands (Lesser Sundas)

❏ *Cynopterus sphinx* Greater Short-nosed Fruit Bat

India and Sri Lanka to S China, Malay Peninsula, Sumatra and near islands

❏ *Cynopterus titthaecheilus*
Indonesian Short-nosed Fruit Bat

Sumatra, Java; Lombok, Timor and near islands (Lesser Sundas)

❏ *Ptenochirus jagori* Greater Musky Fruit Bat

Philippines

❏ *Ptenochirus minor* Lesser Musky Fruit Bat

Philippines

❏ *Megaerops ecaudatus*
Temminck's Tailless Fruit Bat

Thailand, Malaya, Sumatra and Borneo

❏ *Megaerops kusnotoi* Javan Tailless Fruit Bat

Java

❏ *Megaerops niphanae* Ratanaworabhan's Fruit Bat

NE India, Thailand and Vietnam

❏ *Megaerops wetmorei* White-collared Fruit Bat

Malaya, Borneo and Philippines

❏ *Dyacopterus spadiceus* Dyak Fruit Bat

Malaya, Sumatra, Borneo; Luzon and Mindanao (N & SE Philippines)

❏ *Balionycteris maculata* Spotted-winged Fruit Bat

Thailand, Malaya, Riau Is. and Borneo

❏ *Chironax melanocephalus* Black-capped Fruit Bat

Malay Peninsula, Sumatra, Nias (Mentawai Is.), Java, Borneo and Sulawesi

❏ *Thoopterus nigrescens* Swift Fruit Bat

N Sulawesi, Sangihe Is., Morotai (NW & N Moluccas) and Mangole (Sula Is., W Moluccas)

❏ *Sphaerias blanfordi* Blanford's Fruit Bat

N India, Tibet, Bhutan, Myanmar, Thailand and SW China

❏ *Aethalops alecto* Pygmy Fruit Bat

Malaya, Sumatra, Java and Borneo; Lombok (Lesser Sundas)

❏ *Penthetor lucasi* Lucas' Short-nosed Fruit Bat

Malaya, Riau Is. and Borneo

❏ *Latidens salimalii* Salim Ali's Fruit Bat

S India; Critically Endangered

❏ *Alionycteris paucidentata*
Mindanao Pygmy Fruit Bat

Mindanao (SE Philippines)

❏ *Otopteropus cartilagonodus* Luzon Fruit Bat

Luzon (N Philippines)

❏ *Haplonycteris fischeri*
Philippine Pygmy Fruit Bat

Philippines

❏ *Paranyctimene raptor* Unstriped Tube-nosed Bat

Forests of Salawati and New Guinea

❏ *Paranyctimene tenax*
Greater Unstriped Tube-nosed Bat

New Guinea

❏ *Nyctimene aello* Greater Tube-nosed Fruit Bat

Lowland and hill forests of Misoöl, Salawati, and W & N New Guinea

❏ *Nyctimene albiventer*
Common Tube-nosed Fruit Bat

Moluccas incl. Kai Is., New Guinea, Bismarck Archipelago, Solomon Is. and NE Australia

❏ *Nyctimene bougainville*
Solomons Tube-nosed Fruit Bat

Primary forests of Bougainville (New Guinea) and Solomon Is.

❏ *Nyctimene cephalotes*
Pallas' Tube-nosed Fruit Bat

Sulawesi, S Moluccas, Timor and S New Guinea

❏ *Nyctimene certans*
Mountain Tube-nosed Fruit Bat

Hill and montane forests of New Guinea

❏ *Nyctimene cyclotis*
Round-eared Tube-nosed Fruit Bat

Primary rainforests of Numfoor and Biak (Geelvinck Is.), New
Guinea and New Britain (Bismarck Archipelago)

❏ *Nyctimene draconilla* Dragon Tube-nosed Fruit Bat

Lowland forests of C New Guinea

❏ *Nyctimene major* Island Tube-nosed Fruit Bat

Bismarck Archipelago, D'Entrecasteaux Is. and Solomon Is.

❏ *Nyctimene minutus* Lesser Tube-nosed Fruit Bat

Sulawesi and ?C Moluccas

❏ *Nyctimene rabori* Philippine Tube-nosed Fruit Bat

Negros and Cebu (C Philippines); Critically Endangered

❏ *Nyctimene robinsoni*
Queensland Tube-nosed Fruit Bat

E Queensland (Australia)

❏ †*Nyctimene sanctacrucis*
Nendö Tube-nosed Fruit Bat

Formerly Nendö (Santa Cruz Group, Solomon Is.); Extinct
(only known from holotype collected in 1892)

❏ *Eonycteris major* Greater Dawn Bat

Borneo and Philippines

❏ *Eonycteris spelaea* Lesser Dawn Bat

N India and Andaman Is. (India) to S China, Palawan (SW
Philippines), Sulawesi, N Moluccas and Lesser Sundas

❏ *Megaloglossus woermanni* Woermann's Bat

Forests and forest edges of Liberia to Uganda, S to N Angola
and S D.R. Congo; Bioko (Equatorial Guinea)

❏ *Macroglossus minimus*
Lesser Long-tongued Nectar Bat
(Northern Blossom Bat)

Thailand, Malay Peninsula, Java, Borneo, Philippines, New
Guinea, Bismarck Archipelago, Solomon Is. and N Australia

❏ *Macroglossus sobrinus*
Greater Long-tongued Nectar Bat

Indochina, Sumatra, Java and near islands

❏ *Syconycteris australis* Common Blossom Bat

Moluccas, New Guinea, Bismarck Archipelago,
D'Entrecasteaux Is. and coastal E Australia

❏ *Syconycteris carolinae* Halmahera Blossom Bat

Halmahera and Bacan (C Moluccas)

❏ *Syconycteris hobbit* Moss-forest Blossom Bat

Undisturbed montane moss forests of C New Guinea

Melonycteris (Melonycteris)

❏ *Melonycteris melanops* Bismarck Blossom Bat

Bismarck Archipelago

Melonycteris (Nesonycteris)

❏ *Melonycteris fardoulisi* Fardoulis' Blossom Bat

Solomon Is.

❏ *Melonycteris woodfordi* Woodford's Blossom Bat

Buka and Bougainville (New Guinea) and Solomon Is.

❏ *Notopteris macdonaldi* Fijian Blossom Bat

Vanuatu and Fiji

❏ *Notopteris neocaledonica*
New Caledonia Blossom Bat

New Caledonia

Family: EMBALLONURIDAE
(Sheath-tailed Bats—51)

Taphozous (Liponycteris)

❏ *Taphozous hamiltoni* Hamilton's Tomb Bat

Chad, S Sudan and Kenya

❏ *Taphozous nudiventris* Naked-rumped Tomb Bat

N & E Africa, E to Myanmar

Taphozous (Taphozous)

❏ *Taphozous achates* Brown-bearded Tomb Bat

Savu, Roti, Semau and ?Timor (E Lesser Sundas) and Kai Is.
(SE Moluccas)

❏ *Taphozous australis* Coastal Tomb Bat

Coasts of SE New Guinea, Torres Strait islands and NE
Queensland (Australia)

❏ *Taphozous georgianus* Sharp-nosed Tomb Bat

N Australia

❏ *Taphozous hildegardeae* Hildegarde's Tomb Bat

Kenya and NE Tanzania incl. Zanzibar I.

❏ *Taphozous hilli* Hill's Tomb Bat

Western Australia, Northern Territory and South Australia

❏ *Taphozous kapalgensis* Arnhem Tomb Bat

Northern Territory (Australia)

❏ *Taphozous longimanus* Long-winged Tomb Bat

India and Sri Lanka to Indochina, E to Borneo and Flores (Lesser Sundas)

❏ *Taphozous mauritianus* Mauritian Tomb Bat

Savannas and built areas of subsaharan Africa; Madagascar; Assumption, Aldabra Is., Mauritius and Réunion

❏ *Taphozous melanopogon*
Black-bearded Tomb Bat

India and Sri Lanka to S China and Indochina, E to Sulawesi, Timor and Kai Is. (SE Moluccas)

❏ *Taphozous perforatus* Egyptian Tomb Bat

Savannas and built areas of Senegal to Somalia, S to Botswana and Mozambique; Egypt to NW India

❏ *Taphozous philippinensis* Philippine Tomb Bat

Philippines

❏ *Taphozous theobaldi* Theobald's Tomb Bat

C India to Vietnam; Java, Borneo and Sulawesi

❏ *Taphozous troughtoni* Troughton's Tomb Bat

Mt Isa and Cloncurry regions (NW Queensland, Australia); Critically Endangered

❏ *Saccolaimus flaviventris*
Yellow-bellied Pouched Bat

SE New Guinea and Australia

❏ *Saccolaimus mixtus* Troughton's Pouched Bat

SE New Guinea and N Cape York Peninsula (N Queensland, Australia)

❏ *Saccolaimus peli* Pel's Pouched Bat

Rainforests and forest edges of Liberia to W Kenya, S to Angola

❏ *Saccolaimus pluto* Philippine Pouched Bat

Philippines

❏ *Saccolaimus saccolaimus*
Naked-rumped Pouched Bat

India and Sri Lanka to Indochina, E to New Guinea, Solomon Is. and NC & NE Australia

❏ *Mosia nigrescens* Dark Sheath-tailed Bat

Sulawesi and Moluccas to New Guinea, Bismarck Archipelago and Solomon Is.

❏ *Emballonura alecto* Small Asian Sheath-tailed Bat

Borneo, Philippines, Sulawesi, Tanimbar Is. (Lesser Sundas) and S Moluccas incl. Kai Is.

❏ *Emballonura atrata* Madagascar Sheath-tailed Bat

Coastal N, EC & S Madagascar

❏ *Emballonura beccarii* Beccari's Sheath-tailed Bat

Kai Is. (SE Moluccas), Biak, Japen, New Guinea, New Ireland (Bismarck Archipelago) and Trobriand Is.

❏ *Emballonura dianae* Large-eared Sheath-tailed Bat

New Guinea, New Ireland (Bismarck Archipelago) and E Solomon Is.

❏ *Emballonura furax* New Guinea Sheath-tailed Bat

Few scattered localities in lowlands of New Guinea

❏ *Emballonura monticola* Lesser Sheath-tailed Bat

Malay Peninsula, Sumatra, Java, Borneo, Sulawesi and near islands

❏ *Emballonura raffrayana* Raffray's Sheath-tailed Bat

Sulawesi, Moluccas, Geelvinck Is., New Guinea, Bismarck Archipelago and Solomon Is.

❏ *Emballonura semicaudata*
Polynesian Sheath-tailed Bat

Mariana Is., Caroline Is. (Palau and Fed. States of Micronesia), Vanuatu, Fiji and Samoa; Endangered

❏ *Emballonura serii* Seri's Sheath-tailed Bat

New Ireland (Bismarck Archipelago)

❏ *Coleura afra* African Sheath-tailed Bat

Woods and savannas of Guinea-Bissau to Somalia, S to Angola, D.R. Congo and Mozambique; Yemen

❏ *Coleura seychellensis*
Seychelles Sheath-tailed Bat

Seychelles; Critically Endangered

❏ *Rhynchonycteris naso* Proboscis Bat

Near water in lowland rainforests of EC Mexico to French Guiana, S to Peru, Bolivia and S Brazil; Trinidad

❏ *Centronycteris maximiliani* Shaggy Bat

Few records from lowland forests and regrowth areas of EC Mexico to French Guiana, S to Peru and C Brazil

❏ *Balantiopteryx infusca*
Ecuadorian Sac-winged Bat

Lowlands of NW Ecuador; Endangered

❏ *Balantiopteryx io* Least Sac-winged Bat

Lowland forests of EC Mexico to Guatemala and Belize

❏ *Balantiopteryx plicata* Grey Sac-winged Bat

Lowland deciduous forests and scrub of NW & E Mexico to NW Costa Rica; N Colombia

❏ *Saccopteryx bilineata* Greater White-lined Bat

Lowland forests and forest edge of C Mexico to French Guiana, S to Peru, Bolivia and S Brazil; Trinidad & Tobago

❏ *Saccopteryx canescens* Frosted White-lined Bat

Lowland forests of Colombia to French Guiana, S to Peru and E Amazonian Brazil

❏ *Saccopteryx gymnura*
Amazonian White-lined Bat

E Amazonian Brazil

❏ *Saccopteryx leptura* Lesser White-lined Bat

Lowland forests of Pacific slope of S Mexico to French Guiana, S to Peru and S Brazil; Margarita; Trinidad & Tobago

❏ *Cormura brevirostris* Chestnut Sac-winged Bat

Lowland rainforests of Nicaragua to French Guiana, and E of Andes S to Peru and Amazonian Brazil

Peropteryx (Peronymus)

❏ *Peropteryx leucoptera*
White-winged Dog-like Bat

Lowland forests of Colombia to French Guiana, S to Peru and E Brazil

Peropteryx (Peropteryx)

❏ *Peropteryx kappleri* Greater Dog-like Bat

Lowland forests and forest edge of EC Mexico to Suriname, S to Peru and E Brazil

❏ *Peropteryx macrotis* Lesser Dog-like Bat

Caves, rock piles and built areas in lowlands of S Mexico to French Guiana, S to Peru and N Paraguay

❏ *Peropteryx trinitatis* Trinidad Dog-like Bat

Paraguay and SE Brazil; Aruba; Trinidad & Tobago; Grenada

❏ *Cyttarops alecto* Smoky Sheath-tailed Bat

Few records from lowland forest clearings of E Nicaragua, E Costa Rica, Guyana and near Belém (E Amazonian Brazil)

Diclidurus (Depanycteris)

❏ *Diclidurus isabellus* Isabelle's Ghost Bat

Lowland forests and riparian habitats of S Venezuela (and NW Brazil?)

Diclidurus (Diclidurus)

❏ *Diclidurus albus* Northern Ghost Bat

Lowland and mid-elevation clearings and built areas of WC Mexico to Venezuela, S to Ecuador (and E Brazil?); Trinidad

❏ *Diclidurus ingens* Greater Ghost Bat

Lowland rainforests of SE Colombia, Venezuela, Guyana (and NW Brazil?)

❏ *Diclidurus scutatus* Lesser Ghost Bat

Lowland riparian habitats and clearings of Amazon basin of S Venezuela to Suriname, S to Peru (and C Brazil?)

Family: RHINOPOMATIDAE (Mouse-tailed Bats—3)

❏ *Rhinopoma hardwickei* Lesser Mouse-tailed Bat

Deserts and semi-deserts of N Africa, S to Mauritania, Nigeria and Kenya; Arabian Peninsula, E to Myanmar

❏ *Rhinopoma microphyllum* Greater Mouse-tailed Bat

Deserts and semi-deserts of Morocco and Senegal to Thailand and Sumatra

❏ *Rhinopoma muscatellum* Small Mouse-tailed Bat

Yemen, Oman, W Iran and S Afghanistan

Family: CRASEONYCTERIDAE (Bumblebee Bat—1)

❏ *Craseonycteris thonglongyai*
Bumblebee Bat (Kitti's Hog-nosed Bat)

SE Myanmar and C Thailand; Endangered

Family: MEGADERMATIDAE (False Vampire Bats—5)

❏ *Lavia frons* Yellow-winged Bat

Open woods and savannas of Senegal to Somalia, S to Zambia and Malawi, incl. Zanzibar I. (Tanzania)

❏ *Cardioderma cor* Heart-nosed Bat

Dry savannas and coastal thickets of E Sudan, Ethiopia and Somalia, S to Tanzania incl. Zanzibar I.

❏ *Megaderma lyra* Greater False Vampire Bat

Afghanistan to S China, S to Sri Lanka and Malay Peninsula

❏ *Megaderma spasma* Lesser False Vampire Bat

India and Sri Lanka to Indochina, Philippines, Moluccas and Lesser Sundas

❏ *Macroderma gigas*
Australian False Vampire Bat

N & C Australia

Family: NYCTERIDAE (Slit-faced Bats—13)

❏ *Nycteris arge* Bates' Slit-faced Bat

Sierra Leone to SW Sudan, S to NE Angola, S D.R. Congo and W Kenya; Bioko (Equatorial Guinea)

❏ *Nycteris gambiensis* Gambian Slit-faced Bat

Senegal and Gambia to Burkina Faso and Benin

❏ *Nycteris grandis* Large Slit-faced Bat

Riparian woods of Senegal to Kenya, S to E Zimbabwe and Mozambique, incl. near islands

❏ *Nycteris hispida* Hairy Slit-faced Bat

Senegal to Somalia, S to Angola and E South Africa, incl. near islands

❏ *Nycteris intermedia* Intermediate Slit-faced Bat

Liberia to W Tanzania, S to Angola

❏ *Nycteris javanica* Javan Slit-faced Bat

Kangean Is. (NE of Java), Java and Bali

❏ *Nycteris macrotis* Large-eared Slit-faced Bat

Riparian woods of Senegal to Ethiopia, S to Zimbabwe and Mozambique; Zanzibar I. (Tanzania)

❏ *Nycteris madagascariensis*
Madagascar Slit-faced Bat

N Madagascar

❏ *Nycteris major* Ja Slit-faced Bat

Liberia to Zambia

❏ *Nycteris nana* Dwarf Slit-faced Bat

Ivory Coast to SW Sudan, S to NE Angola and W Tanzania

❏ *Nycteris thebaica* Egyptian Slit-faced Bat

N and subsaharan Africa, incl. Pemba and Zanzibar Is. (Tanzania); Middle East and Arabian Peninsula

❏ *Nycteris tragata* Malayan Slit-faced Bat

Myanmar, Thailand, Malay Peninsula, Sumatra and Borneo

❏ *Nycteris woodi* Wood's Slit-faced Bat

Locally in Cameroon; Ethiopia and Somalia; SW Tanzania and Zambia to NE South Africa

Family: RHINOLOPHIDAE (Horseshoe Bats—145)

❏ *Rhinolophus acuminatus*
Acuminate Horseshoe Bat

Thailand and Indochina to Borneo, Palawan (SW Philippines) and Lombok (Lesser Sundas)

❏ *Rhinolophus adami* Adam's Horseshoe Bat

Rep. Congo

❏ *Rhinolophus affinis*
Intermediate Horseshoe Bat

India incl. Andaman Is. to S China, Malay Peninsula, Borneo and Lesser Sundas

❏ *Rhinolophus alcyone* Halcyon Horseshoe Bat

Senegal to SW Sudan, S to Gabon, N D.R. Congo and Uganda; Bioko (Equatorial Guinea)

❏ *Rhinolophus anderseni*
Andersen's Horseshoe Bat

Luzon and Palawan (N & SW Philippines)

❏ *Rhinolophus arcuatus* Arcuate Horseshoe Bat

Sumatra, Borneo, Philippines, Lesser Sundas, S Moluccas and New Guinea

❑ *Rhinolophus beddomei*
Indian Woolly Horseshoe Bat

S India

❑ *Rhinolophus blasii* Blasius' Horseshoe Bat

Forests and woods of SE Europe; NW, E & S Africa; Turkey
and Arabian Peninsula to Turkmenistan and Pakistan

❑ *Rhinolophus borneensis* Bornean Horseshoe Bat

Indochina, Java, Borneo and near islands

❑ *Rhinolophus canuti* Canut's Horseshoe Bat

Java, and Timor (Lesser Sundas)

❑ *Rhinolophus capensis* Cape Horseshoe Bat

Cape Prov. (South Africa)

❑ *Rhinolophus celebensis*
Sulawesi Horseshoe Bat

Java, Madura, Bali, Sulawesi, Sangihe and Talaud Is. (N
Moluccas), and Timor (Lesser Sundas)

❑ *Rhinolophus clivosus* Geoffroy's Horseshoe Bat

N and subsaharan Africa; Arabian Peninsula, E to
Turkmenistan and Afghanistan

❑ *Rhinolophus coelophyllus* Croslet Horseshoe Bat

Myanmar, Thailand and Malaya

❑ *Rhinolophus cognatus* Andaman Horseshoe Bat

Andaman Is. (India)

❑ *Rhinolophus convexus* Convex Horseshoe Bat

Cameron Highlands (Malaysia); Critically Endangered

❑ *Rhinolophus cornutus*
Little Japanese Horseshoe Bat

Japan, incl. Ryukyu Is.

❑ *Rhinolophus creaghi* Creagh's Horseshoe Bat

Java, Madura, Borneo and Timor (Lesser Sundas)

❑ *Rhinolophus darlingi* Darling's Horseshoe Bat

S Angola and Tanzania, S to South Africa

❑ *Rhinolophus deckenii* Decken's Horseshoe Bat

Uganda, Kenya and Tanzania, incl. Pemba and Zanzibar Is.
(Tanzania)

❑ *Rhinolophus denti* Dent's Horseshoe Bat

Guinea, Ivory Coast and Ghana; Namibia to Mozambique, S to
NW South Africa

❑ *Rhinolophus eloquens* Eloquent Horseshoe Bat

NE D.R. Congo to S Somalia, S to Rwanda and N Tanzania,
incl. Pemba and Zanzibar Is. (Tanzania)

❑ *Rhinolophus euryale*
Mediterranean Horseshoe Bat

S Europe, Mediterranean islands and NW Africa; Israel to Iran
and Turkmenistan

❑ *Rhinolophus euryotis*
Broad-eared Horseshoe Bat

Sulawesi, Moluccas incl. Kai and Aru Is., New Guinea and
Bismarck Archipelago

❑ *Rhinolophus ferrumequinum*
Greater Horseshoe Bat

Deciduous woods and pastures of WC & S Europe to China
and Japan, S to NW Africa, Iran and Himalayas

❑ *Rhinolophus fumigatus* Rüppell's Horseshoe Bat

Subsaharan Africa, S to S Namibia and Zimbabwe

❑ *Rhinolophus guineensis* Guinean Horseshoe Bat

Guinea, Sierra Leone and Liberia

❑ *Rhinolophus hildebrandti*
Hildebrandt's Horseshoe Bat

Nigeria; S Sudan, Ethiopia and NE D.R. Congo, S to NE South
Africa and Mozambique

❑ *Rhinolophus hillorum* Hills' Horseshoe Bat

Guinea

❑ *Rhinolophus hipposideros*
Lesser Horseshoe Bat

Deciduous woodland and riparian habitats of C & S Europe
and N Africa, E to Kyrgyzstan and Kashmir (NW India);
Sudan and Ethiopia

❑ *Rhinolophus imaizumii*
Imaizumi's Horseshoe Bat

Iriomote I. (Ryukyu Is., Japan); Endangered

❑ *Rhinolophus inops*
Philippine Forest Horseshoe Bat

Mindanao (SE Philippines)

❏ *Rhinolophus keyensis* Insular Horseshoe Bat

C, S & SE Moluccas incl. Kai Is., and Wetar (Lesser Sundas);
Endangered

❏ *Rhinolophus landeri* Lander's Horseshoe Bat

Subsaharan Africa, incl. Bioko (Equatorial Guinea) and
Zanzibar I. (Tanzania)

❏ *Rhinolophus lepidus* Blyth's Horseshoe Bat

Afghanistan and N India to C & SW China, Malay Peninsula
and Sumatra

❏ *Rhinolophus luctus* Woolly Horseshoe Bat

Nepal, India and Sri Lanka to S China, Taiwan, Indochina,
Sumatra, Java, Bali and Borneo

❏ *Rhinolophus maclaudi* Maclaud's Horseshoe Bat

Guinea, Liberia, E D.R. Congo, W Uganda and Rwanda

❏ *Rhinolophus macrotis* Big-eared Horseshoe Bat

N India to S China, Indochina, Malay Peninsula, Sumatra and
Philippines

❏ *Rhinolophus maendeleo*
Maendeleo Horseshoe Bat

Tanzania

❏ *Rhinolophus malayanus* Malayan Horseshoe Bat

Thailand, Malaya, Laos and Vietnam

❏ *Rhinolophus marshalli* Marshall's Horseshoe Bat

Thailand

❏ *Rhinolophus megaphyllus* Eastern Horseshoe Bat

Lowlands of C & E New Guinea, D'Entrecasteaux Is.,
Bismarck Archipelago and coastal E Australia

❏ *Rhinolophus mehelyi* Mehely's Horseshoe Bat

S Europe incl. Mediterranean islands and N Africa to Iran and
Afghanistan

❏ *Rhinolophus mitratus* Mitred Horseshoe Bat

N India

❏ *Rhinolophus monoceros*
Formosan Lesser Horseshoe Bat

Taiwan

❏ *Rhinolophus nereis* Nereid Horseshoe Bat

Anambas and Natuna Is. (between Malaya and Borneo)

❏ *Rhinolophus osgoodi* Osgood's Horseshoe Bat

Yunnan (SW China)

❏ *Rhinolophus paradoxolophus*
Bourret's Horseshoe Bat

Thailand and Vietnam

❏ *Rhinolophus pearsonii*
Pearson's Horseshoe Bat

N India to E China, S to Myanmar, Thailand, Malaya and
Vietnam

❏ *Rhinolophus philippinensis*
Large-eared Horseshoe Bat

Borneo and Philippines to Timor (Lesser Sundas), Kai Is.
(SE Moluccas), New Guinea and NE Queensland (Australia)

❏ *Rhinolophus pusillus* Least Horseshoe Bat

India, Thailand, Malay Peninsula, Mentawai Is. (Sumatra),
Java and Lesser Sundas

❏ *Rhinolophus rex* King Horseshoe Bat

C & SW China

❏ *Rhinolophus robinsoni* Peninsular Horseshoe Bat

Thailand and Malaya

❏ *Rhinolophus rouxii* Rufous Horseshoe Bat

India and Sri Lanka to S China and Vietnam

❏ *Rhinolophus rufus* Large Rufous Horseshoe Bat

Philippines

❏ *Rhinolophus sakejiensis* Sakeji Horseshoe Bat

NW Zambia

❏ *Rhinolophus sedulus*
Lesser Woolly Horseshoe Bat

Malaya and Borneo

❏ *Rhinolophus shameli* Shamel's Horseshoe Bat

Myanmar, Thailand, Cambodia and Malaya

❏ *Rhinolophus silvestris* Forest Horseshoe Bat

Gabon and Rep. Congo

❏ *Rhinolophus simplex* Lombok Horseshoe Bat

Lombok, Komodo and Sumbawa (Lesser Sundas)

❏ *Rhinolophus simulator* Bushveld Horseshoe Bat

Guinea; Nigeria and Cameroon; S Sudan and Ethiopia, S to NE South Africa

❏ *Rhinolophus stheno* Lesser Brown Horseshoe Bat

Thailand, Malaya, Sumatra and Java

❏ *Rhinolophus subbadius*
Little Nepalese Horseshoe Bat

Nepal, Assam (NE India), Myanmar and Vietnam

❏ *Rhinolophus subrufus*
Small Rufous Horseshoe Bat

Philippines

❏ *Rhinolophus swinnyi* Swinny's Horseshoe Bat

S D.R. Congo and Tanzania to E South Africa

❏ *Rhinolophus thomasi* Thomas' Horseshoe Bat

Myanmar, Yunnan (SW China), Thailand and Vietnam

❏ *Rhinolophus trifoliatus* Trefoil Horseshoe Bat

NE India and Myanmar to Sumatra, Java, Borneo and near islands

❏ *Rhinolophus virgo* Yellow-faced Horseshoe Bat

Philippines

❏ *Rhinolophus yunanensis*
Dobson's Horseshoe Bat

NE India, Yunnan (SW China) and Thailand

❏ *Rhinolophus ziama* Ziama Horseshoe Bat

Highlands of Guinea and Liberia

❏ *Hipposideros abae* Aba Roundleaf Bat

Forests of Guinea-Bissau to SW Sudan and Uganda

❏ *Hipposideros armiger* Great Roundleaf Bat

Nepal and N India to S China, Taiwan, Indochina and Malay Peninsula

❏ *Hipposideros ater* Dusky Roundleaf Bat

India and Sri Lanka to Philippines, Moluccas, New Guinea, Bismarck Archipelago, D'Entrecasteaux Is. and N Australia

❏ *Hipposideros beatus* Benito Roundleaf Bat

Guinea-Bissau to Cameroon, Gabon and N D.R. Congo

❏ *Hipposideros bicolor* Bicoloured Roundleaf Bat

Malaya to Philippines and Timor (Lesser Sundas)

❏ *Hipposideros breviceps*
Short-headed Roundleaf Bat

Mentawai Is. (Sumatra)

❏ *Hipposideros caffer* Sundevall's Roundleaf Bat

Savannas and built areas of Morocco; subsaharan Africa, incl. Pemba and Zanzibar Is. (Tanzania); Yemen

❏ *Hipposideros calcaratus* Spurred Roundleaf Bat

Lowlands of New Guinea and near islands, Bismarck Archipelago and Solomon Is.

❏ *Hipposideros camerunensis*
Greater Roundleaf Bat

Forests of Cameroon, E D.R. Congo and W Kenya

❏ *Hipposideros cervinus* Fawn Roundleaf Bat

Forests of Malaya, Sumatra and Philippines to New Guinea, Bismarck Archipelago, Solomon Is., Vanuatu and NE Australia

❏ *Hipposideros cineraceus* Ashy Roundleaf Bat

Pakistan to Borneo and Philippines

❏ *Hipposideros commersoni*
Commerson's Roundleaf Bat

Subsaharan Africa and near islands, incl. São Tomé (Equatorial Guinea) S to N Namibia, N Botswana, Zimbabwe and C Mozambique; Madagascar

❏ *Hipposideros coronatus*
Large Mindanao Roundleaf Bat

NE Mindanao (SE Philippines)

❏ *Hipposideros corynophyllus*
Telefomin Roundleaf Bat

Few localities in mid-elevation forests of C New Guinea

❏ *Hipposideros coxi* Cox's Roundleaf Bat

Sarawak (NW Borneo)

❏ *Hipposideros crumeniferus* Timor Roundleaf Bat

Timor (Lesser Sundas)

❏ *Hipposideros curtus* Short-tailed Roundleaf Bat

Bioko (Equatorial Guinea) and Cameroon

❏ *Hipposideros cyclops* Cyclops Roundleaf Bat

Forests of Senegal and Guinea-Bissau to S Sudan and Kenya; Bioko (Equatorial Guinea)

❏ *Hipposideros demissus* Makira Roundleaf Bat

San Cristobal (Solomon Is.)

❏ *Hipposideros diadema* Diadem Roundleaf Bat

Nicobar Is. (India) to Philippines, New Guinea, Bismarck Archipelago, Solomon Is. and NC & NE Australia

❏ *Hipposideros dinops* Fierce Roundleaf Bat

Sulawesi and Peleng; Bougainville (New Guinea) and Solomon Is.

❏ *Hipposideros doriae* Borneo Roundleaf Bat

Borneo

❏ *Hipposideros dyacorum* Dayak Roundleaf Bat

Malaya and Borneo

❏ *Hipposideros edwardshilli*
John Hill's Roundleaf Bat

Imonda region (NC New Guinea)

❏ *Hipposideros fuliginosus* Sooty Roundleaf Bat

Liberia to D.R. Congo and Ethiopia

❏ *Hipposideros fulvus* Fulvous Roundleaf Bat

Pakistan, India and Sri Lanka to Vietnam

❏ *Hipposideros galeritus* Cantor's Roundleaf Bat

India and Sri Lanka to Indochina, Java and Borneo; Sanana (Sula Is., W Moluccas)

❏ *Hipposideros halophyllus* Thailand Roundleaf Bat

Thailand

❏ *Hipposideros hypophyllus* Kolar Roundleaf Bat

India

❏ *Hipposideros inexpectatus* Crested Roundleaf Bat

N Sulawesi

❏ *Hipposideros jonesi* Jones' Roundleaf Bat

Guinea and Sierra Leone to Mali, Bukina Faso and Nigeria

❏ *Hipposideros lamottei* Lamotte's Roundleaf Bat

Mt Nimba (Guinea and Liberia)

❏ *Hipposideros lankadiva* Indian Roundleaf Bat

C & S India and Sri Lanka

❏ *Hipposideros larvatus*
Intermediate Roundleaf Bat

Bangladesh to S China, Hainan, Indochina, Malaya, Sumatra, Java and Borneo

❏ *Hipposideros lekaguli* Large Asian Roundleaf Bat

Thailand

❏ *Hipposideros lylei* Shield-faced Roundleaf Bat

Myanmar, Thailand and Malaya

❏ *Hipposideros macrobullatus*
Big-eared Roundleaf Bat

Kangean Is. (NE of Java), Sulawesi, Seram (S Moluccas)

❏ *Hipposideros madurae* Maduran Roundleaf Bat

C Java and Madura

❏ *Hipposideros maggietaylorae*
Maggie Taylor's Roundleaf Bat

Lowlands of New Guinea and Bismarck Archipelago

❏ *Hipposideros marisae* Aellen's Roundleaf Bat

Guinea, Liberia and Ivory Coast

❏ *Hipposideros megalotis*
Large-eared Roundleaf Bat

Very locally in dry upland areas of Ethiopia, Djibouti, Somalia and Kenya; Saudi Arabia

❏ *Hipposideros muscinus* Fly River Roundleaf Bat

Lowlands of S New Guinea

❏ *Hipposideros nequam* Malayan Roundleaf Bat

Selangor region (Malaya); Critically Endangered

❏ *Hipposideros obscurus* Philippine Roundleaf Bat

Philippines

❏ *Hipposideros orbiculus*
Orb-faced Roundleaf Bat

Malaya and Sumatra

❏ *Hipposideros papua* Biak Roundleaf Bat

N Moluccas and W New Guinea incl. Biak (Geelvinck Is.)

❏ *Hipposideros pomona* Pomona Roundleaf Bat

India to S China and Malaya

❏ *Hipposideros pratti* Pratt's Roundleaf Bat

Myanmar, Thailand, S China, Vietnam and Malaya

❏ *Hipposideros pygmaeus*
Philippine Pygmy Roundleaf Bat

Philippines

❏ *Hipposideros ridleyi* Ridley's Roundleaf Bat

Malaya and Borneo

❏ *Hipposideros rotalis* Laotian Roundleaf Bat

Laos

❏ *Hipposideros ruber* Noack's Roundleaf Bat

Senegal to Ethiopia, S to Angola, Zambia and Mozambique;
Equatorial Guinea islands

❏ *Hipposideros sabanus* Least Roundleaf Bat

Malaya, Sumatra and Borneo

❏ *Hipposideros schistaceus* Split Roundleaf Bat

S India

❏ *Hipposideros semoni* Semon's Roundleaf Bat

SE New Guinea and coastal E Queensland (NE Australia)

❏ *Hipposideros sorenseni*
Pangandaran Roundleaf Bat

Pangandaran region (C Java)

❏ *Hipposideros speoris*
Schneider's Roundleaf Bat

India and Sri Lanka

❏ *Hipposideros stenotis*
Narrow-eared Roundleaf Bat

N Australia

❏ *Hipposideros sumbae* Sumba Roundleaf Bat

Sumba (Lesser Sundas)

❏ *Hipposideros turpis* Lesser Roundleaf Bat

Peninsular Thailand and S Ryukyu Is. (Japan); Endangered

❏ *Hipposideros wollastoni*
Wollaston's Roundleaf Bat

W & C New Guinea

❏ *Anthops ornatus* Flower-faced Bat

Bougainville (New Guinea) and Solomon Is.

❏ *Aselliscus stoliczkanus* Stoliczka's Trident Bat

Myanmar, S China, Thailand, Laos, Vietnam and Malaya

❏ *Aselliscus tricuspidatus* Temminck's Trident Bat

Moluccas, New Guinea, Bismarck Archipelago, Solomon Is.
and Vanuatu

❏ *Asellia patrizii* Patrizi's Trident Leaf-nosed Bat

Deserts and semi-deserts of N Ethiopia and Red Sea islands

❏ *Asellia tridens* Trident Leaf-nosed Bat

Deserts and semi-deserts of Senegal, Chad, Sudan and S Somalia;
N Africa; Middle East and Arabian Peninsula; Socotra I.

❏ *Rhinonycteris aurantia* Orange Leaf-nosed Bat

N Australia

❏ *Cloeotis percivali* Percival's Trident Bat

Locally in lowland woods and savannas of S D.R. Congo, Kenya
and Tanzania, S to Botswana, Swaziland and NE South Africa

❏ *Triaenops furculus* Trouessart's Trident Bat

Coastal WC, NW & N Madagascar; Aldabra Is.

❏ *Triaenops persicus* Persian Trident Bat

Ethiopia and Somalia, S to Angola and Mozambique; Yemen
and Oman; Iran

❏ *Triaenops rufus* Rufous Trident Bat

Coastal WC, NW & E Madagascar

❏ *Paracoelops megalotis* Vietnam Leaf-nosed Bat

Annam region (Vietnam); Critically Endangered

❏ *Coelops frithi* East Asian Tailless Leaf-nosed Bat

NE India to S China and Taiwan, S to Vietnam, Malaya, Java
and Bali

❏ *Coelops hirsutus*
Philippine Tailless Leaf-nosed Bat

Mindoro (WC Philippines)

❑ *Coelops robinsoni*
Malayan Tailless Leaf-nosed Bat

Malaya and Borneo

Family: MYSTACINIDAE
(New Zealand Short-tailed Bats—2)

❑ *†Mystacina robusta*
Greater New Zealand Short-tailed Bat

Formerly Solomon I. and Big South Cape I. (S South Island, New Zealand); Extinct (last specimen collected in 1965)

❑ *Mystacina tuberculata*
Lesser New Zealand Short-tailed Bat

Interiors of indigenous forests of New Zealand

Family: NOCTILIONIDAE (Fishing Bats—2)

❑ *Noctilio albiventris* **Lesser Fishing Bat**

Lowland forest clearings and streams of Chiapas (SE Mexico) to French Guiana, S to Peru, Paraguay, NE Argentina and SW Amazonian Brazil

❑ *Noctilio leporinus* **Greater Fishing Bat**

Lowland forests, lakes, rivers and coasts of W Mexico to Suriname, S to Peru, N Argentina and S Brazil; Trinidad; Greater and Lesser Antilles; S Bahamas

Family: MORMOOPIDAE
(Leaf-chinned Bats—8)

Pteronotus (Chilonycteris)

❑ *Pteronotus macleayii*
MacLeay's Moustached Bat

Cuba and Jamaica

❑ *Pteronotus personatus* **Lesser Moustached Bat**

Lowland forests of Mexico to Peru, Suriname and E Brazil; Trinidad

❑ *Pteronotus quadridens* **Sooty Moustached Bat**

Cuba, Jamaica, Hispaniola and Puerto Rico

Pteronotus (Phyllodia)

❑ *Pteronotus parnellii* **Common Moustached Bat**

Lowland and mid-elevation forests and clearings of Mexico to French Guiana, S to Peru and E Brazil; Trinidad & Tobago; Cuba to Puerto Rico

Pteronotus (Pteronotus)

❑ *Pteronotus davyi* **Davy's Naked-backed Bat**

Locally in forests and clearings of Mexico to Venezuela, S to NW Peru and E Brazil; Trinidad; S Lesser Antilles

❑ *Pteronotus gymnonotus* **Big Naked-backed Bat**

Locally in lowland forests of EC Mexico to Guyana, S to Peru (and SW Brazil?)

❑ *Mormoops blainvillii* **Antillean Ghost-faced Bat**

Greater Antilles

❑ *Mormoops megalophylla* **Ghost-faced Bat**

Dry forests of extreme SW & SC USA to E Honduras; N Colombia and N Venezuela, S to NW Peru; Trinidad; Netherlands Antilles

Family: PHYLLOSTOMIDAE
(American Leaf-nosed Bats—155)

Micronycteris (Glyphonycteris)

❑ *Micronycteris behnii* **Behni's Big-eared Bat**

Forests of S Peru and C Brazil

❑ *Micronycteris daviesi*
Davies' Big-eared Bat (Bartica Bat)

Very locally in lowland primary rainforests of E Honduras to French Guiana, S to Peru and Amazonian Brazil

❑ *Micronycteris sylvestris* **Tricoloured Big-eared Bat**

Locally in lowland forests of C Mexico to French Guiana, S to Peru and SE Brazil; Trinidad

Micronycteris (Lampronycteris)

❑ *Micronycteris brachyotis*
Orange-throated Big-eared Bat

Locally in lowland forests of S Mexico to French Guiana and Amazonian Brazil; Trinidad

Micronycteris (Micronycteris)

❑ *Micronycteris brosseti* **Brosset's Big-eared Bat**

Paracou region (French Guiana)

❑ *Micronycteris hirsuta* **Hairy Big-eared Bat**

Locally in forests and forest edge of E Honduras to French Guiana, S to Peru and Amazonian Brazil; Trinidad

❑ *Micronycteris homezi* **Homez's Big-eared Bat**

Guyana, French Guiana and N Brazil

❏ *Micronycteris matses* Matses' Big-eared Bat

Primary rainforests of Rio Galvez drainage (WC Amazonian Brazil)

❏ *Micronycteris megalotis* Little Big-eared Bat

Lowland forests of Colombia to French Guiana, S to Peru, Bolivia and Brazil; Trinidad & Tobago; Grenada

❏ *Micronycteris microtis* Common Big-eared Bat

Forests and regrowth areas of N Mexico to Colombia, French Guiana and Amazonian Brazil

❏ *Micronycteris minuta* White-bellied Big-eared Bat

Lowland forests and farmland of NW Honduras to French Guiana, S to Peru, Bolivia and S Brazil; Trinidad

❏ *Micronycteris sanborni* Sanborn's Big-eared Bat

Chapada do Araripe plateau (NE Brazil)

❏ *Micronycteris schmidtorum*
Schmidts' Big-eared Bat

Locally in lowland forests and forest edge of E Chiapas and Cozumel I. (S Mexico) to Venezuela, S to NE Peru and E Brazil

Micronycteris (Neonycteris)

❏ *Micronycteris pusilla* Least Big-eared Bat

E Colombia and NW Brazil

Micronycteris (Trinycteris)

❏ *Micronycteris nicefori* Niceforo's Big-eared Bat

Locally in lowland forests of Belize to French Guiana, S to Peru and Amazonian Brazil; Trinidad

❏ *Macrotus californicus* California Leaf-nosed Bat

SW USA and NW Mexico

❏ *Macrotus waterhousii*
Waterhouse's Leaf-nosed Bat

Arid habitats of Mexico (and Guatemala?); Cuba, Cayman Is., Jamaica, Hispaniola incl Beata I.; Bahamas

❏ *Lonchorhina aurita* Common Sword-nosed Bat

Locally in lowland forests and farmland of S Mexico to Venezuela, S to Peru and E Brazil; Trinidad

❏ *Lonchorhina fernandezi*
Fernandez's Sword-nosed Bat

Puerto Ayacucho region (S Venezuela)

❏ *Lonchorhina inusitata* Strange Sword-nosed Bat

Venezuela, French Guiana, Suriname and Brazil

❏ *Lonchorhina marinkellei*
Marinkelle's Sword-nosed Bat

Forests of SE Colombia to French Guiana

❏ *Lonchorhina orinocensis*
Orinoco Sword-nosed Bat

Llanos of EC Colombia and W Venezuela

❏ *Macrophyllum macrophyllum* Long-legged Bat

Near streams in lowland forests of S Mexico to French Guiana, S to Peru, Bolivia, NE Argentina and SE Brazil

❏ *Tonatia bidens* Greater Round-eared Bat

Forests of N Argentina, Paraguay and SE Brazil

❏ *Tonatia brasiliense* Pygmy Round-eared Bat

Lowland forests and regrowth areas of EC Mexico to Suriname, S to Peru, Bolivia and E Brazil; Trinidad

❏ *Tonatia carrikeri* Carriker's Round-eared Bat

Lowland forests of SE Colombia to Suriname, S to Peru, Bolivia and SW Amazonian Brazil

❏ *Tonatia evotis* Davis' Round-eared Bat

Lowland forests of EC Mexico to NE Honduras

❏ *Tonatia saurophila* Stripe-headed Round-eared Bat

Locally in lowland forests of E Chiapas (SE Mexico) to Suriname, S to Peru and E Brazil; Trinidad

❏ *Tonatia schulzi* Schulz's Round-eared Bat

Forests of Guyana, Suriname, French Guiana (and N Brazil?)

❏ *Tonatia silvicola* White-throated Round-eared Bat

Lowland primary rainforests of Honduras to French Guiana, S to Bolivia, Paraguay, NE Argentina and SW & E Brazil

Mimon (Anthorhina)

❏ *Mimon crenulatum* Striped Hairy-nosed Bat

Locally in lowland forests and clearings of S Mexico to French Guiana, S to N Peru, N Bolivia and E Brazil; Trinidad

Mimon (Mimon)

❏ *Mimon bennettii* Golden Bat

Lowland forests of EC Mexico to French Guiana, S to SE Brazil

❏ *Mimon cozumelae* Cozumel Golden Bat

Cozumel I. (Quintana Roo, Mexico)

❏ *Phyllostomus discolor* Pale Spear-nosed Bat

Lowland forests of EC Mexico to French Guiana, S to Peru, N Argentina and E Brazil; Margarita; Trinidad

❏ *Phyllostomus elongatus* Lesser Spear-nosed Bat

Lowland rainforests of C Colombia to French Guiana, S to Peru, Bolivia, and Amazonian Brazil

❏ *Phyllostomus hastatus* Greater Spear-nosed Bat

Lowland forests and clearings of S Belize to French Guiana, S to Peru, Bolivia, Paraguay and S Brazil (and N Argentina?); Margarita; Trinidad & Tobago

❏ *Phyllostomus latifolius* Guianan Spear-nosed Bat

Few records from SE Colombia; Guyana, Suriname and French Guiana

❏ *Phylloderma stenops* Pale-faced Bat

Locally in primary rainforests and clearings of S Mexico to Peru, Bolivia and E Brazil

❏ *Trachops cirrhosus* Fringe-lipped Bat

Forests and farmland of S Mexico to French Guiana, S to Bolivia and SE Brazil; Trinidad

❏ *Chrotopterus auritus*
Big-eared Woolly Bat (Woolly False Vampire Bat)

Forests and regrowth areas of EC Mexico to French Guiana, S to N Argentina, Paraguay and SE Brazil

❏ *Vampyrum spectrum*
Spectral Bat (Great False Vampire Bat)

Forests and grasslands of EC Mexico to French Guiana, S to Peru and SC Brazil; Trinidad

❏ *Brachyphylla cavernarum* Antillean Fruit-eating Bat

Puerto Rico; Leeward Is.; Lesser Antilles, S to St Vincent

❏ *Brachyphylla nana* Cuban Fruit-eating Bat

Cuba, Grand Cayman (Cayman Is.), Jamaica, Hispaniola; Middle Caicos (SE Bahamas)

❏ *Erophylla sezekorni* Buffy Flower Bat

Cuba, Cayman Is., Jamaica, Hispaniola, Puerto Rico; Bahamas

Phyllonycteris (Phyllonycteris)

❏ *Phyllonycteris poeyi* Cuban Flower Bat

Cuba incl. Isla de la Juventud, and Hispaniola

Phyllonycteris (Reithronycteris)

❏ *Phyllonycteris aphylla* Jamaican Flower Bat

Jamaica; Endangered

❏ *Glossophaga commissarisi*
Brown Long-tongued Bat

Locally in forests and clearings from Mexico to W Colombia, S to E Peru and NW Brazil

❏ *Glossophaga leachii* Gray's Long-tongued Bat

Dry forests and scrub of C Mexico to Pacific slope of Costa Rica

❏ *Glossophaga longirostris*
Miller's Long-tongued Bat

Lowland dry forests and llanos of Colombia to Guyana, S to W Ecuador and NC Amazonian Brazil; Margarita; Trinidad & Tobago; Lesser Antilles

❏ *Glossophaga morenoi*
Western Long-tongued Bat

Dry forests and scrub of Michoacán to E Chiapas (Mexico)

❏ *Glossophaga soricina* Common Long-tongued Bat

Forests and clearings of Mexico to French Guiana, S to Peru, Paraguay, N Argentina and SE Brazil; Trinidad; Jamaica; Grenada

❏ *Monophyllus plethodon* Insular Single-leaf Bat

Anguilla to St Vincent (Lesser Antilles)

❏ *Monophyllus redmani* Leach's Single-leaf Bat

Cuba, Jamaica, Hispaniola and Puerto Rico; S Bahamas

❏ *Lichonycteris obscura* Dark Long-tongued Bat

Few records from lowland forests and plantations of Chiapas (SE Mexico) to Suriname, S to N Peru (also Bolivia and Amazonian Brazil?)

❏ *Leptonycteris curasoae* Southern Long-nosed Bat

Forests of N Colombia and N Venezuela; Margarita; Netherlands Antilles

❏ *Leptonycteris nivalis* Mexican Long-nosed Bat

Desert scrub and dry forests of Mexico (and Guatemala?), and summer migrant N to SW Texas (USA); Endangered

❏ *Leptonycteris yerbabuenae*
North American Long-nosed Bat

Mexico to El Salvador, and summer migrant N to SE Arizona and SW New Mexico (USA)

158

❏ *Anoura caudifera* Tailed Tailless Bat

Near streams in rainforests of Colombia to French Guiana, S to Peru, Bolivia, Salta Prov. (NW Argentina) and SE Brazil

❏ *Anoura cultrata* Handley's Tailless Bat

Lowland and mid-elevation forests and forest edge of Costa Rica to Venezuela, S in Andes to Peru and Bolivia

❏ *Anoura geoffroyi* Geoffroy's Tailless Bat

Locally in rainforests, fruit groves and clearings of Mexico to French Guiana, S to Peru, Bolivia and SE Brazil; Trinidad; Grenada

❏ *Anoura latidens* Broad-toothed Tailless Bat

Rainforests of Venezuela (also Colombia and Peru?)

❏ *Anoura luismanueli* Luis Manuel's Tailless Bat

Venezuela

❏ *Hylonycteris underwoodi*
Underwood's Long-tongued Bat

Lowland and mid-elevation forests and clearings of C Mexico to W Panama

❏ *Scleronycteris ega* Ega Long-tongued Bat

Very few records from lowland rainforests of S Venezuela

❏ *Choeroniscus godmani*
Godman's Whiskered Long-nosed Bat

Locally in forests and plantations of NW Mexico to Suriname

❏ *Choeroniscus minor*
Lesser Whiskered Long-nosed Bat

Rainforests of C Colombia to French Guiana, S to Peru, C Bolivia and Amazonian Brazil; Trinidad

❏ *Choeroniscus periosus*
Greater Whiskered Long-nosed Bat

Extreme SW Colombia and W Ecuador

❏ *Choeronycteris mexicana*
Mexican Long-tongued Bat

Desert scrub and dry forests of extreme SW & SC USA and Pacific slope of Mexico to S Honduras

❏ *Musonycteris harrisoni* Banana Bat

Jalisco to Guerrero (SW Mexico)

❏ *Lionycteris spurrelli* Chestnut Long-tongued Bat

Rainforests and savannas of E Panama to French Guiana, S to E Peru and Amazonian Brazil

❏ *Lonchophylla bokermanni*
Bokermann's Nectar Bat

Serra do Cipo (Minas Gerais, SE Brazil)

❏ *Lonchophylla dekeyseri* Dekeyser's Nectar Bat

Brasília region (Distrito Federal, E Brazil)

❏ *Lonchophylla handleyi* Handley's Nectar Bat

S Colombia, Ecuador and Peru

❏ *Lonchophylla hesperia* Western Nectar Bat

Zorritos region (Tumbes, N Peru)

❏ *Lonchophylla mordax* Goldman's Nectar Bat

Forests and banana groves of Costa Rica to W Ecuador; EC Brazil

❏ *Lonchophylla robusta* Orange Nectar Bat

Lowland and mid-elevation rainforests of SE Nicaragua to NW Venezuela and Ecuador

❏ *Lonchophylla thomasi* Thomas' Nectar Bat

Lowland forests and clearings of E Panama to French Guiana, S to Peru, Bolivia and Amazonian Brazil

❏ *Platalina genovensium* Long-snouted Bat

Cactus deserts of coastal W Peru

❏ *Carollia brevicauda* Silky Short-tailed Bat

Forest clearings, regrowth areas and plantations of C Mexico to Suriname, S to Peru, Bolivia and SC Brazil

❏ *Carollia castanea* Chestnut Short-tailed Bat

Rainforests and regrowth areas of W Honduras to French Guiana, S to Peru, Bolivia and SW Brazil

❏ *Carollia perspicillata* Seba's Short-tailed Bat

Forest clearings, regrowth areas and plantations of EC Mexico to French Guiana, S to Bolivia, Paraguay, NE Argentina and SE Brazil; Trinidad & Tobago; Grenada

❏ *Carollia sowelli* Sowell's Short-tailed Bat

EC Mexico to W Panama

❏ *Carollia subrufa* Grey Short-tailed Bat

Lowland and mid-elevation dry forests and regrowth areas of C Mexico to NW Costa Rica

❏ *Rhinophylla alethina* Hairy Little Fruit Bat

W of Andes in W Colombia and W Ecuador (also NW Peru and Brazil?)

❑ *Rhinophylla fischerae* Fischer's Little Fruit Bat

Lowland rainforests of Amazon basin of SE Colombia to E Peru and E Amazonian Brazil

❑ *Rhinophylla pumilio* Dwarf Little Fruit Bat

Rainforests of S Colombia to French Guiana, S to Peru, Bolivia and C Brazil

Sturnira (Corvira)

❑ *Sturnira bidens* Bidentate Yellow-shouldered Bat

Montane cloud forests of Colombia, W Venezuela, Ecuador and Peru

❑ *Sturnira nana* Lesser Yellow-shouldered Bat

Huanhuachayo region (Ayacucho, S Peru)

Sturnira (Sturnira)

❑ *Sturnira aratathomasi*
Arata and Thomas' Yellow-shouldered Bat

Few records from montane cloud forests of W Colombia, NW Venezuela, and Ecuador (and Peru?)

❑ *Sturnira erythromos* Hairy Yellow-shouldered Bat

Montane forests of Colombia and N Venezuela to Bolivia and NW Argentina

❑ *Sturnira lilium* Little Yellow-shouldered Bat

Forests and fruit groves of Mexico to French Guiana, S to N Argentina, Uruguay and SE Brazil; Trinidad & Tobago; Lesser Antilles

❑ *Sturnira ludovici* Highland Yellow-shouldered Bat

Lowland and mid-elevation forests and forest edge of Mexico to Guyana, and in Andes S to Peru

❑ *Sturnira luisi* Luis' Yellow-shouldered Bat

Lowland forests and forest edge of Caribbean slope of Costa Rica and Panama; (also Ecuador and NW Peru?)

❑ *Sturnira magna* Greater Yellow-shouldered Bat

Lowland and mid-elevation forests E of the Andes in extreme SE Colombia, Ecuador, Peru and Bolivia

❑ *Sturnira mistratensis*
Mistrató Yellow-shouldered Bat

W flank of Cordillera Occidental (W Colombia)

❑ *Sturnira mordax* Talamancan Yellow-shouldered Bat

Mid- and high-elevation forests and forest edge of mountains of Costa Rica and W Panama

❑ *Sturnira oporaphilum*
Bogotá Yellow-shouldered Bat

Montane forests of Colombia, W Venezuela, Ecuador, Peru and Bolivia

❑ *Sturnira thomasi* Thomas' Yellow-shouldered Bat

Guadeloupe (Lesser Antilles); Endangered

❑ *Sturnira tildae* Tilda Yellow-shouldered Bat

Lowland rainforests of SE Colombia to French Guiana, S to Peru, Bolivia and C Brazil; Trinidad

❑ *Uroderma bilobatum* Common Tent-making Bat

Lowland and mid-elevation forests and fruit groves of EC Mexico to Suriname, S to Peru, Bolivia and E Brazil; Trinidad

❑ *Uroderma magnirostrum* Brown Tent-making Bat

Lowland forests and wetlands of SW Mexico to SW Nicaragua; E Panama to Venezuela, S to Peru, Bolivia and E Brazil

❑ *Platyrrhinus aurarius* Eldorado Broad-nosed Bat

Colombia, S Venezuela and Suriname

❑ *Platyrrhinus brachycephalus*
Short-headed Broad-nosed Bat

Lowland rainforests of Colombia to French Guiana, S to Peru, Bolivia and SW Amazonian Brazil

❑ *Platyrrhinus chocoensis* Choco Broad-nosed Bat

W Colombia

❑ *Platyrrhinus dorsalis* Thomas' Broad-nosed Bat

Mid-elevation forests and forest edge of E Panama to C Colombia, S in Andes to Peru, Bolivia and Paraguay

❑ *Platyrrhinus helleri* Heller's Broad-nosed Bat

Lowland and mid-elevation forests and fruit groves of S Mexico to French Guiana, S to Peru, Bolivia and S Brazil; Trinidad

❑ *Platyrrhinus infuscus* Buffy Broad-nosed Bat

Forests of C Colombia to Peru, Bolivia and SW Amazonian Brazil

❑ *Platyrrhinus lineatus* White-lined Broad-nosed Bat

Forests of Colombia to French Guiana, S to E Paraguay, NE Argentina, NW Uruguay and S Brazil

❑ *Platyrrhinus recifinus* Recife Broad-nosed Bat

Coastal forests of Guyana and E & SE Brazil

❏ *Platyrrhinus umbratus* Shadowy Broad-nosed Bat

Forests of E Panama, NW Colombia and N Venezuela

❏ *Platyrrhinus vittatus* Greater Broad-nosed Bat

Mid-elevation and montane forests and regrowth areas of Costa Rica to Venezuela, S to Peru and Bolivia

❏ *Vampyrodes caraccioli* Great Stripe-faced Bat

Lowland primary rainforests of S Mexico to Peru, N Bolivia and Amazonian Brazil; Trinidad & Tobago

Vampyressa (Metavampyressa)

❏ *Vampyressa brocki* Brock's Yellow-eared Bat

SE Colombia, Guyana, Suriname (and Amazonian Brazil?)

❏ *Vampyressa nymphaea* Striped Yellow-eared Bat

Lowland primary rainforests of SE Nicaragua to W Colombia and NW Ecuador

Vampyressa (Vampyressa)

❏ *Vampyressa melissa* Melissa's Yellow-eared Bat

Few records from S Colombia; French Guiana; Peru

❏ *Vampyressa pusilla* Little Yellow-eared Bat

Lowland and mid-elevation forests of EC Mexico to French Guiana, S to Bolivia, SE Paraguay and SE Brazil

Vampyressa (Vampyriscus)

❏ *Vampyressa bidens* Bidentate Yellow-eared Bat

Lowland rainforests of Colombia to French Guiana, S to N Bolivia and Amazonian Brazil

❏ *Mesophylla macconnelli* MacConnell's Bat

Lowland primary rainforests of SE Nicaragua to French Guiana, S to Peru, Bolivia and SC Brazil; Trinidad

❏ *Ectophylla alba* Honduran White Bat

Lowland rainforests and regrowth areas of E Honduras to NW Panama

❏ *Chiroderma doriae* Brazilian Big-eyed Bat

Minas Gerais and São Paulo (EC & SE Brazil)

❏ *Chiroderma improvisum* Guadeloupe Big-eyed Bat

Guadeloupe and Montserrat (Lesser Antilles); Endangered

❏ *Chiroderma salvini* Salvin's Big-eyed Bat

Locally in mid-elevation forests and clearings of Mexico to Venezuela and S in Andes to Bolivia

❏ *Chiroderma trinitatum* Little Big-eyed Bat

Lowland forests and clearings of Panama to Suriname, S to Peru, Bolivia and SC Brazil; Trinidad & Tobago

❏ *Chiroderma villosum* Hairy Big-eyed Bat

Lowland forests and fruit groves of C Mexico to Suriname, S to Peru, Bolivia and SC Brazil; Trinidad & Tobago

Artibeus (Artibeus)

❏ *Artibeus amplus* Large Fruit-eating Bat

Foothill rainforests of N Colombia and Venezuela

❏ *Artibeus fimbriatus* Fringed Fruit-eating Bat

Lowlands of E Paraguay and SE Brazil (and NE Argentina?)

❏ *Artibeus fraterculus* Fraternal Fruit-eating Bat

Lowland rainforests of W Ecuador and NW Peru

❏ *Artibeus glaucus* Silver Fruit-eating Bat

Venezuela and Suriname, S to Peru, Bolivia and SC & E Brazil; Trinidad & Tobago; Grenada

❏ *Artibeus hirsutus* Hairy Fruit-eating Bat

Sonora to Guerrero (Mexico)

❏ *Artibeus inopinatus* Honduran Fruit-eating Bat

Dry scrub and dry forests of Pacific slope of El Salvador, S Honduras and SW Nicaragua

❏ *Artibeus intermedius*
Intermediate Fruit-eating Bat

Forests and regrowth areas of Mexico to N South America

❏ *Artibeus jamaicensis* Jamaican Fruit-eating Bat

Forests and plantations of Mexico to French Guiana, S to Bolivia, N Argentina and S Brazil; Trinidad & Tobago; Greater and Lesser Antilles

❏ *Artibeus lituratus* Great Fruit-eating Bat

Rainforests and clearings of Mexico incl. Tres Marías Is. to French Guiana, S to Paraguay, NE Argentina and S Brazil; Trinidad & Tobago; S Lesser Antilles

❏ *Artibeus obscurus* Dark Fruit-eating Bat

Lowland rainforests of Colombia to French Guiana, S to Peru, Bolivia and S Brazil

❏ *Artibeus phaeotis* Pygmy Fruit-eating Bat

Forests and fruit groves of Mexico to Colombia and Ecuador (and E to Guyana?)

❏ *Artibeus planirostris* Flat-faced Fruit-eating Bat

Lowland rainforests of S Colombia and S Venezuela, S to Peru, N Argentina, Paraguay and E Brazil

❏ *Artibeus watsoni* Thomas' Fruit-eating Bat

Forests and fruit groves of EC Mexico to Colombia

Artibeus (Dermanura)

❏ *Artibeus anderseni* Andersen's Fruit-eating Bat

Ecuador, Peru, Bolivia and W Brazil

❏ *Artibeus aztecus* Aztec Fruit-eating Bat

Mid- and high-elevation cloud forests and fruit groves of Mexico to W Panama

❏ *Artibeus cinereus* Gervais' Fruit-eating Bat

Mid-elevation forests of Venezuela to French Guiana, S to Bolivia and SC Brazil

❏ *Artibeus incomitatus* Solitary Fruit-eating Bat

Isla Escudo de Veraguas (Bocas del Toro Is., Panama)

❏ *Artibeus toltecus* Toltec Fruit-eating Bat

Mid-elevation forests and fruit groves of NE Mexico to W Colombia

Artibeus (Enchisthenes)

❏ *Artibeus hartii* Velvety Fruit-eating Bat

Forests and forest edge of N Mexico to Venezuela, S in Andes to Peru and Bolivia

Artibeus (Koopmania)

❏ *Artibeus concolor* Brown Fruit-eating Bat

Lowland forests of Colombia to French Guiana, S to Amazonian and E Brazil (and Peru?)

❏ *Ardops nichollsi* Tree Bat

St Eustatius to St Vincent (Lesser Antilles)

❏ *Phyllops falcatus* Cuban Fig-eating Bat

Cuba and Hispaniola

❏ *Ariteus flavescens* Jamaican Fig-eating Bat

Jamaica

❏ *Stenoderma rufum* Red Fruit Bat

Puerto Rico and Virgin Is. (Leeward Is.)

❏ *Pygoderma bilabiatum* Ipanema Bat

Forests and fruit groves of Suriname; SC Bolivia, Paraguay, N Argentina and SE Brazil

❏ *Ametrida centurio* Little White-shouldered Bat

Locally in lowland rainforests of Panama to French Guiana, S to C & E Amazonian & SC Brazil; Bonaire I. (Netherlands Antilles); Trinidad

❏ *Sphaeronycteris toxophyllum* Visored Bat

Forests, plantations and gardens of NE Colombia and Venezuela, S to Amazon basin of E Peru, N Bolivia and NW Amazonian Brazil

❏ *Centurio senex* Wrinkle-faced Bat

Lowland and mid-elevation regrowth areas and seasonally flooded forests of Mexico to N Colombia and N Venezuela; Trinidad & Tobago

❏ *Diphylla ecaudata* Hairy-legged Vampire Bat

Lowland and mid-elevation forests and farmland of E Mexico to Venezuela, S to Peru, Bolivia and SE Brazil; vagrant S Texas (USA)

❏ *Diaemus youngi* White-winged Vampire Bat

Lowland forests of E Mexico to Venezuela, S to N Argentina, Paraguay and S Brazil; Margarita; Trinidad

❏ *Desmodus rotundus* Common Vampire Bat

Livestock farmland, gardens and forests of Mexico to Suriname, S to C Chile, N Argentina and Uruguay; Trinidad

Family: MYZOPODIDAE
(Sucker-footed Bat—1)

❏ *Myzopoda aurita* Sucker-footed Bat

Rainforests of coastal E Madagascar

Family: THYROPTERIDAE
(Disk-winged Bats—3)

❏ *Thyroptera discifera* Peters' Disk-winged Bat

Lowland forests and fruit groves of SE Nicaragua; C Panama to French Guiana, S to Peru, Bolivia and E Amazonian Brazil

❏ *Thyroptera lavali* LaVal's Disk-winged Bat

Upper Amazon basin of NE Peru

❏ *Thyroptera tricolor* Spix's Disk-winged Bat

Lowland and mid-elevation forests and regrowth areas of EC Mexico to French Guiana, S to Peru, Bolivia and Amazonian Brazil; Trinidad

Family: FURIPTERIDAE
(Thumbless Bats—2)

❏ *Furipterus horrens* Thumbless Bat

Locally in lowland forests of Costa Rica to French Guiana, S to Peru and E Brazil; Trinidad

❏ *Amorphochilus schnablii* Smoky Thumbless Bat

Coasts and estuaries of W Ecuador, W Peru and extreme NW Chile

Family: NATALIDAE
(Funnel-eared Bats—5)

Natalus (Chilonatalus)

❏ *Natalus micropus* Cuban Funnel-eared Bat

Providencia I. (Colombia); Cuba, Jamaica and Hispaniola

❏ *Natalus tumidifrons* Bahaman Funnel-eared Bat

Bahamas

Natalus (Natalus)

❏ *Natalus stramineus* Mexican Funnel-eared Bat

Lowland and mid-elevation forests of Mexico to Panama; E Brazil; Jamaica and Hispaniola; Lesser Antilles

❏ *Natalus tumidirostris* Trinidadian Funnel-eared Bat

Lowland dry forests of N Colombia to Suriname; Netherlands Antilles; Trinidad & Tobago

Natalus (Nyctiellus)

❏ *Natalus lepidus* Gervais' Funnel-eared Bat

Cuba and Bahamas

Family: MOLOSSIDAE
(Free-tailed Bats—94)

❏ *Tomopeas ravus* Blunt-eared Bat

Rocky semiarid areas of coastal NW Peru

Mormopterus (Mormopterus)

❏ *Mormopterus acetabulosus* Natal Free-tailed Bat

Ethiopia; South Africa; Madagascar?; Réunion and Mauritius

❏ *Mormopterus beccarii* Beccari's Mastiff Bat

Moluccas, New Guinea and near islands, and N Australia

❏ *Mormopterus doriae* Sumatran Mastiff Bat

Sumatra

❏ *Mormopterus jugularis* Peters' Wrinkle-lipped Bat

Madagascar

❏ *Mormopterus kalinowskii* Kalinowski's Mastiff Bat

C Peru and N Chile

❏ *Mormopterus loriae* Little Northern Free-tailed Bat

C New Guinea and N Australia

❏ *Mormopterus minutus* Little Goblin Bat

Cuba

❏ *Mormopterus norfolkensis*
East-coast Free-tailed Bat

SE Queensland and E New South Wales (SE Australia); Norfolk I.? (type locality uncertain)

❏ *Mormopterus phrudus* Incan Little Mastiff Bat

Riparian habitats in Machu Picchu region (Cusco, Peru); Endangered

❏ *Mormopterus planiceps* Southern Free-tailed Bat

S Australia

Mormopterus (Platymops)

❏ *Mormopterus setiger* Peters' Flat-headed Bat

Arid rocky habitats of S Sudan, Ethiopia and Kenya

Mormopterus (Sauromys)

❏ *Mormopterus petrophilus*
Roberts' Flat-headed Bat

Namibia, Botswana, Zimbabwe, Mozambique and South Africa

Molossops (Cabreramops)

❏ *Molossops aequatorianus*
Equatorial Dog-faced Bat

Babahoyo region (Los Ríos, Ecuador) and Peru

Molossops (Cynomops)

❏ *Molossops abrasus* Cinnamon Dog-faced Bat

Lowland rainforests and thorn scrub of Venezuela to Suriname, S to Peru, Bolivia, Paraguay, NE Argentina and E Brazil

163

❏ *Molossops greenhalli* Greenhall's Dog-faced Bat

Forests and clearings, usually near water, of WC Mexico to Venezuela, S to Ecuador and E Brazil; Trinidad

❏ *Molossops mexicanus* Mexican Dog-faced Bat

WC Mexico

❏ *Molossops paranus* Brown-bellied Dog-faced Bat

Venezuela and French Guiana

❏ *Molossops planirostris* Southern Dog-faced Bat

Lowland forests and savannas of C Panama to Suriname, S to Peru, N Argentina, Paraguay and S Brazil

Molossops (Molossops)

❏ *Molossops neglectus* Rufous Dog-faced Bat

Suriname; Peru and Amazonian Brazil

❏ *Molossops temminckii* Dwarf Dog-faced Bat

Thorn scrub and chaco of Colombia and Venezuela, S to Peru, Bolivia, N Argentina, Paraguay, Uruguay and S Brazil

Molossops (Neoplatymops)

❏ *Molossops matogrossensis*
Mato Grosso Dog-faced Bat

Lowland rainforests of Venezuela and W Guyana; locally in arid rocky habitats and caatinga of E Amazonian & E Brazil

❏ *Myopterus daubentonii*
Daubenton's Free-tailed Bat

Forests and forest-savanna mosaics of Senegal, Ivory Coast and NE D.R. Congo

❏ *Myopterus whitleyi* Bini Free-tailed Bat

Forests of Ghana, Nigeria, Cameroon, D.R. Congo and Uganda

❏ *Cheiromeles parvidens* Lesser Hairless Bat

Negros and Mindanao (C & SE Philippines) and Sanana (Sula Is.)

❏ *Cheiromeles torquatus* Hairless Bat

Malaya, Sumatra, Java, Borneo, SW Philippines and Sulawesi

❏ *Tadarida aegyptiaca* Egyptian Free-tailed Bat

N and subsaharan Africa; Arabian Peninsula, E to India and Sri Lanka

❏ *Tadarida australis* White-striped Free-tailed Bat

C & S Australia

❏ *Tadarida brasiliensis* Mexican Free-tailed Bat

Locally in dry open areas and forests of S USA to Venezuela, S to C Chile, C Argentina and S Brazil; Trinidad; Greater and Lesser Antilles

❏ *Tadarida fulminans* Large Free-tailed Bat

E D.R. Congo, Rwanda and Kenya, S to NE South Africa; Madagascar

❏ *Tadarida kuboriensis* New Guinea Mastiff Bat

Montane forests of C New Guinea

❏ *Tadarida latouchei* Latouche's Free-tailed Bat

Thailand and N China

❏ *Tadarida lobata*
Kenyan Big-eared Free-tailed Bat

Kenya and Zimbabwe

❏ *Tadarida teniotis* European Free-tailed Bat

Open rocky and built areas of S Europe and NW Africa to S China, Taiwan and Japan; Madeira; Canary Is.

❏ *Tadarida ventralis* African Giant Free-tailed Bat

Ethiopia to South Africa

❏ *Chaerephon aloysiisabaudiae*
Duke of Abruzzi's Free-tailed Bat

Ghana, Gabon, D.R. Congo and Uganda

❏ *Chaerephon ansorgei* Ansorge's Free-tailed Bat

Cameroon to Ethiopia, S to Angola and KwaZulu-Natal (SE South Africa)

❏ *Chaerephon bemmeleni*
Gland-tailed Free-tailed Bat

Liberia, Cameroon, D.R. Congo, Sudan, Uganda, Kenya and Tanzania

❏ *Chaerephon bivittata* Spotted Free-tailed Bat

Sudan and Ethiopia, S to Zambia, Zimbabwe and Mozambique

❏ *Chaerephon bregullae* Fijian Mastiff Bat

Locally on islands of Vanuatu and Fiji

❏ *Chaerephon chapini* Chapin's Free-tailed Bat

Ethiopia, D.R. Congo and Uganda, S to Namibia, Botswana and Zimbabwe

❏ *Chaerephon gallagheri* Gallagher's Free-tailed Bat

D.R. Congo; Critically Endangered

❏ *Chaerephon jobensis* **Northern Mastiff Bat**

Lowlands of Seram (S Moluccas), Japen, S & E New Guinea and N Australia

❏ *Chaerephon johorensis* **Northern Free-tailed Bat**

Malaya and Sumatra

❏ *Chaerephon leucogaster*
Pale-bellied Free-tailed Bat

W Madagascar

❏ *Chaerephon major* **Lappet-eared Free-tailed Bat**

Liberia and Mali to Sudan, Uganda and Tanzania

❏ *Chaerephon nigeriae* **Nigerian Free-tailed Bat**

Ghana and Niger to Ethiopia, S to Namibia, Botswana and Zimbabwe; Saudi Arabia

❏ *Chaerephon plicata* **Wrinkle-lipped Free-tailed Bat**

India and Sri Lanka to S China, Indochina, Borneo, Philippines and Lesser Sundas

❏ *Chaerephon pumila* **Little Free-tailed Bat**

Subsaharan Africa and near islands; Yemen; E Madagascar and Comoro Is.; Aldabra Is; Seychelles

❏ *Chaerephon russata* **Russet Free-tailed Bat**

Ghana, Cameroon, D.R. Congo and Kenya

❏ *Chaerephon solomonis* **Solomons Mastiff Bat**

Choiseul and Santa Isabel (Solomon Is.)

❏ *Chaerephon tomensis* **São Tomé Free-tailed Bat**

São Tomé (Equatorial Guinea)

Mops (Mops)

❏ *Mops condylurus* **Angolan Free-tailed Bat**

Subsaharan Africa

❏ *Mops congicus* **Medje Free-tailed Bat**

Ghana, Nigeria, Cameron, D.R. Congo and Uganda

❏ *Mops demonstrator* **Mongalla Free-tailed Bat**

Burkina Faso, D.R. Congo, Sudan, and Uganda

❏ *Mops leucostigma* **Pale-marked Free-tailed Bat**

Madagascar

❏ *Mops midas* **Midas Free-tailed Bat**

Subsaharan Africa; Madagascar; Saudi Arabia

❏ *Mops mops* **Malayan Free-tailed Bat**

Malaya, Sumatra and Borneo

❏ *Mops niangarae* **Niangara Free-tailed Bat**

Niangara region (D.R. Congo); Critically Endangered

❏ *Mops niveiventer* **White-bellied Free-tailed Bat**

D.R. Congo, Rwanda, Burundi, Tanzania, Angola, Zambia and Mozambique

❏ *Mops sarasinorum* **Sulawesi Free-tailed Bat**

Sulawesi and near islands, and Philippines

❏ *Mops trevori* **Trevor's Free-tailed Bat**

NW D.R. Congo and Uganda

Mops (Xiphonycteris)

❏ *Mops brachypterus* **Sierra Leone Free-tailed Bat**

Gambia to Kenya, Tanzania incl. Zanzibar I. and Mozambique

❏ *Mops nanulus* **Dwarf Free-tailed Bat**

Gambia to Ethiopia and Kenya

❏ *Mops petersoni* **Peterson's Free-tailed Bat**

Ghana and Cameroon

❏ *Mops spurrelli* **Spurrell's Free-tailed Bat**

Liberia and Ivory Coast to D.R. Congo

❏ *Mops thersites* **Railer Bat**

Sierra Leone to Rwanda; Bioko (Equatorial Guinea)

❏ *Otomops formosus* **Javan Mastiff Bat**

Java

❏ *Otomops johnstonei* **Alor Mastiff Bat**

Alor (Lesser Sundas)

❏ *Otomops martiensseni* **Large-eared Free-tailed Bat**

Locally in Central African Republic and Djibouti, S to Angola and Natal (SE South Africa); Madagascar

❏ *Otomops papuensis* Papuan Mastiff Bat

SE New Guinea

❏ *Otomops secundus* Mantled Mastiff Bat

NE New Guinea

❏ *Otomops wroughtoni* Wroughton's Free-tailed Bat

S India; Critically Endangered

❏ *Nyctinomops aurispinosus* Peale's Free-tailed Bat

Mexico; Colombia and Venezuela, S to Peru, Bolivia and SE Brazil

❏ *Nyctinomops femorosaccus*
Pocketed Free-tailed Bat

SW USA to Guerrero (SW Mexico)

❏ *Nyctinomops laticaudatus* Broad-eared Bat

Dry forests and scrub of Mexico to Honduras; E Panama to Suriname, S to NW Peru, N Argentina, Paraguay and S Brazil; Trinidad; Cuba

❏ *Nyctinomops macrotis* Big Free-tailed Bat

Rocky canyons of SW USA to Venezuela, S to NW Argentina, Paraguay and Uruguay; Cuba, Jamaica and Hispaniola

❏ *Eumops auripendulus* Black Bonneted Bat

Forests and savannas of S Mexico to French Guiana, S to Peru, Bolivia, N Argentina and S Brazil; Trinidad

❏ *Eumops bonariensis* Dwarf Bonneted Bat

Lowland dry deciduous forests and scrub of EC Mexico to NW Venezuela, S to NW Peru, Bolivia, N Argentina, Uruguay and S Brazil

❏ *Eumops dabbenei* Big Bonneted Bat

Dry deciduous thorn forests of Colombia and Venezuela; chaco of N Argentina and Paraguay

❏ *Eumops glaucinus* Wagner's Bonneted Bat

Lowland forests, scrub and built areas of S Florida (USA); WC Mexico to Venezuela, S to N Argentina, Paraguay and SE Brazil; Cuba and Jamaica

❏ *Eumops hansae* Sanborn's Bonneted Bat

Rainforests and forest edge of Chiapas (SE Mexico), NW Honduras and S Costa Rica to French Guiana, S to Peru, Bolivia and Amazonian Brazil

❏ *Eumops maurus* Guyanan Bonneted Bat

Guyana and Suriname

❏ *Eumops patagonicus*
Argentine Dwarf Bonneted Bat

NW Argentina

❏ *Eumops perotis* Greater Bonneted Bat

Rocky canyons of SW & SC USA to C Mexico; Colombia to N Argentina, Paraguay and S Brazil; Cuba

❏ *Eumops underwoodi* Underwood's Bonneted Bat

Dry forests and deserts of extreme S Arizona (USA) to W Nicaragua

❏ *Promops centralis* Big Crested Mastiff Bat

Locally in forests, clearings and built areas of WC Mexico to Suriname, S to Peru, N Argentina and Paraguay; Trinidad

❏ *Promops nasutus* Brown Mastiff Bat

Forests of Venezuela to Suriname, S to Peru, Bolivia, N Argentina and S Brazil; Trinidad

❏ *Molossus ater* Black Mastiff Bat

Forests, scrub and built areas of Mexico to French Guiana, S to Peru, N Argentina, Paraguay and S Brazil; Trinidad

❏ *Molossus barnesi* Barnes' Mastiff Bat

French Guiana and Brazil

❏ *Molossus currentium* Bonda Mastiff Bat

Forests of Cozumel I. (Quintana Roo, Mexico) and Caribbean slope of E Honduras to Panama; Venezuela, Colombia and Ecuador

❏ *Molossus molossus* Pallas' Mastiff Bat

Forests and built areas of Florida Keys (USA) and Mexico to French Guiana, S to Uruguay and S Brazil; Trinidad & Tobago; Greater and Lesser Antilles

❏ *Molossus pretiosus* Miller's Mastiff Bat

Lowland dry and semideciduous forests of S Mexico to Guyana

❏ *Molossus sinaloae* Sinaloan Mastiff Bat

Forests, farmland and built areas of WC Mexico to Suriname; Trinidad

Family: ANTROZOIDAE
(Antrozoid Bats—2)

❏ *Bauerus dubiaquercus* Van Gelder's Bat

Forest interiors of S Mexico incl. Tres Marías Is. (Nayarit), Belize, Honduras and Costa Rica

❏ *Antrozous pallidus* Pallid Bat

Arid and semiarid habitats of extreme SW Canada and W USA to C Mexico; Cuba

Family: VESPERTILIONIDAE
(Vesper Bats—364)

❏ *Barbastella barbastellus* Western Barbastelle

Forest edge, parks and built areas of Europe, Mediterranean islands and NW Africa to Ukraine and Caucasus Mtns; Canary Is.

❏ *Barbastella leucomelas* Eastern Barbastelle

Caucasus Mtns and Middle East to W China, Taiwan and Japan; N Ethiopia

❏ *Euderma maculatum* Spotted Bat

Montane habitats of interior SW Canada and W USA to C Mexico

❏ *Idionycteris phyllotis* Allen's Big-eared Bat

Arizona and W New Mexico (USA) to C Mexico

Plecotus (Corynorhinus)

❏ *Plecotus mexicanus* Mexican Big-eared Bat

Highland pine-oak forests and arid lowlands of Mexico incl. Cozumel I. (Quintana Roo)

❏ *Plecotus rafinesquii* Rafinesque's Big-eared Bat

SC & SE USA

❏ *Plecotus townsendii* Townsend's Big-eared Bat

SW Canada, W & locally in EC USA, to S Mexico

Plecotus (Plecotus)

❏ *Plecotus alpinus* Alpine Long-eared Bat

Alps of C Europe

❏ *Plecotus auritus* Brown Long-eared Bat

Woods and parks of Europe to Nepal, E Siberia incl. Sakhalin I. (Russia), N China and Japan

❏ *Plecotus austriacus* Grey Long-eared Bat

Built and rocky areas of C & S Europe and NW Africa to Mongolia and W China; Senegal; Canary Is.; Cape Verde Is.

❏ *Plecotus balensis* Bale Long-eared Bat

Bale Mtns (Ethiopia)

❏ *Plecotus kolombatovici*
Kolombatovic's Long-eared Bat

S Austria, Bosnia-Herzegovina, Croatia; Greece

❏ *Plecotus sardus* Sardinian Long-eared Bat

Sardinia

❏ *Plecotus taivanus* Taiwan Long-eared Bat

Taiwan

❏ *Plecotus teneriffae* Canary Long-eared Bat

Forests and scrub of Tenerife, La Palma and El Hierro (W Canary Is.)

❏ *Eudiscopus denticulus* Disk-footed Bat

C Myanmar and Laos

Pipistrellus (Arielulus)

❏ *Pipistrellus aureocollaris*
Gold-collared Gilded Pipistrelle

Chiang Mai Prov. (Thailand) and Vietnam

❏ *Pipistrellus circumdatus* Black Gilded Pipistrelle

India and Nepal to Yunnan (China), Malaya and Java

❏ *Pipistrellus cuprosus* Coppery Gilded Pipistrelle

Borneo

❏ *Pipistrellus societatis* Social Gilded Pipistrelle

Malaya

❏ *Pipistrellus torquatus* Taiwan Gilded Pipistrelle

Highlands of C Taiwan

Pipistrellus (Falsistrellus)

❏ *Pipistrellus affinis* Chocolate Pipistrelle

N India, NE Myanmar and Yunnan (SW China)

❏ *Pipistrellus mackenziei* Western False Pipistrelle

SW Western Australia (Australia)

❏ *Pipistrellus mordax* Pungent Pipistrelle

Java

❏ *Pipistrellus petersi* Peters' Pipistrelle

Borneo, Philippines, Sulawesi, Buru and Ambon (S Moluccas)

❏ *Pipistrellus tasmaniensis* **Eastern False Pipistrelle**

SE Australia incl. Tasmania

Pipistrellus (Hypsugo)

❏ *Pipistrellus anchietai* D'Anchieta's Pipistrelle

S D.R. Congo, Angola and Zambia; E South Africa

❏ *Pipistrellus anthonyi* Anthony's Pipistrelle

Changyinku region (Myanmar); Critically Endangered

❏ *Pipistrellus arabicus* Arabian Pipistrelle

Oman

❏ *Pipistrellus ariel* Desert Pipistrelle

Egypt and N Sudan

❏ *Pipistrellus bodenheimeri* Bodenheimer's Pipistrelle

Israel, S Yemen and Oman

❏ *Pipistrellus cadornae* Cadorna's Pipistrelle

NE India, Myanmar and Thailand

❏ *Pipistrellus eisentrauti* Eisentraut's Pipistrelle

Liberia to Somalia and Kenya

❏ *Pipistrellus hesperus* Western Pipistrelle

Arid rocky habitats and grasslands of W & SC USA to SW Mexico

❏ *Pipistrellus imbricatus* Brown Pipistrelle

Java, Kangean Is. (NE of Java), Bali, Borneo and Lesser Sundas

❏ *Pipistrellus inexspectatus* Aellen's Pipistrelle

Benin, Cameroon, D.R. Congo, Uganda and Kenya

❏ *Pipistrellus joffrei* Joffre's Pipistrelle

N Myanmar; Critically Endangered

❏ *Pipistrellus kitcheneri* Red-brown Pipistrelle

Borneo

❏ *Pipistrellus lophurus* Burma Pipistrelle

Peninsular Myanmar

❏ *Pipistrellus musculus* Mouse-like Pipistrelle

Cameroon, Gabon and D.R. Congo

❏ *Pipistrellus pulveratus* Chinese Pipistrelle

Thailand; C & S China incl. Hong Kong

❏ *Pipistrellus savii* Savi's Pipistrelle

Rocky and built areas of S Europe and NW Africa to Korea, NE China and Japan, S to NE India and Myanmar; Canary Is.; Cape Verde Is.

❏ *Pipistrellus stenopterus* Narrow-winged Pipistrelle

Malaya, Sumatra, Riau Is., N Borneo and Mindanao (SE Philippines)

Pipistrellus (Neoromicia)

❏ *Pipistrellus brunneus* Dark-brown Serotine

Liberia to D.R. Congo

❏ *Pipistrellus capensis* Cape Serotine

Guinea to Ethiopia, S to South Africa

❏ *Pipistrellus flavescens* Yellow Serotine

Angola; Burundi

❏ *Pipistrellus guineensis* Tiny Serotine

Senegal and Guinea to Ethiopia and NE D.R. Congo

❏ *Pipistrellus matroka* Madagascar Serotine

Madagascar

❏ *Pipistrellus melckorum* Melck's House Bat

Tanzania, Zambia, Mozambique and Cape Prov. (South Africa)

❏ *Pipistrellus rendalli* Rendall's Serotine

Gambia to Somalia, S to Botswana, Malawi and Mozambique

❏ *Pipistrellus somalicus* Somali Serotine

Guinea-Bissau to Somalia, S to NE South Africa; SW Madagascar

❏ *Pipistrellus tenuipinnis* White-winged Serotine

Senegal to Kenya, S to Angola and D.R. Congo

Pipistrellus (Perimyotis)

❏ *Pipistrellus subflavus* Eastern Pipistrelle

Forest clearings and forest edge of SE Canada and C & E USA to E Honduras

Pipistrellus (Pipistrellus)

❏ *Pipistrellus adamsi* Cape York Pipistrelle

Coastal E Cape York Peninsula (NE Queensland, Australia)

❏ *Pipistrellus aegyptius* Egyptian Pipistrelle

Algeria, Libya, Egypt and N Sudan; Burkina Faso

❏ *Pipistrellus aero* Mount Gargues Pipistrelle

NW Kenya

❏ *Pipistrellus angulatus* New Guinea Pipistrelle

Aru Is. (SE Moluccas), New Guinea, D'Entrecasteaux Is., Bismarck Archipelago and Solomon Is.

❏ *Pipistrellus babu* Himalayan Pipistrelle

Afghanistan and Pakistan to Myanmar and SW China

❏ *Pipistrellus ceylonicus* Kelaart's Pipistrelle

Pakistan, India and Sri Lanka to S China, Hainan, Vietnam and Borneo

❏ *Pipistrellus collinus* Mountain Pipistrelle

Montane forests of Central Cordillera (New Guinea)

❏ *Pipistrellus coromandra* Indian Pipistrelle

Afghanistan, Pakistan, India incl. Nicobar Is. and Sri Lanka to S China, Thailand and Vietnam

❏ *Pipistrellus crassulus* Broad-headed Pipistrelle

S Sudan; Cameroon, D.R. Congo and N Angola

❏ *Pipistrellus endoi* Endo's Pipistrelle

Honshu (C Japan); Endangered

❏ *Pipistrellus javanicus* Javan Pipistrelle

E Siberia, Korea, China and Japan, S to Indochina, Philippines, Talaud Is. and Lesser Sundas

❏ *Pipistrellus kuhlii* Kuhl's Pipistrelle

Built areas and forest edge of WC & S Europe to Middle East, Caucasus Mtns, Kazakhstan and Pakistan; subsaharan Africa; Canary Is.

❏ *Pipistrellus macrotis* Big-eared Pipistrelle

Malaya, Sumatra, Borneo, Bali, and near islands

❏ *Pipistrellus maderensis* Madeira Pipistrelle

Forests and built areas of Madeira; Tenerife, La Gomera, La Palma and El Hierro (W Canary Is.)

❏ *Pipistrellus mimus* Indian Pygmy Pipistrelle

Afghanistan, Pakistan, India and Sri Lanka to Thailand and Vietnam

❏ *Pipistrellus minahassae* Minahassa Pipistrelle

Sulawesi

❏ *Pipistrellus nanulus* Tiny Pipistrelle

Sierra Leone to Kenya; Bioko (Equatorial Guinea)

❏ *Pipistrellus nanus* Banana Pipistrelle

Subsaharan Africa incl. Pemba and Zanzibar Is. (Tanzania) S to E South Africa; Madagascar

❏ *Pipistrellus nathusii* Nathusius' Pipistrelle

Forest edge, parks and riparian habitats of C Europe to Ural Mtns and Caucasus Mtns

❏ *Pipistrellus papuanus* Papuan Pipistrelle

S & SE Moluccas, Misoöl, Salawati, Biak, New Guinea, D'Entrecasteaux Is. and New Ireland (Bismarck Arch.)

❏ *Pipistrellus paterculus* Mount Popa Pipistrelle

N India, Myanmar, Thailand and SW China

❏ *Pipistrellus peguensis* Pegu Pipistrelle

Myanmar

❏ *Pipistrellus permixtus* Dar es Salaam Pipistrelle

Tanzania

❏ *Pipistrellus pipistrellus* Common Pipistrelle

Woods, farmland and built areas of Europe and NW Africa to Middle East, C Siberia, Kazakhstan and NW China

❏ *Pipistrellus pygmaeus* Soprano Pipistrelle

Europe

❏ *Pipistrellus rueppelli* Rüppell's Pipistrelle

Algeria, Egypt and subsaharan Africa incl. Zanzibar I. (Tanzania) S to NE South Africa; Iraq

❏ *Pipistrellus rusticus* Rusty Pipistrelle

Liberia and Ethiopia, S to NE South Africa

❏ †*Pipistrellus sturdeei* Sturdee's Pipistrelle

Formerly Bonin Is. (Japan); Extinct (not recorded since its original description in 1915)

❏ *Pipistrellus tenuis* **Least Pipistrelle**

Thailand to Borneo, Philippines and Sulawesi; Cocos Keeling and Christmas Is.

❏ *Pipistrellus vordermanni* **White-winged Pipistrelle**

Belitung I. (between Sumatra and Borneo) and W Borneo

❏ *Pipistrellus wattsi* **Watts' Pipistrelle**

Lowlands of SE New Guinea and Samarai I.

❏ *Pipistrellus westralis* **Mangrove Pipistrelle**

Mangrove forests and swamps of coastal NC Australia

Pipistrellus (Scotozous)

❏ *Pipistrellus dormeri* **Dormer's Pipistrelle**

Pakistan and India

Pipistrellus (Vespadelus)

❏ *Pipistrellus baverstocki* **Inland Forest Bat**

Arid and semiarid areas of interior Australia

❏ *Pipistrellus caurinus* **Northern Cave Bat**

NE Western Australia and N Northern Territory (Australia)

❏ *Pipistrellus darlingtoni* **Large Forest Bat**

Forests and woods of SE Australia incl. Kangaroo I. and Tasmania

❏ *Pipistrellus douglasorum* **Yellow-lipped Bat**

Tropical woodlands in W Kimberley region (NE Western Australia)

❏ *Pipistrellus finlaysoni* **Inland Cave Bat**

Arid regions of W & C Australia

❏ *Pipistrellus pumilus* **Eastern Forest Bat**

Wet forests in scattered localities of E Australia incl. Lord Howe I.

❏ *Pipistrellus regulus* **Southern Forest Bat**

Woodland and scrub in coastal and subcoastal S Australia incl. Tasmania

❏ *Pipistrellus troughtoni* **Eastern Cave Bat**

Forests of coastal E Queensland and NE New South Wales (NE Australia)

❏ *Pipistrellus vulturnus* **Little Forest Bat**

Forests of SE Australia and Tasmania

❏ *Nyctalus aviator* **Bird-like Noctule**

Korea, E China and Japan

❏ *Nyctalus azoreum* **Azores Noctule**

Built areas of C & E Azores

❏ *Nyctalus lasiopterus* **Greater Noctule**

Locally in deciduous forests and parks of C & S Europe and NW Africa to Ural Mtns and Caucasus Mtns, S to Middle East and Iran

❏ *Nyctalus leisleri* **Lesser Noctule (Leisler's Bat)**

Forest clearings and built areas of C & S Europe and NW Africa to Ural Mtns, Caucasus Mtns and E Afghanistan; Madeira; Azores

❏ *Nyctalus montanus* **Mountain Noctule**

Afghanistan, Pakistan, Nepal and N India

❏ *Nyctalus noctula* **Common Noctule**

Forest edge, wetlands and built areas of Europe and NW Africa to SW Siberia, China, Taiwan, Japan and Malaya

❏ *Ia io* **Great Evening Bat**

NE India, Thailand, S China, Laos and Vietnam

❏ *Glischropus javanus* **Javan Thick-thumbed Bat**

Java; Endangered

❏ *Glischropus tylopus* **Common Thick-thumbed Bat**

Myanmar, Thailand, Malaya, Sumatra, Borneo, SW Philippines and N Moluccas

Eptesicus (Eptesicus)

❏ *Eptesicus bobrinskoi* **Bobrinski's Serotine**

N Caucasus Mtns, Turkmenistan, Uzbekistan and Kazakhstan

❏ *Eptesicus bottae* **Botta's Serotine**

Rocky and built areas of Turkey, Egypt and Yemen, E to Pakistan and Mongolia

❏ *Eptesicus brasiliensis* **Brazilian Brown Bat**

Forests and forest edge of E & S Mexico; Costa Rica to E Venezuela, S to N Argentina and Uruguay; Trinidad & Tobago

❏ *Eptesicus demissus* **Surat Serotine**

Peninsular Thailand

❏ *Eptesicus diminutus* **Diminutive Serotine**

Forests of Venezuela; SE Paraguay, N Argentina, Uruguay and E Brazil

❑ *Eptesicus furinalis* Argentine Brown Bat

Forests and forest clearings of Mexico to French Guiana, S to N Argentina, Paraguay and S Brazil

❑ *Eptesicus fuscus* Big Brown Bat

Forest edge of S Canada and USA to montane rain and cloud forests of Venezuela and N Brazil; Greater Antilles; Dominica, Barbados (Lesser Antilles); Bahamas

❑ *Eptesicus guadeloupensis*
Guadeloupe Big Brown Bat

Guadeloupe (Lesser Antilles); Endangered

❑ *Eptesicus hottentotus* Long-tailed House Bat

Locally in Angola and Kenya, S to South Africa

❑ *Eptesicus innoxius* Harmless Serotine

W Ecuador and NW Peru

❑ *Eptesicus kobayashii* Kobayashi's Serotine

Korea

❑ *Eptesicus nasutus* Sind Bat

Iraq, Iran, Arabian Peninsula, Afghanistan and Pakistan

❑ *Eptesicus nilssoni* Northern Bat

Forest edge, wetlands and built areas of C & N Europe to E Siberia incl Sakhalin I. (Russia), S to Iraq, N Iran, Nepal, W China and Japan

❑ *Eptesicus pachyotis* Thick-eared Bat

Assam (NE India), N Myanmar and N Thailand

❑ *Eptesicus platyops* Lagos Serotine

Senegal; Nigeria; Bioko (Equatorial Guinea)

❑ *Eptesicus serotinus* Common Serotine

Forest edge, parks and built areas of C & S Europe incl. Mediterranean islands and N Africa to Russia, China, Korea, Taiwan and Thailand

❑ *Eptesicus tatei* Sombre Bat

NE India

Eptesicus (Rhinopterus)

❑ *Eptesicus floweri* Horn-skinned Bat

Mali; Sudan

❑ *Vespertilio murinus* Particoloured Bat

Built areas and woods of C & E Europe to E Siberia and China, S to Afghanistan

❑ *Vespertilio sinensis* Asian Particoloured Bat

E SIberia, Korea, China, Taiwan and Japan

❑ *Laephotis angolensis* Angolan Long-eared Bat

Locally in savannas of D.R. Congo and Angola

❑ *Laephotis botswanae* Botswanan Long-eared Bat

Riparian woods of D.R. Congo, Zambia, Malawi, Botswana, Zimbabwe and NE South Africa

❑ *Laephotis namibensis* Namib Long-eared Bat

Riparian woods of Kuiseb R. region (WC Namibia); Endangered

❑ *Laephotis wintoni* De Winton's Long-eared Bat

Locally in savannas of Ethiopia and Kenya; SW Cape Prov. (South Africa)

❑ *Histiotus alienus* Strange Big-eared Brown Bat

SE Brazil and Uruguay

❑ *Histiotus humboldti*
Humboldt's Big-eared Brown Bat

Highlands of Colombia and Venezuela

❑ *Histiotus macrotus* Andean Big-eared Brown Bat

Andes of S Peru, S Bolivia, Chile and W Argentina

❑ *Histiotus montanus* Small Big-eared Brown Bat

Montane forests, clearings and savannas of Colombia and Venezuela, S in Andes to Bolivia, S Chile, S Argentina and Uruguay

❑ *Histiotus velatus* Tropical Big-eared Brown Bat

SE Paraguay, Misiones Prov. (NE Argentina), S Uruguay and E Brazil

❑ *Philetor brachypterus* Rohu's Bat

Nepal; Malaya, Sumatra, Java, Borneo, Philippines, Sulawesi, New Guinea and New Britain (Bismarck Archipelago)

❑ *Tylonycteris pachypus* Lesser Bamboo Bat

India incl. Andaman Is. to S China, Philippines, Lesser Sundas and Ambon (S Moluccas)

❑ *Tylonycteris robustula* Greater Bamboo Bat

S China to Philippines, Sulawesi and Lesser Sundas

❑ *Mimetillus moloneyi* Moloney's Flat-headed Bat

Forests and forest-savanna mosaics of Sierra Leone to
Ethiopia, S to Angola and Zambia; Bioko (Equatorial Guinea)

Hesperoptenus (Hesperoptenus)

❑ *Hesperoptenus doriae* Peters' False Serotine

Malay Peninsula and Borneo; Endangered

Hesperoptenus (Milithronycteris)

❑ *Hesperoptenus blanfordi* Blanford's False Serotine

Myanmar, Thailand, Malaya and Borneo

❑ *Hesperoptenus gaskelli* Gaskell's False Serotine

Sulawesi

❑ *Hesperoptenus tickelli* Tickell's False Serotine

Nepal, India incl. Andaman Is. and Sri Lanka to Myanmar and
Thailand

❑ *Hesperoptenus tomesi* Large False Serotine

Malay Peninsula and Borneo

Chalinolobus (Chalinolobus)

❑ *Chalinolobus dwyeri* Large-eared Pied Bat

Eucalyptus forests of SE Queensland and E New South Wales
(Australia)

❑ *Chalinolobus gouldii* Gould's Wattled Bat

Australia incl. Tasmania; Norfolk I. (extinct?)

❑ *Chalinolobus morio* Chocolate Wattled Bat

S Australia incl. Tasmania

❑ *Chalinolobus neocaledonicus*
New Caledonia Wattled Bat

New Caledonia; Endangered

❑ *Chalinolobus nigrogriseus* Hoary Wattled Bat

SE New Guinea and near islands, Fergusson (D'Entrecasteaux
Is.) and N & NE Australia

❑ *Chalinolobus picatus* Little Pied Bat

Dry forests of interior S Queensland, NE South Australia and
NW New South Wales (Australia)

❑ *Chalinolobus tuberculatus*
New Zealand Wattled Bat

New Zealand and near islands

Chalinolobus (Glauconycteris)

❑ *Chalinolobus alboguttatus* Allen's Striped Bat

Cameroon and D.R. Congo

❑ *Chalinolobus argentatus* Silvered Bat

Cameroon to Kenya, S to Angola and Tanzania

❑ *Chalinolobus beatrix* Beatrix's Bat

Ivory Coast, Rep. Congo and Kenya

❑ *Chalinolobus curryi* Curry's Bat

W Africa

❑ *Chalinolobus egeria* Bibundi Bat

Cameroon and Uganda

❑ *Chalinolobus gleni* Glen's Wattled Bat

Cameroon and Uganda

❑ *Chalinolobus kenyacola* Kenyan Wattled Bat

Kenya

❑ *Chalinolobus poensis* Abo Bat

Senegal to Uganda; Bioko (Equatorial Guinea)

❑ *Chalinolobus superbus* African Pied Bat

Ivory Coast, Ghana and NE D.R. Congo

❑ *Chalinolobus variegatus* Butterfly Bat

Savannas of subsaharan Africa S to E South Africa

Nycticeius (Nycticeinops)

❑ *Nycticeius schlieffeni* Schlieffen's Bat

Dry savannas of Mauritania and Ghana to Sudan, S to N
Namibia, N Botswana and E South Arica; Egypt and Arabian
Peninsula

Nycticeius (Nycticeius)

❑ *Nycticeius humeralis* Evening Bat

Built areas of EC Mexico, and summer migrant N to C & E
USA; Cuba

Nycticeius (Scoteanax)

❏ *Nycticeius rueppellii* Rüppell's Broad-nosed Bat

Coastal E Queensland and E New South Wales (Australia)

Nycticeius (Scotorepens)

❏ *Nycticeius balstoni* Inland Broad-nosed Bat

Arid and semiarid interior Australia

❏ *Nycticeius greyii* Little Broad-nosed Bat

N & E Australia

❏ *Nycticeius orion* Eastern Broad-nosed Bat

Sclerophyll forests of coastal plains of SE Australia

❏ *Nycticeius sanborni* Northern Broad-nosed Bat

SE New Guinea, and coastal NW & NE Australia

Rhogeesa (Baeodon)

❏ *Rhogeessa alleni* Allen's Yellow Bat

Zacatecas to Oaxaca (C & S Mexico); Endangered

Rhogeessa (Rhogeessa)

❏ *Rhogeessa genowaysi* Genoways' Yellow Bat

Pacific slope lowlands of S Chiapas (S Mexico)

❏ *Rhogeessa gracilis* Slender Yellow Bat

Jalisco and Zacatecas to Oaxaca (C & S Mexico)

❏ *Rhogeessa hussoni* Husson's Yellow Bat

Suriname

❏ *Rhogeessa minutilla* Tiny Yellow Bat

Forests of Cordillera Oriental (Colombia) and N Venezuela; Margarita; Trinidad

❏ *Rhogeessa mira* Least Yellow Bat

S Michoacán (SW Mexico); Endangered

❏ *Rhogeessa parvula* Little Yellow Bat

Mexico incl. Tres Marías Is. (Nayarit)

❏ *Rhogeessa tumida*
Central American Yellow Bat

Forests, clearings and built areas of NE Mexico to Venezuela, S to Bolivia and E Brazil; Trinidad & Tobago

❏ *Scotoecus albofuscus*
Light-winged Lesser House Bat

Woods and dry savannas of Gambia to Kenya, S to Mozambique and extreme E South Africa

❏ *Scotoecus hirundo*
Dark-winged Lesser House Bat

Woods and dry savannas of Senegal to Ethiopia, S to Angola, Zambia and Malawi

❏ *Scotoecus pallidus* Desert Yellow Bat

Pakistan and N India

❏ *Scotomanes emarginatus*
Emarginate Harlequin Bat

India

❏ *Scotomanes ornatus* Harlequin Bat

NE India, Myanmar, Thailand, S China and Vietnam

❏ *Scotophilus borbonicus* Lesser Yellow Bat

Coastal forests of Madagascar; Réunion (probably extinct); Critically Endangered

❏ *Scotophilus celebensis* Sulawesi Yellow Bat

Sulawesi

❏ *Scotophilus collinus* Small Asiatic Yellow Bat

Java and Lesser Sundas

❏ *Scotophilus dinganii* African Yellow Bat

Savannas and built areas of Senegal and Sierra Leone to Somalia, S to E South Africa; Yemen

❏ *Scotophilus heathi* Greater Asiatic Yellow Bat

Afghanistan to S China incl. Hainan, S to Sri Lanka and Vietnam

❏ *Scotophilus kuhlii* Lesser Asiatic Yellow Bat

Pakistan to Taiwan, S to Sri Lanka, Malaya, Philippines and Aru Is. (SE Moluccas)

❏ *Scotophilus leucogaster*
White-bellied Yellow Bat

Mauritania and Senegal to Ethiopia and N Kenya

❏ *Scotophilus nigrita* Schreber's Yellow Bat

Savannas of Senegal to Sudan and Kenya, S to Zimbabwe and Mozambique

❑ *Scotophilus nux* Nut-coloured Yellow Bat

Sierra Leone to Kenya

❑ *Scotophilus robustus* Robust Yellow Bat

Coastal forests of Madagascar

❑ *Scotophilus viridis* Greenish Yellow Bat

Savannas and riparian habitats of subsaharan Africa

❑ *Otonycteris hemprichii* Desert Long-eared Bat

Rocky areas and buildings in deserts and semi-desert grasslands of Sahara Desert; Arabian Peninsula to Afghanistan, Tajikistan and Kashmir

Lasiurus (Dasypterus)

❑ *Lasiurus ega* Southern Yellow Bat

Lowland scrub and foothill forests of extreme S Texas (USA) to Guyana, S to C Argentina, Uruguay and S Brazil; Trinidad

❑ *Lasiurus intermedius* Northern Yellow Bat

Dry forests and scrub of SC & SE USA to Honduras; Cuba

❑ *Lasiurus xanthinus* Western Yellow Bat

SW USA and W Mexico

Lasiurus (Lasiurus)

❑ *Lasiurus atratus* Mourning Bat

Guyana, Suriname and French Guiana

❑ *Lasiurus blossevillii* Western Red Bat

Forests of extreme SW Canada and W USA to Suriname, S to S Chile and S Argentina; Galápagos Is.; Greater Antilles; Bahamas; Bermuda

❑ *Lasiurus borealis* Eastern Red Bat

SC & SE Canada, C & E USA and NE Mexico

❑ *Lasiurus castaneus* Tacarcuna Bat

Very few records from rainforests of Costa Rica and Panama

❑ *Lasiurus cinereus* Hoary Bat

Forests of Canada to Guatemala; Colombia to S Chile, C Argentina and Uruguay; Galápagos Is.; Bermuda; vagrant Greater Antilles and NW Europe

❑ *Lasiurus ebenus* Ebony Bat

Parque Estadual da Ilha do Cardoso (São Paulo, SE Brazil)

❑ *Lasiurus egregius* Big Red Bat

Very few records from forests of E Panama, French Guiana and S Brazil

❑ *Lasiurus seminolus* Seminole Bat

SC & E USA; Bermuda

❑ *Nyctophilus arnhemensis* Arnhem Nyctophilus

Coastal and subcoastal NW & NC Australia

❑ *Nyctophilus bifax* Northern Nyctophilus

Forests and woods of W New Guinea and coastal N & EC Australia

❑ *Nyctophilus geoffroyi* Lesser Nyctophilus

Australia (except NE Queensland) incl. Tasmania

❑ *Nyctophilus gouldi* Gould's Nyctophilus

New Guinea and extreme SW & E Australia

❑ *Nyctophilus heran* Sunda Nyctophilus

Lomblen (Lesser Sundas); Endangered

❑ *Nyctophilus microdon*
Small-toothed Nyctophilus

Montane forests of EC New Guinea

❑ *Nyctophilus microtis* Small-eared Nyctophilus

Forests of Salawati and New Guinea

❑ *Nyctophilus nebulosus*
New Caledonia Nyctophilus

Rainforests of Noumea region (New Caledonia)

❑ *Nyctophilus timoriensis* Greater Nyctophilus

Timor (Lesser Sundas), E New Guinea; S & E Australia incl. Tasmania

❑ *Nyctophilus walkeri* Pygmy Nyctophilus

Coastal NC Australia

❑ *Pharotis imogene* Large-eared Nyctophilus

Kamali region (SE New Guinea); unrecorded since type series collected in 1890, Critically Endangered

❑ *Miniopterus australis* Little Bent-winged Bat

Forests of Java, Borneo and Philippines to Vanuatu and E Australia

❏ *Miniopterus fraterculus*
Lesser Long-fingered Bat

Angola, Zambia, Malawi, Mozambique and South Africa; E & S Madagascar

❏ *Miniopterus fuscus* **Ryukyu Bent-winged Bat**

Ryukyu Is. (Japan)

❏ *Miniopterus gleni* **Glen's Long-fingered Bat**

N, E & S Madagascar

❏ *Miniopterus inflatus* **Greater Long-fingered Bat**

Liberia; Cameroon to Ethiopia and Somalia, S to Gabon, Zimbabwe and Mozambique

❏ *Miniopterus macrocneme*
Small Melanesian Bent-winged Bat

Sulawesi, Moluccas, New Guinea, D'Entrecasteaux Is., Bismarck Archipelago, Solomon Is., Vanuatu and New Caledonia

❏ *Miniopterus magnater* **Western Bent-winged Bat**

NE India and SE China to Timor (Lesser Sundas) and lowlands of New Guinea

❏ *Miniopterus majori* **Major's Long-fingered Bat**

E Madagascar

❏ *Miniopterus manavi*
Malagasy Least Long-fingered Bat

Madagascar

❏ *Miniopterus medius* **Medium Bent-winged Bat**

Thailand, SE China, Malaya, Java, Borneo, Philippines, Sulawesi and New Guinea

❏ *Miniopterus minor* **Least Long-fingered Bat**

Rep. Congo, D.R. Congo, Kenya and Tanzania; São Tomé (Equatorial Guinea)

❏ *Miniopterus propritristis*
Large Melanesian Bent-winged Bat

Biak, Japen, C & E New Guinea, D'Entrecasteaux Is., Bismarck Archipelago, Solomon Is. and Vanuatu

❏ *Miniopterus pusillus* **Small Bent-winged Bat**

India to Philippines, Moluccas, New Guinea, Solomon Is. and New Caledonia

❏ *Miniopterus robustior* **Loyalty Bent-winged Bat**

Loyalty Is. (E of New Caledonia); Endangered

❏ *Miniopterus schreibersi*
Common Bent-winged Bat (Schreibers' Bat)

Open rocky areas of S Europe, NW and subsaharan Africa; S Asia E to Japan, New Guinea, Solomon Is. and Australia

❏ *Miniopterus tristis* **Great Bent-winged Bat**

Sulawesi, Moluccas and Philippines

Myotis (Cistugo)

❏ *Myotis lesueuri* **Lesueur's Hairy Bat**

SW Cape Prov. (South Africa)

❏ *Myotis seabrai* **Angolan Hairy Bat**

SW Angola, W Namibia and NW Cape Prov. (South Africa)

Myotis (Leuconoe)

❏ *Myotis abei* **Sakhalin Myotis**

Shirutoru region (Sakhalin I., E Siberia, Russia)

❏ *Myotis adversus* **Large-footed Myotis**

Malaya and Taiwan to New Guinea, Bismarck Archipelago, Solomon Is. and NC & E Australia

❏ *Myotis aelleni* **Southern Myotis**

SW Argentina

❏ *Myotis albescens* **Silver-haired Myotis**

Lowland and mid-elevation forests and forest edge of EC Mexico to Suriname, S to Peru, N Argentina and Uruguay

❏ *Myotis annamiticus* **Annamite Myotis**

Annamite Mtns (Vietnam)

❏ *Myotis austroriparius* **Southeastern Myotis**

Wetland areas of SC & SE USA

❏ *Myotis bocagei* **Rufous Mouse-eared Bat**

Subsaharan Africa; Yemen

❏ *Myotis capaccinii* **Long-fingered Myotis**

Wetlands of S Europe incl. Mediterranean islands, Turkey, Israel, Iraq, Iran and Uzbekistan

❏ *Myotis chiloensis* **Chilean Myotis**

Forests and savannas, usually near fresh water, of C & S Chile and WC & S Argentina

❏ *Myotis cobanensis* **Guatemalan Myotis**

Cobán region (C Guatemala); only known from holotype specimen, Critically Endangered

❏ *Myotis csorbai* **Csorba's Myotis**

Nepal

❏ *Myotis dasycneme* **Pond Bat**

Wetlands of C Europe to E Siberia, S to Ukraine, NW Kazakhstan and NE China

❏ *Myotis daubentoni* **Daubenton's Bat**

Forests and wetlands of Europe to E Siberia incl. Sakhalin and Kuril Is. (Russia), S to NE India, S China and Japan

❏ *Myotis fortidens* **Cinnamon Myotis**

Lowland dry and semideciduous forests and forests edge of Mexico and Guatemala

❏ *Myotis grisescens* **Grey Myotis**

Interior SE USA; Endangered

❏ *Myotis hasseltii* **Hasselt's Myotis**

Sri Lanka, Myanmar, Indochina, Malaya, Sumatra, Java, Borneo and near islands

❏ *Myotis horsfieldii* **Horsfield's Myotis**

India incl. Andaman Is. to SE China, Malaya, Java, Bali, Borneo, Sulawesi and Philippines

❏ *Myotis levis* **Yellowish Myotis**

Open areas and forest clearings of SC Bolivia, N & C Argentina, Uruguay and SE Brazil

❏ *Myotis longipes* **Kashmir Cave Myotis**

Afghanistan and Kashmir (NW India)

❏ *Myotis lucifugus* **Little Brown Myotis**

S Alaska, Canada and USA to C Mexico

❏ *Myotis macrodactylus* **Big-footed Myotis**

C Asia and S China to Japan and Kuril Is.

❏ *Myotis macrotarsus* **Pallid Large-footed Myotis**

N Borneo and Philippines

❏ *Myotis montivagus* **Large Brown Myotis**

India, Myanmar, S China, Malaya and Borneo

❏ *Myotis oxyotus* **Montane Myotis**

Montane rainforests and forest edge of Costa Rica to SE Venezuela, and S in Andes to Peru (and Bolivia?)

❏ *Myotis pruinosus* **Frosted Myotis**

Honshu and Shikoku (C & S Japan); Endangered

❏ *Myotis ricketti* **Rickett's Big-footed Myotis**

S & SE China incl. Hong Kong

❏ *Myotis riparius* **Riparian Myotis**

Lowland forests and forest clearings of E Honduras to E Venezuela, S to Paraguay, NE Argentina, Uruguay and SE Brazil; Trinidad

❏ *Myotis ruber* **Red Myotis**

Paraguay, NE Argentina and SE Brazil

❏ *Myotis simus* **Velvety Myotis**

Lowland rainforests of extreme SE Colombia and Amazonian Brazil, S to Peru, Bolivia, Paraguay and NE Argentina

❏ *Myotis stalkeri* **Kai Myotis**

Gebe and Kai Is. (E & SE Moluccas); Endangered

❏ *Myotis velifer* **Cave Myotis**

Riparian habitat near desert scrub of SW & SC USA and in mid- to high-elevation forests of Mexico to Honduras

❏ *Myotis vivesi* **Fish-eating Myotis**

Coasts and islands of Baja California and Sonora (NW Mexico)

❏ *Myotis volans* **Long-legged Myotis**

Forests, woods and deserts of SE Alaska, SW Canada and W USA to C Mexico

❏ *Myotis yumanensis* **Yuma Myotis**

Wetlands in valleys and deserts of British Columbia (SW Canada) and W USA to C Mexico

Myotis (Myotis)

❏ *Myotis annectans* **Hairy-faced Bat**

NE India to Thailand

❏ *Myotis auriculus* **Southwestern Myotis**

Dry forests and desert scrub of SE Arizona and SW New Mexico (USA) to C Mexico; ?vagrant to S Guatemala

❏ *Myotis bechsteini* **Bechstein's Bat**

Deciduous forests and parks of C & S Europe to Caucasus Mtns and Iran

❏ *Myotis blythii* **Lesser Mouse-eared Bat**

Scrub and built areas of S Europe and NW Africa to Himalayas

❏ *Myotis bombinus* **Far Eastern Myotis**

SE Siberia, NE China, Korea and Japan

❏ *Myotis chinensis* **Large Myotis**

N Thailand to C & SE China incl. Hong Kong

❏ *Myotis emarginatus* **Geoffroy's Bat**

Parks and built areas of C & S Europe and NW Africa to Uzbekistan, E Iran and Afghanistan

❏ *Myotis evotis* **Long-eared Myotis**

SW Canada, W USA and locally in E Baja California Norte (NW Mexico)

❏ *Myotis formosus* **Hodgson's Myotis**

Afghanistan to Korea, China, Taiwan, Philippines and Sulawesi, S to Sumatra, Java and Bali

❏ *Myotis goudoti* **Malagasy Mouse-eared Bat**

Coastal Madagascar and Anjouan I. (Comoro Is.)

❏ *Myotis keenii* **Keen's Myotis**

Dense coastal forests of SE Alaska, W British Columbia (Canada) and W Washington (NW USA)

❏ *Myotis milleri* **Miller's Myotis**

N Baja California (Mexico); Endangered

❏ *Myotis morrisi* **Morris' Bat**

Nigeria; Ethiopia

❏ *Myotis myotis* **Greater Mouse-eared Bat**

Forest edges, farmland and built areas of C & S Europe incl. Mediterranean islands to Ukraine, Israel and Lebanon; Azores

❏ *Myotis nattereri* **Natterer's Bat**

Forest edge, parks and roadsides of Europe and NW Africa to Caucasus Mtns and Turkmenistan

❏ *Myotis peninsularis* **Peninsular Myotis**

S Baja California (Mexico)

❏ *Myotis pequinius* **Peking Myotis**

NC & E China

❏ *Myotis schaubi* **Schaub's Myotis**

Caucasus Mtns and W Iran; Endangered

❏ *Myotis septentrionalis* **Northern Myotis**

Forests of C & E Canada and NE & EC USA

❏ *Myotis sicarius* **Mandelli's Mouse-eared Bat**

Sikkim (NE India)

❏ *Myotis thysanodes* **Fringed Myotis**

Dry forests, grassland and desert scrub of SC British Columbia (Canada) and W USA to S Mexico

❏ *Myotis tricolor* **Cape Hairy Bat**

D.R. Congo and Ethiopia, S to South Africa

❏ *Myotis welwitschii* **Welwitsch's Bat**

Ethiopia, S to E South Africa

Myotis (Selysius)

❏ *Myotis alcathoe* **Alcathoe's Myotis**

Mountains of Hungary; Greece

❏ *Myotis altarium* **Sichuan Myotis**

Thailand to C China

❏ *Myotis atacamensis* **Atacama Myotis**

Coastal deserts and mountains of S Peru and N Chile

❏ *Myotis ater* **Small Black Myotis**

Borneo, Sulawesi and Moluccas

❏ *Myotis aurascens* **Golden Myotis**

SE Europe and W Asia

❏ *Myotis australis* **Australian Myotis**

New South Wales (Australia)

❏ *Myotis brandti* **Brandt's Bat**

Woods and wetlands of WC & E Europe to E Siberia incl. Sakhalin and Kuril Is. (Russia), Mongolia, Korea and Japan

❏ *Myotis californicus* **California Myotis**

Desert scrub and dry forests of SE Alaska and W USA to Guatemala

❏ *Myotis ciliolabrum* Western Small-footed Myotis

Rocky habitats of SW Canada, W USA and N Mexico

❏ *Myotis dominicensis* Dominican Myotis

N Lesser Antilles

❏ *Myotis elegans* Elegant Myotis

Lowland forests and clearings of Mexico to Costa Rica

❏ *Myotis findleyi* Findley's Myotis

Tres Marías Is. (Nayarit, Mexico); Endangered

❏ *Myotis frater* Fraternal Myotis

Uzbekistan and Afghanistan to Korea, SE China and Japan

❏ *Myotis gomantongensis* Gomantong Myotis

N Borneo

❏ *Myotis hajastanicus* Armenian Myotis

Sevan Lake basin (Armenia)

❏ *Myotis hosonoi* Hosono's Myotis

Honshu (C Japan)

❏ *Myotis ikonnikovi* Ikonnikov's Myotis

C & E Siberia incl. Sakhalin I. (Russia), Mongolia, NE China, N Korea and Hokkaido (N Japan)

❏ *Myotis insularum* Insular Myotis

Samoa

❏ *Myotis keaysi* Hairy-legged Myotis

Forests, regrowth areas and forest edge of NE Mexico to N Venezuela, S in foothills of Andes to N Argentina; Trinidad

❏ *Myotis leibii* Eastern Small-footed Myotis

Forests of SE Canada and NE USA

❏ *Myotis martiniquensis* Schwartz's Myotis

Martinique and Barbados (Lesser Antilles)

❏ *Myotis muricola* Wall-roosting Myotis

Afghanistan to Taiwan; Sumatra, Java, Philippines, Sulawesi, Lesser Sundas and Ambon (S Moluccas)

❏ *Myotis mystacinus* European Whiskered Bat

Parks, gardens and riparian habitats of Europe and NW Africa

❏ *Myotis nesopolus* Curaçao Myotis

Arid habitats of NE Colombia and N Venezuela; Curaçao and Bonaire (Netherlands Antilles)

❏ *Myotis nigricans* Black Myotis

Forests and built areas of Mexico to French Guiana, S to Peru, N Argentina, Paraguay and S Brazil; Trinidad & Tobago; Grenada (Lesser Antilles)

❏ *Myotis nipalensis* Asian Whiskered Bat

Caucasus Mtns to E Siberia and China

❏ *Myotis oreias* Singapore Whiskered Bat

Singapore

❏ *Myotis ozensis* Honshu Myotis

Honshu (C Japan); Endangered

❏ *Myotis planiceps* Flat-headed Myotis

Coahuila, Nuevo Leon and Zacatecas (NC Mexico); Critically Endangered

❏ *Myotis ridleyi* Ridley's Myotis

Malaya, Sumatra and Borneo

❏ *Myotis rosseti* Thick-thumbed Myotis

Thailand and Cambodia

❏ *Myotis scotti* Scott's Mouse-eared Bat

Ethiopia

❏ *Myotis siligorensis* Small-toothed Myotis

N India to S China, S to Vietnam, Malaya and Borneo

❏ *Myotis sodalis* Indiana Myotis

Interior C & NE USA; Endangered

❏ *Myotis yanbarensis* Okinawa Myotis

Okinawa (Ryukyu Is., Japan)

❏ *Myotis yesoensis* Yoshiyuki's Myotis

Hokkaido (N Japan)

❏ *Lasionycteris noctivagans* Silver-haired Bat

SE Alaska, S Canada, USA and NE Mexico; Bermuda

Murina (Harpiola)

❏ *Murina grisea* Peters' Tube-nosed Bat

Himalayas of NW India; Endangered

Murina (Murina)

❏ *Murina aenea* Bronze Tube-nosed Bat

Malaya and Borneo

❏ *Murina aurata* Little Tube-nosed Bat

Nepal to Myanmar and SW China

❏ *Murina cyclotis* Round-eared Tube-nosed Bat

India and Sri Lanka to S China and Hainan, S to Indochina, Malaya, Borneo, Philippines and Lesser Sundas

❏ *Murina florium* Flores Tube-nosed Bat

Rainforests of Sulawesi, Lesser Sundas, Moluccas, New Guinea and coastal NE Queensland (Australia)

❏ *Murina fusca* Dusky Tube-nosed Bat

Manchuria (NE China)

❏ *Murina huttoni* Hutton's Tube-nosed Bat

NW India and Tibet to SE China, S to Thailand and Malaya

❏ *Murina leucogaster* Greater Tube-nosed Bat

Kazakhstan and C Siberia to Sakhalin I. (Russia), Korea, China and Japan

❏ *Murina puta* Taiwan Tube-nosed Bat

Taiwan

❏ *Murina rozendaali* Gilded Tube-nosed Bat

Borneo

❏ *Murina ryukyuana* Ryukyu Tube-nosed Bat

Ryukyu Is. (Japan)

❏ *Murina silvatica* Forest Tube-nosed Bat

Japan incl. Tsushima Is.

❏ *Murina suilla* Brown Tube-nosed Bat

Malaya, Java, Sumatra, Borneo and near islands

❏ *Murina tenebrosa* Gloomy Tube-nosed Bat

Tsushima Is. (Japan); Critically Endangered

❏ *Murina tubinaris* Scully's Tube-nosed Bat

Pakistan, N India, Myanmar, Thailand, Laos and Vietnam

❏ *Murina ussuriensis* Ussuri Tube-nosed Bat

E Siberia incl. Sakhalin and Kuril Is. (Russia), and Korea; Endangered

❏ *Harpiocephalus harpia* Hairy-winged Bat

India to Taiwan, S to Indochina, Java, Lesser Sundas and Moluccas

Kerivoula (Kerivoula)

❏ *Kerivoula africana* Tanzanian Woolly Bat

Tanzania

❏ *Kerivoula agnella* St Aignan's Trumpet-eared Bat

Fergusson, Woodlark, Misima and Sudest (D'Entrecasteaux Is., New Guinea)

❏ *Kerivoula argentata* Damara Woolly Bat

Savannas of Uganda and S Kenya, S to Namibia and KwaZulu-Natal (SE South Africa)

❏ *Kerivoula cuprosa* Copper Woolly Bat

S Cameroon, N D.R. Congo and Kenya

❏ *Kerivoula eriophora* Ethiopian Woolly Bat

Ethiopia

❏ *Kerivoula flora* Flores Woolly Bat

Borneo and Lesser Sundas

❏ *Kerivoula hardwickii* Hardwicke's Woolly Bat

India and Sri Lanka to China, S to Indochina, Sumatra, Java, Borneo, Philippines, Sulawesi and Lesser Sundas

❏ *Kerivoula intermedia* Small Woolly Bat

Malaya and Borneo

❏ *Kerivoula lanosa* Lesser Woolly Bat

Savannas of Liberia to Ethiopia, S to South Africa

❏ *Kerivoula minuta* Least Woolly Bat

S Thailand, Malaya and Borneo

❏ *Kerivoula muscina* Fly River Trumpet-eared Bat

C New Guinea

❏ *Kerivoula myrella* Bismarck Trumpet-eared Bat

Wetar (Lesser Sundas); Bismarck Archipelago

❏ *Kerivoula papillosa* Papillose Woolly Bat

NE India, Vietnam, Malaya, Sumatra, Java, Borneo and Sulawesi

❏ *Kerivoula pellucida* Clear-winged Woolly Bat

Malaya, Sumatra, Java, Borneo and Philippines

❏ *Kerivoula phalaena* Spurrell's Woolly Bat

Liberia, Ghana, Cameroon, Rep. Congo and D.R. Congo

❏ *Kerivoula picta* Painted Bat

India and Sri Lanka to S China, S to Indochina, Malaya, Sumatra, Java, Borneo, W Lesser Sundas and Moluccas

❏ *Kerivoula smithii* Smith's Woolly Bat

Liberia, Ivory Coast, Nigeria, Cameroon, N & E D.R. Congo and Kenya

❏ *Kerivoula whiteheadi* Whitehead's Woolly Bat

S Thailand, Malaya, Borneo and Philippines

Kerivoula (Phoniscus)

❏ *Kerivoula aerosa* Dubious Trumpet-eared Bat

"E coast of South Africa" (type locality doubtful)

❏ *Kerivoula atrox* Groove-toothed Bat

S Thailand, Malaya, Sumatra and Borneo

❏ *Kerivoula jagorii* Peters' Trumpet-eared Bat

Java, Bali, Borneo, Samar (EC Philippines), Sulawesi and Lesser Sundas

❏ *Kerivoula papuensis* Golden-tipped Bat

Forests of Biak, SE New Guinea and coastal E Australia

Order: PRIMATES
Family: GALEOPITHECIDAE (Flying Lemurs—2)

❏ *Cynocephalus variegatus* Malayan Flying Lemur

Indochina, Malaya, Sumatra, Java and Borneo

❏ *Cynocephalus volans* Philippine Flying Lemur

C & S Philippines

Family: DAUBENTONIIDAE (Aye-aye—1)

❏ *Daubentonia madagascariensis* Aye-aye

Scattered localities in Madagascar; Endangered

Family: LEMURIDAE (Lemurs—26)

❏ *Lepilemur dorsalis* Grey-backed Sportive Lemur

NW Madagascar and near islands

❏ *Lepilemur edwardsi*
Milne-Edwards' Sportive Lemur

Coastal forests of WC Madagascar

❏ *Lepilemur leucopus* White-footed Sportive Lemur

Xerophytic bush zone of S Madagascar

❏ *Lepilemur microdon*
Small-toothed Sportive Lemur

Coastal forests of EC & SE Madagascar

❏ *Lepilemur mustelinus* Weasel Sportive Lemur

Coastal forests of NE & EC Madagascar

❏ *Lepilemur ruficaudatus* Red-tailed Sportive Lemur

Coastal forests of SW Madagascar

❏ *Lepilemur septentrionalis* Northern Sportive Lemur

Extreme N Madagascar

❏ *Hapalemur alaotrensis* Alaotran Bamboo Lemur

Reedbeds of Lake Alaotra (EC Madagascar)

❏ *Hapalemur aureus* Golden Bamboo Lemur

Namorona R. to Bevoahazo (SE Madagascar); Critically Endangered

❏ *Hapalemur griseus* Grey Bamboo Lemur

Forests of E Madagascar

❏ *Hapalemur occidentalis*
Sambirano Bamboo Lemur

Sambirano region (W Madagascar)

❏ *Prolemur simus* Greater Bamboo Lemur

Rainforests of Ranomafana region (E Madagascar); Critically Endangered

❏ *Varecia rubra* Red Ruffed Lemur

Masoala Peninsula (NE Madagascar); Endangered

❏ *Varecia variegata*
Black-and-white Ruffed Lemur

Coastal E Madagascar; Endangered

❏ *Lemur catta* Ring-tailed Lemur

S Madagascar

❏ *Eulemur albifrons* White-fronted Lemur

Coastal NE & EC Madagascar

❏ *Eulemur albocollaris* White-collared Lemur

Forest between Manampatrana and Mananara Rs. (SE Madagascar)

❏ *Eulemur cinereiceps* Grey-headed Lemur

Farafangana region? (SE Madagascar); status uncertain, only known from two late 19th Century specimens

❏ *Eulemur collaris* Red-collared Lemur

Coastal SE Madagascar

❏ *Eulemur coronatus* Crowned Lemur

NW & extreme NE Madagascar

❏ *Eulemur fulvus* Brown Lemur

Coastal W, NE & E Madagascar; feral Mayotte (Comoro Is.)

❏ *Eulemur macaco* Black Lemur

Between Mahavavy and Sandrakota Rs. (W Madagascar)

❏ *Eulemur mongoz* Mongoose Lemur

Antsohihy to Betsiboka R. (NW Madagascar); feral Anjouan and Mohéli (Comoro Is.)

❏ *Eulemur rubriventer* Red-bellied Lemur

Mid- to high-altitude rainforests of E Madagascar

❏ *Eulemur rufus* Red-fronted Lemur

Coastal NW, SW & E Madagascar

❏ *Eulemur sanfordi* Sanford's Lemur

Coastal NW & extreme NE Madagascar

Family: LORIDAE
(Galagos and Lorises—29)

❏ *Pseudopotto martini* Martin's False Potto

Cameroon

Galago (Galago)

❏ *Galago alleni* Allen's Squirrel Galago

Primary rainforests of Bioko (Equatorial Guinea)

❏ *Galago cameronensis* Cross River Squirrel Galago

Primary rainforests of SE Nigeria and NW Cameroon

❏ *Galago gabonensis* Gabon Squirrel Galago

Primary rainforests of S Cameroon, Gabon and Rep. Congo

❏ *Galago gallarum* Somali Galago

Savannas and scrub thickets of Ethiopia, S Somalia and E Kenya

❏ *Galago matschiei* Spectacled Galago

Lowland and mid-elevation forests of E D.R. Congo

❏ *Galago moholi* South African Galago

Miombo (*Brachystegia*) woods, savannas and riparian woods of Angola, E D.R. Congo and SW Tanzania, S to South Africa

❏ *Galago senegalensis* Senegal Galago

Savannas and forest edges of Senegal to Sudan, S to Cameroon and N Tanzania

Galago (Galagoides)

❏ *Galago demidoff* Demidoff's Galago

Dense forest edges and regrowth areas of Senegal to Uganda and Kenya, S to Malawi; Bioko (Equatorial Guinea)

❏ *Galago granti* Mozambique Galago

Forests of S Tanzania to S Mozambique

❏ *Galago nyasae* Malawi Galago

Malawi and C Mozambique

❏ *Galago orinus* Uluguru Galago

Uluguru and Usambara Mtns (Tanzania)

❏ *Galago rondoensis* Rondo Galago

Lowland forest remnants on E edge of Rondo plateau (extreme SE Tanzania); Endangered

❏ *Galago thomasi* Thomas' Galago

Primary forests of Kivu region (E D.R. Congo) and SW Uganda

❏ *Galago udzungwensis* Matundu Galago

Lowland forests SE of Udzungwa Mtns and E of Uluguru Mtns (Tanzania)

❏ *Galago zanzibaricus* Zanzibar Galago

Coastal forests, regrowth areas and gardens of S Somalia, SE Kenya and N Tanzania incl. Zanzibar I.

❏ *Otolemur crassicaudatus* Brown Greater Galago

Dense forests and riparian woods of S Malawi, Zimbabwe and Mozambique to Natal (SE South Africa)

❏ *Otolemur garnettii* Northern Greater Galago

Forests, farmland and built areas of Juba R. (S Somalia) to Kenya and Tanzania incl. Pemba, Zanzibar and Mafia Is.

❏ *Otolemur monteiri* Silvery Greater Galago

Riparian and mid-elevation forests of Kenya, Rwanda and Tanzania, S to Angola, Zambia and Malawi

❏ *Euoticus elegantulus*
Western Needle-clawed Galago

Forests of S Cameroon, Equatorial Guinea, Gabon and Rep. Congo

❏ *Euoticus pallidus* Pallid Needle-clawed Galago

Locally in forests and regrowth areas of SE Nigeria and NW Cameroon; Bioko (Equatorial Guinea)

❏ *Nycticebus bengalensis* Bengal Slow Loris

Rainforests of Assam (NE India) and Bangladesh to Yunnan (SW China), S to C Thailand and Vietnam

❏ *Nycticebus coucang* Sunda Slow Loris

Malaya, Sumatra, Java, Borneo and Tawi-tawi (S Philippines)

❏ *Nycticebus pygmaeus* Pygmy Slow Loris

S Yunnan (SW China), Laos, Vietnam and E Cambodia

❏ *Perodicticus potto* Potto

Lowland and mid-elevation forests and regrowth areas of Guinea to W Kenya, S to Gabon and Rep. Congo

❏ *Arctocebus aureus* Golden Angwantibo

Dense undergrowth in lowland rainforests between Sanaga and Congo Rs. (S Cameroon and S Central African Republic to Rep. Congo)

❏ *Arctocebus calabarensis* Calabar Angwantibo

Dense undergrowth in lowland rainforests, forest edges and regrowth areas between Niger and Sanaga Rs. (SE Nigeria and NW Cameroon)

❏ *Loris lydekkerianus* Grey Slender Loris

S India and Sri Lanka (except SW)

❏ *Loris tardigradus* Red Slender Loris

Wet zone of SW Sri Lanka

Family: CHEIROGALEIDAE
(Dwarf Lemurs and Mouse-Lemurs—20)

❏ *Cheirogaleus adipicaudatus*
Southern Fat-tailed Dwarf Lemur

Xerophytic bush zone of S Madagascar

❏ *Cheirogaleus crossleyi* Furry-eared Dwarf Lemur

E Madagascar, inland from the coast

❏ *Cheirogaleus major* Greater Dwarf Lemur

E coast of Madagascar

❏ *Cheirogaleus medius*
Western Fat-tailed Dwarf Lemur

Dry forests of W Madagascar

❏ *Cheirogaleus minusculus*
Lesser Iron-grey Dwarf Lemur

Ambositra region (Madagascar)

❏ *Cheirogaleus ravus*
Greater Iron-grey Dwarf Lemur

Tamatave to Mahambo (coastal E Madagascar)

❏ *Cheirogaleus sibreei* Sibree's Dwarf Lemur

Ankeramadinika, Imerima and Pasandava regions (coastal E Madagascar)

❏ *Mirza coquereli* Coquerel's Mouse-Lemur

W coast of Madagascar

❏ *Microcebus berthae* Berthe's Mouse-Lemur

W Madagascar

❏ *Microcebus murinus* Grey Mouse-Lemur

Dry forests of W & S Madagascar

❏ *Microcebus myoxinus* Pygmy Mouse-Lemur

Scattered localities in forests of W Madagascar; Endangered

❏ *Microcebus ravelobensis* Golden Mouse-Lemur

Ankarafantsika region (Madagascar); Endangered

❏ *Microcebus rufus* Brown Mouse-Lemur

E coast of Madagascar

❏ *Microcebus sambiranensis*
Sambirano Mouse-Lemur

W Madagascar

❏ *Microcebus tavaratra*
Northern Rufous Mouse-Lemur

W Madagascar

❏ *Allocebus trichotis* Hairy-eared Dwarf Lemur

Lowlands around Mananara and Masoala Peninsula (E Madagascar); Endangered

❏ *Phaner electromontis*
Amber Mountain Fork-crowned Lemur

Montagne d'Ambre (Madagascar)

❏ *Phaner furcifer* Masoala Fork-crowned Lemur

Masoala Peninsula (Madagascar)

❏ *Phaner pallescens*
Western Fork-crowned Lemur

Scattered localities in forests of W Madagascar

❏ *Phaner parienti* Sambirano Fork-crowned Lemur

Sambirano region (Madagascar)

Family: INDRIIDAE
(Indris and Sifakas—11)

❏ *Indri indri* Indri

NE & EC Madagascar; Endangered

❏ *Propithecus coquereli* Coquerel's Sifaka

Ambato-Boéni to Antsohihy (NW Madagascar)

❏ *Propithecus deckenii* Van der Decken's Sifaka

Between Mahavavy and Manambolo Rs. (WC Madagascar)

❏ *Propithecus diadema* Diademed Sifaka

Rainforests of N & E Madagascar; Endangered

❏ *Propithecus edwardsi* Milne-Edwards' Sifaka

Rainforests between Mangoro and Mananara Rs. (EC Madagascar)

❏ *Propithecus perrieri* Perrier's Sifaka

Dry forest from Ankarana Massif to E coast (N Madagascar)

❏ *Propithecus tattersalli* Golden-crowned Sifaka

Ampandrana, Madirabe and Daraina regions (N Madagascar); Critically Endangered

❏ *Propithecus verreauxi* Verreaux's Sifaka

Arid SW Madagascar

❏ *Avahi laniger* Eastern Avahi

E Madagascar

❏ *Avahi occidentalis* Western Avahi

Locally in NW Madagascar

❏ *Avahi unicolor* Unicoloured Avahi

Madagascar

Family: TARSIIDAE (Tarsiers—7)

❏ *Tarsius bancanus* Western Tarsier

Sumatra, Bangka, Belitung, Karimata Strait Is., S Natuna Is. and Borneo

❏ *Tarsius dianae* Dian's Tarsier

C Sulawesi

❏ *Tarsius pelengensis* Peleng Tarsier

Peleng I. (W Moluccas)

❏ *Tarsius pumilus* Pygmy Tarsier

Montane rainforests of Rano Rano and Latimojong Mtns (C Sulawesi)

❏ *Tarsius sangirensis* Sangihe Tarsier

Sangihe Is. (N Moluccas)

❏ *Tarsius spectrum* Spectral Tarsier

Lowlands of Sulawesi and Togian Is.

❏ *Tarsius syrichta* Philippine Tarsier

Samar, Leyte, Bohol and Mindanao (EC & SE Philippines)

Family: CERCOPITHECIDAE
(Old World Monkeys—132)

❏ *Colobus angolensis* Angola Pied Colobus

Forests of E & S D.R. Congo, Rwanda, Burundi, SE Kenya, NE Angola, NE Zambia and E Tanzania

❏ *Colobus guereza* Guereza Colobus

Forests of Nigeria to Ethiopia; Kenya, Uganda and Tanzania

❏ *Colobus polykomos* Western Pied Colobus

Rain and gallery forests of Gambia E to R. Sassandra (Ivory Coast)

❏ *Colobus satanas* Black Colobus

Forests of SW Cameroon, Rio Muni and Bioko (Equatorial Guinea), and SW Gabon

❏ *Colobus vellerosus* Geoffroy's Pied Colobus

Lowland rain and gallery forests of R. Bandama-Nzi (Ivory Coast) E to W Nigeria

❏ *Piliocolobus badius* Western Red Colobus

Forests and woods of Senegal to Ghana; Endangered

❏ *Piliocolobus foai* Central African Red Colobus

Forests of Rep. Congo, N of Congo R. (N & NE D.R. Congo), Central African Republic and S Sudan

❏ *Piliocolobus gordonorum* Udzungwa Red Colobus

Forests of Udzungwa Mtns and between Little Ruaha and Ulanga Rs. (Tanzania)

❏ *Piliocolobus kirkii* Zanzibar Red Colobus

Forest remnants of Zanzibar I. (Tanzania); introd. Pemba I. (Tanzania); Endangered

❏ *Piliocolobus pennantii* Pennant's Red Colobus

Forests of Niger Delta (S Nigeria), Bioko (Equatorial Guinea) and C Rep. Congo; Endangered

❏ *Piliocolobus preussi* Preuss' Red Colobus

Highland forests of Korup and Cross River regions (Nigeria/Cameroon border)

❏ *Piliocolobus rufomitratus* Tana River Red Colobus

Riparian forests on lower Tana R. (SE Kenya); Critically Endangered

❏ *Piliocolobus tephrosceles* Ugandan Red Colobus

Forests of E border of W Rift Valley (W Uganda, Rwanda, Burundi and W Tanzania)

❏ *Piliocolobus tholloni* Thollon's Red Colobus

Rainforests S of Congo R. (C D.R. Congo)

❏ *Procolobus verus* Olive Colobus

Forest edge, swamp forests and dense regrowth areas of Sierra Leone to Togo; S bank of R. Benue (E Nigeria)

❏ *Presbytis chrysomelas* Sarawak Leaf Monkey

Borneo

❏ *Presbytis comata* Grizzled Leaf Monkey

W & C Java; Endangered

❏ *Presbytis femoralis* Banded Leaf Monkey

Malay Peninsula, Singapore and NE Sumatra

❏ *Presbytis frontata* White-fronted Leaf Monkey

NW, C & E Borneo

❏ *Presbytis hosei* Grey Leaf Monkey

N & E Borneo

❏ *Presbytis melalophos* Mitred Leaf Monkey

Sumatra

❏ *Presbytis natunae* Natuna Islands Leaf Monkey

Bunguran I. (Natuna Is., between Malaya and Borneo)

❏ *Presbytis potenziani* Mentawai Leaf Monkey

Mentawai Is. (Sumatra)

❏ *Presbytis rubicunda* Red Leaf Monkey

Borneo and Karimata Strait Is.

❏ *Presbytis siamensis* White-thighed Leaf Monkey

Malaya, E Sumatra and Riau Is.

❏ *Presbytis thomasi* North Sumatran Leaf Monkey

NW Sumatra

❏ *Semnopithecus ajax* Kashmir Grey Langur

Himalayas of NE Pakistan, Kashmir and NW India

❏ *Semnopithecus dussumieri*
Southern Plains Grey Langur

SW & S India

❏ *Semnopithecus entellus*
Northern Plains Grey Langur

Pakistan and lowlands of N India

❏ *Semnopithecus hector* **Tarai Grey Langur**

Himalayan foothills of N India, Nepal and Bhutan

❏ *Semnopithecus hypoleucos*
Black-footed Grey Langur

South Coorg region (Kerala, SW India)

❏ *Semnopithecus priam* **Tufted Grey Langur**

SE India and Sri Lanka

❏ *Semnopithecus schistaceus* **Nepal Grey Langur**

Himalayan slopes of C Nepal, S Tibet and Sikkim (NE India)

❏ *Trachypithecus auratus* **Javan Langur**

Java, Bali and Lombok (Lesser Sundas); Endangered

❏ *Trachypithecus barbei* **Tenasserim Leaf Monkey**

N Peninsular Myanmar and Thailand

❏ *Trachypithecus cristatus* **Silvered Leaf Monkey**

W coast of Malaya, Riau Is., Sumatra, Bangka, Belitung, Natuna Is. and Borneo

❏ *Trachypithecus delacouri* **Delacour's Leaf Monkey**

Scattered localities in NC Vietnam; Critically Endangered

❏ *Trachypithecus ebenus* **Black Leaf Monkey**

Highlands on border of Laos and Vietnam (type locality uncertain)

❏ *Trachypithecus francoisi* **François' Leaf Monkey**

Guangxi (S China), C Laos and N Vietnam

❏ *Trachypithecus geei* **Golden Leaf Monkey**

Forests between Sankosh and Manos Rs. (Assam, NE India) and Bhutan; Endangered

❏ *Trachypithecus germaini*
Indochinese Leaf Monkey

Myanmar and C Thailand to S Vietnam

❏ *Trachypithecus hatinhensis* **Hatinh Leaf Monkey**

C Vietnam

❏ *Trachypithecus johnii*
Hooded Leaf Monkey (Nilgiri Langur)

Forests and farmland of Nilgiri, Annaimalai and Palni hills (Western Ghats, S India)

❏ *Trachypithecus laotum* **Laotian Leaf Monkey**

C Laos

❏ *Trachypithecus obscurus* **Dusky Leaf Monkey**

S Peninsular Myanmar and Thailand, Malaya and near islands

❏ *Trachypithecus phayrei* **Phayre's Leaf Monkey**

S Assam (NE India), N Myanmar, Yunnan (SW China), N & C Thailand, Laos and N Vietnam

❏ *Trachypithecus pileatus* **Capped Leaf Monkey**

Hill forests of Assam (NE India), E Bangladesh, NW Myanmar and S Yunnan (SW China); Endangered

❏ *Trachypithecus poliocephalus*
White-headed Leaf Monkey

Guangxi (SC China) and Cat Ba I. (NE Vietnam); Critically Endangered

❏ *Trachypithecus shortridgei*
Shortridge's Leaf Monkey

Gongshan Dist. (Yunnan, SW China)

❏ *Trachypithecus vetulus*
Purple-faced Leaf Monkey

Sri Lanka; Endangered

❏ *Pygathrix cinerea* **Grey-shanked Douc Langur**

Few localities in C Vietnam

❏ *Pygathrix nemaeus* **Red-shanked Douc Langur**

E Laos and N & C Vietnam; Endangered

❏ *Pygathrix nigripes* **Black-shanked Douc Langur**

S Vietnam and E Cambodia; Endangered

❏ *Rhinopithecus avunculus*
Tonkin Snub-nosed Monkey

Restricted to a few areas of extreme NW Vietnam; Critically Endangered

❏ *Rhinopithecus bieti* Black Snub-nosed Monkey

Ridge of the upper Mekong-Salween Rs. divide (NW Yunnan, SW China); Endangered

❏ *Rhinopithecus brelichi* Grey Snub-nosed Monkey

Fanjin Shan, S of Yangtze R. (Guizhou, SC China); Endangered

❏ *Rhinopithecus roxellana*
Golden Snub-nosed Monkey

Mountains of Gansu, Sichuan, Shaanxi and Hubei (C China)

Nasalis (Nasalis)

❏ *Nasalis larvatus* Proboscis Monkey

Borneo; Endangered

Nasalis (Simias)

❏ *Nasalis concolor* Pig-tailed Langur

Mentawai Is. (Sumatra); Endangered

❏ *Macaca arctoides* Stump-tailed Macaque

Assam (NE India) and S China to Malaya

❏ *Macaca assamensis* Assam Macaque

Hill forests and mangrove swamps of Nepal to S China and N Vietnam

❏ *Macaca cyclopis* Taiwan Macaque

Taiwan

❏ *Macaca fascicularis* Long-tailed Macaque

Nicobar Is. (India), Myanmar and Indochina to Borneo, Philippines and Timor (Lesser Sundas)

❏ *Macaca fuscata* Japanese Macaque

Honshu, Shikoku and Kyushu (C & S Japan) and Yaku I. (Ryukyu Is., Japan)

❏ *Macaca hecki* Heck's Macaque

N Sulawesi

❏ *Macaca leonina* Northern Pig-tailed Macaque

Myanmar, Yunnan (SW China), Thailand and Laos

❏ *Macaca maura* Moor Macaque

S Sulawesi; Endangered

❏ *Macaca mulatta* Rhesus Macaque

Forest edges and built areas of Afghanistan and India to N Thailand and China incl. Hainan

❏ *Macaca nemestrina* Sunda Pig-tailed Macaque

Malaya, Sumatra, Bangka and Borneo

❏ *Macaca nigra* Sulawesi Crested Macaque

Sungai Onggak Dumoga and Mt Padang E to tip of NE Sulawesi; feral Bacan (N Moluccas); Endangered

❏ *Macaca nigrescens* Gorontalo Macaque

Gorontalo to Sungai Onggak Dumoga and Mt Padang (N Sulawesi)

❏ *Macaca ochreata* Booted Macaque

SE Sulawesi incl. Muna and Buton Is.

❏ *Macaca pagensis* Mentawai Macaque

Sipora and S Pagai Is. (Mentawai Is., Sumatra); Critically Endangered

❏ *Macaca radiata* Bonnet Macaque

Forests, farmland and built areas of S India

❏ *Macaca siberu* Siberut Macaque

Siberut (Mentawai Is., Sumatra); Critically Endangered

❏ *Macaca silenus* Lion-tailed Macaque

Evergreen hill forests and plantations of Western Ghats (SW India); Endangered

❏ *Macaca sinica* Toque Macaque

Sri Lanka

❏ *Macaca sylvanus* Barbary Macaque

Locally in montane forests, scrub and rocky slopes of Morocco and Algeria, formerly E to Libya; feral Gibraltar

❏ *Macaca thibetana* Père David's Macaque

E Tibet, SW & SC China

❏ *Macaca tonkeana* Tonkean Macaque

C Sulawesi incl. Togian Is.

❏ *Theropithecus gelada* Gelada Baboon

Montane grassland plateaus of N Ethiopia

❏ *Papio anubis* Olive Baboon

Forests and woods of Mali to C Ethiopia, S to E D.R. Congo, Uganda and NW Tanzania

❏ *Papio cynocephalus* Yellow Baboon

Miombo (*Brachystegia*) woods and thickets of S Ethiopia and S Somalia, S to C Angola, N Zambia and N Mozambique

❏ *Papio hamadryas* Hamadryas Baboon

Arid semi-deserts and rocky slopes of NE Sudan, Eritrea, N & E Ethiopia and N Somalia, formerly N to Egypt; S Arabian Peninsula

❏ *Papio papio* Guinea Baboon

Gallery forests, woods and savannas of Senegal and S Mauritania to Guinea

❏ *Papio ursinus* Chacma Baboon

Woods, savannas, scrub and grasslands of S Angola, S Zambia and C Mozambique, S to South Africa

❏ *Mandrillus leucophaeus* Drill

Rainforests and forest-savanna mosaics of SE Nigeria, NW Cameroon and Bioko (Equatorial Guinea); Endangered

❏ *Mandrillus sphinx* Mandrill

Primary rainforests of S Cameroon, Rio Muni (Equatorial Guinea), Gabon and Rep. Congo

❏ *Cercocebus agilis* Agile Mangabey

Lowland swamp forests of Rio Muni (Equatorial Guinea) and NE Gabon to N of Congo R. (N & NE D.R. Congo)

❏ *Cercocebus atys* Sooty Mangabey

Lowland swamp and valley forests of Senegal to Volta R. (Ghana)

❏ *Cercocebus chrysogaster*
Golden-bellied Mangabey

Lowland swamp forests S of Congo R. (C D.R. Congo)

❏ *Cercocebus galeritus* Tana River Mangabey

Gallery forests along lower Tana R. and its former courses (SE Kenya)

❏ *Cercocebus sanjei* Sanje Mangabey

Valley forests of Mwanihana Forest Reserve (E slopes of Udzungwa Mtns, Tanzania)

❏ *Cercocebus torquatus* Red-capped Mangabey

Lowland swamp and valley forests of W Nigeria to C Gabon

❏ *Lophocebus albigena* Grey-cheeked Mangabey

Rainforests of SE Nigeria to W Kenya, S to NE Angola, N of Congo R. (N D.R. Congo) and W Tanzania

❏ *Lophocebus atterimus* Black Mangabey

Rainforests of C D.R. Congo, S of Congo R.

❏ *Lophocebus opdenboschi*
Opdenbosch's Mangabey

Forests along Kwilu, Wamba and Kwango Rs. (SW D.R. Congo and N Angola)

❏ *Cercopithecus albogularis* Sykes' Monkey

Forests, regrowth areas and bamboo thickets of Somalia, Kenya incl. Patta and Witu Is. and Tanzania incl. Zanzibar, S to E South Africa

❏ *Cercopithecus ascanius* Red-tailed Monkey

Forests and regrowth areas of Central African Republic, D.R. Congo, Uganda, W Kenya, Angola and Zambia

❏ *Cercopithecus campbelli* Campbell's Monkey

Lowland rainforests, mangrove swamps and regrowth areas of Gambia SE to R. Cavally (Liberia/Ivory Coast border)

❏ *Cercopithecus cephus* Moustached Monkey

Lowland rainforests of S Cameroon and SW Central African Republic to NW Angola

❏ *Cercopithecus denti* Dent's Monkey

Lowland rainforests of S Central African Republic, NE & E D.R. Congo and Rwanda

❏ *Cercopithecus diana* Diana Monkey

Forests of Sierra Leone E to R. Sassandra (Ivory Coast); Endangered

❏ *Cercopithecus doggetti* Silver Monkey

E D.R. Congo, S Uganda, Rwanda, Burundi and NW Tanzania

❏ *Cercopithecus dryas* Dryas Monkey

Riparian forests thickets of Lomela and Wamba regions (C D.R. Congo)

❏ *Cercopithecus erythrogaster* Red-bellied Monkey

Lowland rainforest remnants of Benin and S Nigeria; Endangered

❏ *Cercopithecus erythrotis* Red-eared Monkey

Lowland rainforests of S & E Nigeria, Bioko (Equatorial Guinea) and Cameroon

187

❏ *Cercopithecus hamlyni* Owl-faced Monkey

Forests and bamboo thickets of E D.R. Congo and Rwanda

❏ *Cercopithecus kandti* Golden Monkey

High altitude of Virunga Volcanoes region (E D.R. Congo) and Nyungwe Forest (S Rwanda)

❏ *Cercopithecus lhoesti* L'Hoest's Monkey

Forests of E D.R. Congo, W Uganda, Rwanda and Burundi

❏ *Cercopithecus lowei* Lowe's Monkey

Lowland forests and regrowth areas of R. Cavally (Liberia/Ivory Coast border) E to R. Volta (Ghana)

❏ *Cercopithecus mitis* Blue Monkey

Forests, regrowth areas and bamboo thickets of Sudan and Ethiopia, S to Angola, Zambia and Tanzania

❏ *Cercopithecus mona* Mona Monkey

Lowland rainforests and mangrove swamps of Ghana to Cameroon; feral Lesser Antilles

❏ *Cercopithecus neglectus* De Brazza's Monkey

Riparian forests of SE Cameroon to S Sudan and SW Ethiopia, S to N Angola and Uganda

❏ *Cercopithecus nictitans*
Greater Spot-nosed Monkey

Forests and regrowth areas of Liberia and Ivory Coast; Nigeria to Central African Republic and NW D.R. Congo; Rio Muni and Bioko (Equatorial Guinea)

❏ *Cercopithecus petaurista*
Lesser Spot-nosed Monkey

Lowland forests and regrowth areas of Gambia to Benin

❏ *Cercopithecus pogonias* Crowned Monkey

Lowland forests and regrowth areas of SE Nigeria to Rep. Congo and W D.R. Congo; Rio Muni and Bioko (Equatorial Guinea)

❏ *Cercopithecus preussi* Preuss' Monkey

Forests of highlands of Nigeria/Cameroon border region; Bioko (Equatorial Guinea); Endangered

❏ *Cercopithecus roloway* Roloway Monkey

Forests of R. Sassandra (Ivory Coast) E to R. Pra (Ghana)

❏ *Cercopithecus sclateri* Sclater's Monkey

Forest remnants and swamps of SE Nigeria; Endangered

❏ *Cercopithecus solatus* Sun-tailed Monkey

Rainforests of Fôret des Abeilles (C Gabon)

❏ *Cercopithecus wolfi* Wolf's Monkey

Lowland rainforests of D.R. Congo, S of Congo R.

❏ *Chlorocebus aethiops* Grivet Monkey

Savannas, woods and farmland of E Sudan, Eritrea and W Ethiopia

❏ *Chlorocebus cynosuros* Malbrouck Monkey

Angola, N Namibia, S D.R. Congo and Zambia

❏ *Chlorocebus djamdjamensis* Bale Monkey

Montane forest edges and bamboo thickets of Bale Mtns (C Ethiopia)

❏ *Chlorocebus pygerythrus* Vervet Monkey

Woods and savannas of Ethiopia, Somalia, Kenya incl. Witu Is., Uganda and Tanzania incl. Pemba and Mafia Is., S to South Africa

❏ *Chlorocebus sabaeus* Green Monkey

Forest edge, woods and mangrove swamps of Senegal to Volta R. (Burkina Faso and Ghana), N to Niger R. in SE Mali

❏ *Chlorocebus tantalus* Tantalus Monkey

Woods and savannas of Volta R. (Burkina Faso and Ghana) to Sudan and Kenya, S to Cameroon, Central African Republic and N D.R. Congo

❏ *Miopithecus ogouensis* Northern Talapoin

Riparian forests of Cameroon, Rep. Congo, Rio Muni (Equatorial Guinea) and Gabon

❏ *Miopithecus talapoin* Southern Talapoin

Riparian forests of Angola and SW D.R. Congo

❏ *Allenopithecus nigroviridis*
Allen's Swamp Monkey

Seasonally flooded swamp and riparian forests of middle and lower Congo R. basin (Rep. Congo, W D.R. Congo and N Angola)

❏ *Erythrocebus patas* Patas Monkey

Savannas and open grasslands of Gambia to Ethiopia, S to Tanzania

Family: HOMINIDAE (Apes—21)

❏ *Pongo abelii* Sumatran Orangutan

Forests of NW Sumatra; Critically Endangered

❏ *Pongo pygmaeus* Bornean Orangutan

Forests of Borneo; Endangered

❏ *Pan paniscus* Bonobo (Pygmy Chimpanzee)

Forests of Congo Basin, S of Congo R. (NW D.R. Congo);
Endangered

❏ *Pan troglodytes* Chimpanzee

Forests and savanna woods of Senegal to W Nigeria; S
Cameroon to S Sudan and Uganda, S to Rep. Congo, C D.R.
Congo and W Tanzania; Endangered

❏ *Homo sapiens* Human

Cosmopolitan

❏ *Gorilla beringei* Eastern Gorilla

Forests and dense regrowth areas of E D.R. Congo, SW
Uganda and Rwanda; Endangered

❏ *Gorilla gorilla* Western Gorilla

Lowland rainforests of SE Nigeria to Rep. Congo;
Endangered

Hylobates (Bunopithecus)

❏ *Hylobates hoolock* Hoolock Gibbon

Hill forests of Assam (NE India), Bangladesh, Myanmar and
Yunnan (SW China); Endangered

Hylobates (Hylobates)

❏ *Hylobates agilis* Agile Gibbon

Malay Peninsula and Sumatra

❏ *Hylobates albibarbis*
White-bearded Bornean Gibbon

SW Borneo

❏ *Hylobates klossii* Kloss' Gibbon

Mentawai Is. (Sumatra)

❏ *Hylobates lar* White-handed Gibbon

E & S Myanmar, S Yunnan (SW China), Thailand, Malay
Peninsula and NW Sumatra

❏ *Hylobates moloch* Silvery Gibbon

Java; Critically Endangered

❏ *Hylobates muelleri* Müller's Bornean Gibbon

Borneo (except for SW)

❏ *Hylobates pileatus* Pileated Gibbon

SE Thailand, SW Laos and W Cambodia

Hylobates (Nomascus)

❏ *Hylobates concolor* Black Crested Gibbon

S Yunnan (SW China), Laos and Vietnam; Endangered

❏ *Hylobates gabriellae* Buff-cheeked Gibbon

S Laos, S Vietnam and E Cambodia

❏ *Hylobates hainanus* Hainan Gibbon

Hainan I. (S China)

❏ *Hylobates leucogenys*
Northern White-cheeked Gibbon

SW Yunnan (SW China), N Laos and N Vietnam

❏ *Hylobates siki* Southern White-cheeked Gibbon

C Laos and C Vietnam

Hylobates (Symphalangus)

❏ *Hylobates syndactylus* Siamang

Highlands of Malay Peninsula and Barisan Mtns (W Sumatra)

Family: CALLITRICHIDAE
(Tamarins and Marmosets—43)

❏ *Saguinus bicolor* Pied Bare-faced Tamarin

Rainforests of Manaus region, E of lower R. Negro and N of
R. Amazonas (C Amazonian Brazil); Critically Endangered

❏ *Saguinus fuscicollis* Saddle-backed Tamarin

Dense forests of S Colombia, E Ecuador, E Peru, N Bolivia
and N & W Brazil

❏ *Saguinus geoffroyi* Geoffroy's Tamarin

Dense regrowth areas and forest remnants of C Panama to NW
Colombia

❏ *Saguinus graellsi* Graells' Black-mantled Tamarin

Amazon basin of Colombia, E Ecuador and E Peru

❑ *Saguinus imperator* Emperor Tamarin

Dense lowland rainforests of Amazon basin of E Peru and W Brazil

❑ *Saguinus inustus* Mottle-faced Tamarin

S of R. Guaviare (SE Colombia) and between upper Rs. Japurá and Negro (NW Brazil)

❑ *Saguinus labiatus*
Red-chested Moustached Tamarin

Rainforests and regrowth areas between Rs. Solimões and Japurá and between Rs. Purus and Madeira (W Amazonian Brazil and N Bolivia)

❑ *Saguinus leucopus*
Silvery-brown Bare-faced Tamarin

Rainforests between lower Rs. Cauca and Magdalena, S to Mariquita (N Colombia)

❑ *Saguinus martinsi*
Martins' Bare-faced Tamarin

Rainforests between Rs. Uatumã and Erepecurú, N of R. Amazonas (NC Amazonian Brazil)

❑ *Saguinus melanoleucus*
White-mantled Tamarin

Between upper Rs. Juruá and Tarauacá (Acre Prov., W Amazonian Brazil)

❑ *Saguinus midas* Red-handed Tamarin

Terra firme rainforests of Guyana to French Guiana and E of R. Negro (NE Amazonian Brazil)

❑ *Saguinus mystax*
Black-chested Moustached Tamarin

Rainforests of Amazon basin of NE Peru and between lower Rs. Huallaga and Madeira, S of R. Solimões (W Amazonian Brazil)

❑ *Saguinus niger* Black Tamarin

Between R. Xingu and the coast, S of R. Amazonas, and on Marajó I. (E Amazonian Brazil)

❑ *Saguinus nigricollis* Black-mantled Tamarin

Rainforests of Amazon basin of Colombia, E Ecuador, E Peru and NW Brazil

❑ *Saguinus oedipus* Cotton-top Tamarin

Lowland rainforests of Panama (extinct?) and between lower Rs. Cauca and Magdalena, W to Atlantic coast (N Colombia); Endangered

❑ *Saguinus pileatus*
Red-capped Moustached Tamarin

Between Rs. Tefé and Purus, S of R. Solimões to Rs. Pauiní or Marmoriá (W Amazonian Brazil)

❑ *Saguinus tripartitus* Golden-mantled Tamarin

Lowland rainforests of Amazon basin of E Ecuador and NE Peru

❑ *Leontopithecus caissara*
Black-faced Lion Tamarin

Rs. Sebuí and dos Patos, Superagui I. and Cananéia region (Paraná and São Paulo Provs., SE Brazil); Critically Endangered

❑ *Leontopithecus chrysomelas*
Golden-headed Lion Tamarin

Very locally between Rs. Contas and Jequitinhonha (S Bahia and NE Minas Gerais, EC Brazil); Endangered

❑ *Leontopithecus chrysopygus* Black Lion Tamarin

Two small remnant forest patches in São Paulo Prov. (SE Brazil); Critically Endangered

❑ *Leontopithecus rosalia* Golden Lion Tamarin

A few isolated forest patches in C & S Rio de Janeiro Prov. (SE Brazil); Critically Endangered

❑ *Callithrix acariensis* Rio Acarí Marmoset

Lower R. Acari to between Rs. Aripuaná and Juruena and between Rs. Acari and Sucundurí (C Amazonian Brazil)

❑ *Callithrix argentata* Silvery Marmoset

Forests between Rs. Tapajós and Tocantins (Pará, E Amazonian Brazil)

❑ *Callithrix aurita* Buffy-tufted Marmoset

Minas Gerais, Rio de Janeiro and São Paulo Provs. (SE Brazil); Endangered

❑ *Callithrix chrysoleuca* Gold-and-white Marmoset

Right bank of lower R. Madeira-Aripuanã, N to R. Amazonas (C Amazonian Brazil)

❑ *Callithrix emiliae* Emilia's Marmoset

Locally in forests between upper Rs. Jamari and Xingu (Rondônia and N Mato Grosso, S Amazonian Brazil)

❑ *Callithrix flaviceps* Buffy-headed Marmoset

Locally in Minas Gerais, S Espírito Santo and Rio de Janeiro Provs. (EC Brazil); Endangered

❑ *Callithrix geoffroyi* Geoffroy's Marmoset

Between Rs. Itaunas and Jucú (Minas Gerais and Espírito Santo, EC Brazil)

❑ *Callithrix humeralifera* Santarém Marmoset

Rainforests between Rs. Maués and Tapajós, S of R. Amazonas (Amazonas and Pará, C Amazonian Brazil)

❑ *Callithrix humilis*
Dwarf Marmoset

Between lower Rs. Madeira and Aripuanã, and E of R. Atininga (SC Amazonian Brazil)

❑ *Callithrix intermedia* Hershkovitz's Marmoset

Near mouth of R. Guariba, a tributary of R. Aripuanã (SC Amazonian Brazil)

❑ *Callithrix jacchus* White-tufted Marmoset

Piauí, Ceará, Paraíba, Pernambuco, Alagoas and N Bahia Provs. (NE Brazil)

❑ *Callithrix kuhlii* Wied's Black-tufted Marmoset

Between Rs. de Contas and Jequitinhonha (S Bahia and extreme NW Minas Gerais, EC Brazil)

❑ *Callithrix leucippe* White Marmoset

Between Rs. Jamanxim and Cupari on right bank of R. Tapajós (Pará, C Amazonian Brazil)

❑ *Callithrix manicorensis* Rio Manicoré Marmoset

Rio Manicoré region (C Amazonian Brazil)

❑ *Callithrix marcai* Marca's Marmoset

Mouth of R. Castanho, a tributary of R. Roosevelt (SC Amazonian Brazil)

❑ *Callithrix mauesi* Rio Maués Marmoset

Dense primary rainforest between Rs. Urariá-Abacaxis and Rio Maués-Açú (C Amazonian Brazil)

❑ *Callithrix melanura* Black-tailed Marmoset

Forests of W & C Brazil, N Bolivia and N Paraguay

❑ *Callithrix nigriceps* Black-headed Marmoset

Very locally E of R. Madeira-Jiparaná, probably as far as R. dos Marmelos (SC Amazonian Brazil)

❑ *Callithrix penicillata* Black-tufted Marmoset

Maranhão to Goiás and Bahia Provs. (NE & EC Brazil)

❑ *Callithrix pygmaea* Pygmy Marmoset

Seasonally flooded and riparian forests of E Ecuador and NE Peru, E to Rs. Caquetá and Madeira (C Amazonian Brazil)

❑ *Callithrix saterei* Saterê-Maués' Marmoset

Between Rs. Canumã and Abacaxis, S of R. Paraná-Urariá (C Amazonian Brazil)

❑ *Callimico goeldii* Goeldi's Marmoset

Scrubby terra firme forests of Amazon basin of SE Colombia, E Peru, NW Bolivia and W Brazil

Family: ATELIDAE
(New World Monkeys—82)

❑ *Saimiri boliviensis* Bolivian Squirrel Monkey

Amazon basin of E Peru, Bolivia and SW Brazil

❑ *Saimiri oerstedii*
Central American Squirrel Monkey

Lowland regrowth areas of Pacific slope of C Costa Rica and W Panama; Endangered

❑ *Saimiri sciureus* Common Squirrel Monkey

Lowland riparian rainforests and mangroves of Colombia to French Guiana, S to NE Peru and N Brazil

❑ *Saimiri ustus* Bare-eared Squirrel Monkey

Amazonas, Rondônia and Pará Provs., S of R. Amazonas-Solimões (C Amazonian Brazil)

❑ *Saimiri vanzolinii* Black Squirrel Monkey

Between Rs. Solimões and Japurá (W Amazonian Brazil)

❑ *Cebus albifrons* White-fronted Capuchin

Forests of Colombia, Venezuela, NW Brazil, Ecuador and N Peru; probably feral Trinidad

❑ *Cebus apella* Brown Tufted Capuchin

Forests and regrowth areas of Colombia to French Guiana, S to Peru and C Brazil

❑ *Cebus capucinus* White-faced Capuchin

Forests and regrowth areas of W Honduras to NW Colombia, Ecuador and N Venezuela

❑ *Cebus kaapori* Ka'apor Capuchin

Between Rs. Gurupi and Pindaré (Maranhão, NE Brazil)

❏ *Cebus libidinosus* Black-striped Tufted Capuchin

Forests of Bolivia, N Argentina, Paraguay and S Brazil

❏ *Cebus nigritus* Black Tufted Capuchin

Atlantic forests of SE Brazil

❏ *Cebus olivaceus* Weeping Capuchin

Forests of N Colombia to French Guiana, S to Rs. Negro and
Amazonas (NE Amazonian Brazil)

❏ *Cebus xanthosternos*
Golden-bellied Tufted Capuchin

Now very locally in forest remnants of S Bahia and N Minas
Gerais (EC Brazil); Critically Endangered

❏ *Aotus azarai* Azara's Night Monkey

Between Rs. Tapajós-Juruena and Tocantins (E Amazonian
Brazil), S to Bolivia, NE Argentina and Paraguay

❏ *Aotus hershkovitzi* Hershkovitz's Night Monkey

Boyacá Dept. (E side Cordillera Oriental, C Colombia)

❏ *Aotus lemurinus* Lemurine Night Monkey

Lowland forests and regrowth areas of Costa Rica?, Panama,
Colombia, NW Venezuela and NW Ecuador

❏ *Aotus miconax* Peruvian Night Monkey

Between Rs. Marañón and Ucayali (NC Peru)

❏ *Aotus nancymaae* Nancy Ma's Night Monkey

NE Peru and region NW of R. Juruá (W Amazonian Brazil)

❏ *Aotus nigriceps* Black-headed Night Monkey

E Peru and region S of R. Solimões and W of R. Tapajós-
Juruena (W & C Amazonian Brazil)

❏ *Aotus trivirgatus* Northern Night Monkey

E Colombia, S Venezuela, NC Brazil and Guyana

❏ *Aotus vociferans* Noisy Night Monkey

Colombia, E Ecuador, N Peru and region N of R. Solimões and
W of R. Negro (NW Amazonian Brazil)

❏ *Callicebus baptista* Baptista Lake Titi

Between lower Rs. Madeira and Uíra-Curupá, S of R.
Amazonas (C Amazonian Brazil)

❏ *Callicebus barbarabrownae* Blond Titi

Locally in remnant forests of N Bahia, between Rs. São
Francisco and Paraguaçu (EC Brazil); Critically Endangered

❏ *Callicebus bernhardi* Prince Bernhard's Titi

Between Rs. Madeira-Jiparaná and Aripuanã-Roosevelt (SC
Amazonian Brazil)

❏ *Callicebus brunneus* Brown Titi

Between upper Rs. Madeira and Jiparaná (SC Amazonian
Brazil)

❏ *Callicebus caligatus* Booted Titi

Between lower Rs. Purus-Ipixuna and Madeira, S of R.
Solimões (C Amazonian Brazil)

❏ *Callicebus cinerascens* Ashy-black Titi

E of lower Rs. Aripuanã and Roosevelt, S of lower R. Madeira
(C Amazonian Brazil)

❏ *Callicebus coimbrai* Coimbra-Filho's Titi

Atlantic coastal forests between lower Rs. São Francisco and
Itapicuru (Sergipe and extreme NE Bahia, NE Brazil);
Critically Endangered

❏ *Callicebus cupreus* Coppery Titi

Between Rs. Ucayali and Purus, S of R. Solimões (N Peru and
WC Amazonian Brazil)

❏ *Callicebus discolor* Double-browed Titi

Amazon basin of extreme S Colombia, E Ecuador, N Peru and
between Rs. Ucayali and Solimões (WC Amazonian Brazil)

❏ *Callicebus donacophilus* White-eared Titi

Upper basins of Rs. Mamoré-Grande and San Miguel (Beni
and Santa Cruz Provs., WC Bolivia)

❏ *Callicebus dubius* Dubious Titi

Between upper Rs. Purus and Madeira-Madre de Dios (WC
Amazonian Brazil)

❏ *Callicebus hoffmannsi* Hoffmanns' Titi

Between Rs. Paraná do Urariá and lower Tapajós (C
Amazonian Brazil)

❏ *Callicebus lucifer* Lucifer Titi

Between Rs. Japurá-Caquetá and Solimões-Napo (S Colombia,
E Ecuador, NE Peru and W Amazonian Brazil)

❏ *Callicebus lugens* Mourning Titi

Amazon basin of E Colombia and S Venezuela, S to Rs.
Caquetá and Negro-Uaupés (NW Amazonian Brazil)

❏ *Callicebus medemi* Medem's Titi

Between upper Rs. Caquetá and Aguarico (SC Colombia and
NE Ecuador)

❏ *Callicebus melanochir* Black-handed Titi

Atlantic coastal forests of SE Bahia and N Espírito Santo (EC Brazil)

❏ *Callicebus modestus* Rio Beni Titi

Upper R. Beni basin (NW Bolivia)

❏ *Callicebus moloch* Dusky Titi

Between Rs. Tapajós and Tocantins-Araguaia, S of lower R. Amazonas (EC Amazonian Brazil)

❏ *Callicebus nigrifrons* Black-fronted Titi

Atlantic coastal forests of S Minas Gerias, São Paulo and SW Rio de Janeiro (SE Brazil)

❏ *Callicebus oenanthe* Rio Mayo Titi

Upper R. Mayo valley (San Martín Dept., N Peru)

❏ *Callicebus olallae* Olalla Brothers' Titi

Upper R. Beni basin (NW Bolivia)

❏ *Callicebus ornatus* Ornate Titi

Amazon basin of E Colombia

❏ *Callicebus pallescens* Pallid Titi

Pantanal region (Mato Grosso do Sul, SC Brazil) and W Paraguay

❏ *Callicebus personatus* Masked Titi

Atlantic coastal forests of EC Minas Gerias, S Espírito Santo and NE Rio de Janeiro (EC Brazil)

❏ *Callicebus purinus* Red-crowned Titi

Between Rs. Juruá and Purus-Tapauá, S of R. Solimões (C Amazonian Brazil)

❏ *Callicebus regulus* Kinglet Titi

Between Rs. Javarí and Juruá, S of R. Solimões (C Amazonian Brazil)

❏ *Callicebus stephennashi* Stephen Nash's Titi

Probably between middle Rs. Purus and Madeira (WC Amazonian Brazil) (type locality uncertain)

❏ *Callicebus torquatus* Collared Titi

Between Rs. Negro-Uaupés and Solimões-Japurá-Apaporis (C Amazonian Brazil)

❏ *Pithecia aequatorialis* Equatorial Saki

Lowland primary forests of upper Amazon basin of E Ecuador and N Peru

❏ *Pithecia albicans* Buffy Saki

Primary forests between lower Rs. Juruá and Purus, S of R. Solimões (C Amazonian Brazil)

❏ *Pithecia irrorata* Grey Monk Saki

Amazon basin of S Peru, N Bolivia and between Rs. Juruá and Tapajós (C & SW Amazonian Brazil)

❏ *Pithecia monachus* Monk Saki

Lowland rainforests of upper Amazon basin between Rs. Japurá-Caquetá and Juruá (S Colombia, E Ecuador, N Peru and W Brazil)

❏ *Pithecia pithecia* Guianan Saki

Terra firme rainforests of E Venezuela to French Guiana, S to R. Amazonas (NC Brazil)

❏ *Chiropotes albinasus* White-nosed Bearded Saki

Undisturbed terra firme rainforests between Rs. Madeira and Xingu, S of R. Amazonas, S to Mato Grosso Prov. (C Brazil)

❏ *Chiropotes satanas* Brown Bearded Saki

Lowland rainforests of Amazon basin from S Venezuela to S French Guiana, SE to Maranhão Prov. (NE Brazil); Endangered

❏ *Cacajao calvus* Red Uakari

Primary swamp forests between Rs. Ucayali and Juruá, S of R. Solimões (N Peru and WC Brazil)

❏ *Cacajao melanocephalus* Black Uakari

Lowland rainforests between Rs. Japurá-Apaporis and Negro, N of R. Solimões (SE Colombia, extreme S Venezuela and NW Brazil)

❏ *Ateles belzebuth* White-bellied Spider Monkey

Primary forests of Amazon basin of E Colombia, S Venezuela, E Ecuador, NE Peru and NW Brazil

❏ *Ateles chamek* Peruvian Spider Monkey

E Peru and N Bolivia, E to R. Tapajós and N to R. Solimões (W Amazonian Brazil)

❏ *Ateles fusciceps* Brown-headed Spider Monkey

Primary rainforests and tall regrowth areas of C Panama to W Colombia and C Ecuador

❏ *Ateles geoffroyi* Central American Spider Monkey

Rainforests and riparian habitats of E Mexico to SE Panama

❏ *Ateles hybridus* Brown Spider Monkey

N Colombia and NW Venezuela; Critically Endangered

❏ *Ateles marginatus*
White-whiskered Spider Monkey

Between Rs. Tapajós and Tocantins, S of lower R. Amazonas (EC Amazonian Brazil); Endangered

❏ *Ateles paniscus* **Black Spider Monkey**

Undisturbed primary rainforests of Guyana to French Guiana, S to lower R. Amazonas and W to R. Negro (NE Amazonian Brazil)

❏ *Brachyteles arachnoides* **Southern Muriqui**

Atlantic coastal forest remnants of Rio de Janeiro and São Paulo Provs. (SE Brazil); Endangered

❏ *Brachyteles hypoxanthus* **Northern Muriqui**

Atlantic coastal forest remnants from Bahia to Espírito Santo Provs. (EC Brazil); Critically Endangered

❏ *Oreonax flavicauda* **Yellow-tailed Woolly Monkey**

Montane cloud forests on E Andean slopes of Amazonas, San Martín and La Libertad Depts. (NE Peru); Critically Endangered

❏ *Lagothrix cana* **Grey Woolly Monkey**

Highlands of S Peru and S Amazonian Brazil

❏ *Lagothrix lagotricha* **Brown Woolly Monkey**

Primary rainforests of SE Colombia, NE Ecuador and extreme N Peru, E to R. Negro and S to R. Solimões-Napo (NW Amazonian Brazil)

❏ *Lagothrix lugens* **Colombian Woolly Monkey**

N Colombia and W Venezuela

❏ *Lagothrix poeppigii* **Silvery Woolly Monkey**

Highlands of E Ecuador and N Peru, E to R. Juruá (extreme W Amazonian Brazil)

❏ *Alouatta belzebul* **Red-handed Howler Monkey**

Forests between R. Madeira and Marajó I., S of R. Amazonas, E to Alagoas Prov. (C & NE Brazil)

❏ *Alouatta caraya* **Black Howler Monkey**

Forests of E Bolivia and SC Brazil, S to Paraguay and NE Argentina

❏ *Alouatta coibensis* **Coiba Island Howler Monkey**

Azuero Peninsula and Coiba I. (SC Panama); Endangered

❏ *Alouatta guariba* **Brown Howler Monkey**

Atlantic coastal forests from Bahia to Rio Grande do Sul Provs. (EC & SE Brazil) and Misiones Prov. (NE Argentina)

❏ *Alouatta macconnelli*
Guyanan Red Howler Monkey

Forests of Guyana to French Guiana and NC Amazonian Brazil; Trinidad

❏ *Alouatta nigerrima*
Amazon Black Howler Monkey

Forests both N and S of R. Amazonas, E of Rs. Trombetas and Purus (C & E Amazonian Brazil)

❏ *Alouatta palliata* **Mantled Howler Monkey**

Forests of S Mexico to Colombia, W Ecuador and extreme NW Peru

❏ *Alouatta pigra* **Yucatán Black Howler Monkey**

Lowland forests and tall regrowth areas of Yucatán Peninsula (SE Mexico, Belize and N Guatemala); Endangered

❏ *Alouatta sara* **Bolivian Red Howler Monkey**

Forests of S Peru, Bolivia and SW Brazil

❏ *Alouatta seniculus*
Venezuelan Red Howler Monkey

Forests of Colombia to Venezuela and NW Brazil

Order: SCANDENTIA
Family: TUPAIIDAE (Tree-Shrews—19)

❏ *Ptilocercus lowii* **Pen-tailed Tree-Shrew**

Malay Peninsula, Sumatra, Borneo and near islands

❏ *Tupaia belangeri* **Northern Tree-Shrew**

India and Myanmar to China, Indochina, Malay Peninsula and near islands

❏ *Tupaia chrysogaster* **Golden-bellied Tree-Shrew**

Sipora and Pagai Is. (Mentawai Is., Sumatra)

❏ *Tupaia dorsalis* **Striped Tree-Shrew**

Borneo

❏ *Tupaia glis* **Common Tree-Shrew**

Malay Peninsula, Sumatra, Java and near islands

❏ *Tupaia gracilis* **Slender Tree-Shrew**

Bangka, Belitung, Karimata Strait Is. and Borneo

❏ *Tupaia javanica* **Javan Tree-Shrew**

Sumatra, Nias (Mentawai Is. Sumatra), Java and Bali

❏ *Tupaia longipes* Long-footed Tree-Shrew

Borneo; Endangered

❏ *Tupaia minor* Pygmy Tree-Shrew

Malay Peninsula, Sumatra, Borneo and near islands

❏ *Tupaia montana* Mountain Tree-Shrew

Highlands of Borneo

❏ *Tupaia nicobarica* Nicobar Tree-Shrew

Nicobar Is. (India); Endangered

❏ *Tupaia palawensis* Palawan Tree-Shrew

Culion, Cuyo, Busuanga and Palawan (WC & SW Philippines)

❏ *Tupaia picta* Painted Tree-Shrew

C Borneo

❏ *Tupaia splendidula* Ruddy Tree-Shrew

Karimata Strait Is., Natuna Is. and Borneo

❏ *Tupaia tana* Large Tree-Shrew

Sumatra, Karimata Strait Is. and Borneo

❏ *Anathana ellioti* Madras Tree-Shrew

India

❏ *Dendrogale melanura*
Bornean Smooth-tailed Tree-Shrew

Mountains of N Borneo

❏ *Dendrogale murina*
Northern Smooth-tailed Tree-Shrew

E Thailand, Cambodia and Vietnam

❏ *Urogale everetti* Mindanao Tree-Shrew

Dinagat, Siargao and Mindanao (EC & SE Philippines)

Grandorder: UNGULATA (Ungulates)
Order: TUBULIDENTATA
Family: ORYCTEROPODIDAE (Aardvark—1)

❏ *Orycteropus afer* Aardvark

Locally in savannas and semi-deserts of subsaharan Africa

Mirorder: EPARCTOCYONA
Order: CETE
Family: BALAENOPTERIDAE (Rorquals—8)

❏ *Balaenoptera acutorostrata*
Northern Minke Whale

Northern polar, temperate and subtropical waters

❏ *Balaenoptera bonaerensis* Antarctic Minke Whale

Southern polar, temperate and subtropical waters

❏ *Balaenoptera borealis* Sei Whale

Worldwide in deep cold-temperate to tropical waters; Endangered

❏ *Balaenoptera brydei* Bryde's Whale

Worldwide in warm-temperate, subtropical and tropical waters

❏ *Balaenoptera musculus* Blue Whale

Locally distributed worldwide in polar to tropical waters; Endangered

❏ *Balaenoptera physalus* Fin Whale

Worldwide in polar to tropical waters; Endangered

❏ *Megaptera novaeangliae* Humpback Whale

Worldwide in cold-temperate to tropical waters

❏ *Eschrichtius robustus* Grey Whale

Sea of Okhotsk, Bering Sea and neighbouring Arctic Ocean, wintering S to Korea Strait and Baja California (Mexico)

Family: BALAENIDAE (Right Whales—5)

Balaena (Balaena)

❏ *Balaena mysticetus* Bowhead Whale

Circumpolar in Arctic seas, usually close to pack ice; vagrants S to Japan and NE USA

Balaena (Eubalaena)

❏ *Balaena australis* Southern Right Whale

Antarctic and southern temperate waters, breeding in coastal S South America, S Africa and S Australasia

❏ *Balaena glacialis* North Atlantic Right Whale

N Atlantic Ocean, mainly from E Canada to SE USA and now very rare in European waters; Endangered

❑ *Balaena japonica* North Pacific Right Whale

N Pacific Ocean, Bering Sea and Sea of Okhotsk; Endangered

❑ *Caperea marginata* Pygmy Right Whale

Scattered records in southern cold-temperate waters

Family: PHYSETERIDAE
(Great Sperm Whale—1)

❑ *Physeter catodon* Great Sperm Whale

Worldwide in subpolar to tropical waters

Family: KOGIIDAE
(Pygmy Sperm Whales—2)

❑ *Kogia breviceps* Pygmy Sperm Whale

Scattered records worldwide in temperate and tropical waters

❑ *Kogia sima* Dwarf Sperm Whale

Scattered records worldwide in temperate and tropical waters

Family: PLATANISTIDAE
(Indian River Dolphins—2)

❑ *Platanista gangetica* Ganges River Dolphin

Ganges, Brahmaputra and Meghna R. systems of India, Nepal, Bhutan and Bangladesh; Endangered

❑ *Platanista minor* Indus River Dolphin

Indus R. system of Pakistan; Endangered

Family: HYPEROODONTIDAE
(Beaked Whales—21)

❑ *Mesoplodon bidens* Sowerby's Beaked Whale

Temperate waters of N Atlantic Ocean and North Sea; vagrant to Florida (USA)

❑ *Mesoplodon bowdoini* Andrews' Beaked Whale

Probably circumpolar in cold-temperate waters, mainly S Australia, New Zealand and subantarctic islands

❑ *Mesoplodon carlhubbsi* Hubbs' Beaked Whale

Temperate coastal waters off NE Honshu (C Japan) and S Alaska to S California (USA)

❑ *Mesoplodon densirostris*
Blainville's Beaked Whale

Scattered records worldwide in temperate and tropical waters

❑ *Mesoplodon europaeus* Gervais' Beaked Whale

Warm-temperate and tropical waters of W & E Atlantic Ocean and Caribbean Sea; vagrant NE Atlantic Ocean

❑ *Mesoplodon ginkgodens*
Ginkgo-toothed Beaked Whale

Warm-temperate waters near Japan and Taiwan; other scattered records widely in Indian and Pacific Oceans

❑ *Mesoplodon grayi* Gray's Beaked Whale

Southern cold-temperate waters; one vagrant in Netherlands

❑ *Mesoplodon hectori* Hector's Beaked Whale

Southern cold-temperate waters

❑ *Mesoplodon layardii* Strap-toothed Whale

Probably circumpolar in southern cold-temperate waters

❑ *Mesoplodon mirus* True's Beaked Whale

Scattered records from warm-temperate waters of N Atlantic Ocean; vagrant S Africa and S Australia

❑ *Mesoplodon perrini* Perrin's Beaked Whale

S California (USA)

❑ *Mesoplodon peruvianus* Pygmy Beaked Whale

Cold-temperate and tropical waters of E Pacific Ocean, from S California (USA) to N Chile; vagrant New Zealand

❑ *Mesoplodon stejnegeri*
Stejneger's Beaked Whale

Cold-temperate waters of N Pacific Ocean

❑ *Mesoplodon traversii* Spade-toothed Whale

New Zealand, Chatham Is. and Juan Fernández Archipelago (S Chile)

❑ *Hyperoodon ampullatus*
Northern Bottle-nosed Whale

Subpolar and cold-temperate waters of N Atlantic Ocean and Labrador Sea

❑ *Hyperoodon planifrons*
Southern Bottle-nosed Whale

Circumpolar in southern polar and cold-temperate waters

❑ *Ziphius cavirostris* Cuvier's Beaked Whale

Worldwide in temperate and tropical waters

❏ *Berardius arnuxii* Arnoux's Beaked Whale

Circumpolar in southern polar and cold-temperate waters

❏ *Berardius bairdii* Baird's Beaked Whale

Deep subpolar to temperate waters of N Pacific Ocean

❏ *Tasmacetus shepherdi* Shepherd's Beaked Whale

Few scattered records from southern cold-temperate waters

❏ *Indopacetus pacificus* Longman's Beaked Whale

Very few records from Indian and S Pacific Oceans

Family: INIIDAE
(Amazon River Dolphin—1)

❏ *Inia geoffrensis* Amazon River Dolphin (Boto)

Orinoco and Amazon R. systems (E Colombia to Guyana, S to E Ecuador, NE Peru, N Bolivia and SC Brazil)

Family: LIPOTIDAE
(Yangtze River Dolphin—1)

❏ *Lipotes vexillifer* Yangtze River Dolphin

Yangtze R. system (China); Critically Endangered

Family: PONTOPORIIDAE
(Franciscana—1)

❏ *Pontoporia blainvillei* Franciscana

Inshore waters off coasts of SE Brazil, Uruguay and C & N Argentina

Family: PHOCOENIDAE (Porpoises—6)

❏ *Phocoenoides dalli* Dall's Porpoise

Cold-temperate waters of Sea of Okhotsk, N Pacific Ocean and Bering Sea

❏ *Australophocoena dioptrica* Spectacled Porpoise

Scattered records in southern cold-temperate waters

❏ *Neophocaena phocaenoides* Finless Porpoise

Inshore waters and river systems from Persian Gulf to Japan, S to Lesser Sundas

❏ *Phocoena phocoena* Harbour Porpoise

Discontinuously in northern coastal subpolar and temperate waters

❏ *Phocoena sinus* Vaquita

Sea of Cortez (NW Gulf of California, Mexico); Critically Endangered

❏ *Phocoena spinipinnis* Burmeister's Porpoise

Inshore coastal waters of S South America, S from N Peru and SE Brazil

Family: DELPHINIDAE
(Marine Dolphins—34)

❏ *Orcinus orca* Killer Whale

Worldwide in polar to tropical waters

❏ *Pseudorca crassidens* False Killer Whale

Probably worldwide in warm-temperate and tropical waters

❏ *Delphinus capensis*
Long-beaked Common Dolphin

Coastal warm-temperate and tropical waters of Atlantic, SW Indian and Pacific Oceans

❏ *Delphinus delphis*
Short-beaked Common Dolphin

Probably worldwide in temperate and tropical waters

❏ *Delphinus tropicalis* Arabian Common Dolphin

Coastal waters of Red Sea, N Indian Ocean and South China Sea

❏ *Tursiops aduncus* Indo-Pacific Bottle-nosed Dolphin

Inshore waters off coasts of Indian and SW Pacific Oceans

❏ *Tursiops truncatus* Common Bottle-nosed Dolphin

Worldwide in temperate and tropical waters

❏ *Lagenorhynchus acutus*
Atlantic White-sided Dolphin

Cold-temperate waters of Labrador Sea, N Atlantic Ocean, North and Norwegian Seas

❏ *Lagenorhynchus albirostris* White-beaked Dolphin

Subpolar and cold-temperate waters of Labrador Sea, N Atlantic Ocean, North, Norwegian and Barents Seas

❏ *Lagenorhynchus australis* Peale's Dolphin

Inshore waters off coasts of S Chile, S Argentina and Falkland Is.

❏ *Lagenorhynchus cruciger* Hourglass Dolphin

Circumpolar in southern subpolar waters

❏ *Lagenorhynchus obliquidens*
Pacific White-sided Dolphin

Deep cold-temperate waters of Sea of Japan, S Sea of Okhotsk, N Pacific Ocean and S Bering Sea

❏ *Lagenorhynchus obscurus* **Dusky Dolphin**

Southern coastal temperate waters of S South America, S Africa, S Australasia and subantarctic islands

❏ *Steno bredanensis* **Rough-toothed Dolphin**

Worldwide in deep warm-temperate and tropical waters

❏ *Lissodelphis borealis*
Northern Right-Whale Dolphin

Deep cool-temperate waters of N Pacific Ocean

❏ *Lissodelphis peronii*
Southern Right-Whale Dolphin

Circumpolar in southern cold-temperate waters

❏ *Lagenodelphis hosei* **Fraser's Dolphin**

Worldwide in deep warm-temperate and tropical waters

❏ *Stenella attenuata* **Pantropical Spotted Dolphin**

Worldwide in warm-temperate and tropical waters

❏ *Stenella clymene* **Clymene Dolphin**

Subtropical and tropical waters of Atlantic Ocean and Caribbean Sea

❏ *Stenella coeruleoalba* **Striped Dolphin**

Worldwide in warm-temperate and tropical waters

❏ *Stenella frontalis* **Atlantic Spotted Dolphin**

Warm-temperate and tropical waters of Atlantic Ocean and Caribbean Sea

❏ *Stenella longirostris* **Spinner Dolphin**

Worldwide in subtropical and tropical waters

❏ *Globicephala macrorhynchus*
Short-finned Pilot Whale

Worldwide in warm-temperate and tropical waters

❏ *Globicephala melas* **Long-finned Pilot Whale**

Cold-temperate waters of Southern Ocean and N Atlantic Ocean

❏ *Grampus griseus* **Risso's Dolphin**

Worldwide in warm-temperate and tropical waters

❏ *Peponocephala electra* **Melon-headed Whale**

Worldwide in warm-temperate and tropical waters

❏ *Feresa attenuata* **Pygmy Killer Whale**

Worldwide in warm-temperate and tropical waters

❏ *Cephalorhynchus commersonii*
Commerson's Dolphin

Inshore waters off coasts of S Chile, Argentina, Falkland Is., South Georgia and Kerguelen I.

❏ *Cephalorhynchus eutropia* **Chilean Dolphin**

Inshore waters off coasts of C & S Chile

❏ *Cephalorhynchus heavisidii* **Heaviside's Dolphin**

Inshore waters off coasts of Namibia and Cape Prov. (South Africa)

❏ *Cephalorhynchus hectori* **Hector's Dolphin**

Inshore waters off coasts of New Zealand; Endangered

❏ *Sotalia fluviatilis* **Tucuxi**

Inshore waters off coasts of E Nicaragua to Lesser Antilles and EC Brazil; Amazon and lower Orinoco R. systems (E Colombia to French Guiana and Brazil)

❏ *Sotalia chinensis*
Indo-Pacific Hump-backed Dolphin

Inshore waters off coasts of Indian and SW Pacific Oceans

❏ *Sousa teuszii* **Atlantic Hump-backed Dolphin**

Inshore waters off coasts of tropical W Africa, from Mauritania to Cameroon

Family: MONODONTIDAE
(Single-toothed Whales—3)

❏ *Delphinapterus leucas* **Beluga**

Circumpolar in northern polar and subpolar seasonally ice-covered waters

❏ *Monodon monoceros* **Narwhal**

Discontinuously in Arctic Ocean seas, near pack ice

❏ *Orcaella brevirostris* **Irrawaddy Dolphin**

Inshore waters and river systems from India and Indochina to New Guinea and NE Australia

Order: ARTIODACTYLA
Family: SUIDAE (Pigs—17)

❏ *Sus barbatus* **Bearded Pig**

Malay Peninsula, Riau Is., Sumatra, Bangka, Borneo, and SW Philippines

❏ *Sus bucculentus* **Vietnam Warty Pig**

S Vietnam

❏ *Sus cebifrons* **Visayan Warty Pig**

Negros and Cebu (C Philippines); Critically Endangered

❏ *Sus celebensis* **Sulawesi Wild Boar**

Sulawesi and near islands; feral Simeulue Is. (Mentawai Is., Sumatra) and Halmahera (N Moluccas)

❏ *Sus domesticus* **Domestic Pig**

Domesticated worldwide, feral South Africa, Australia, Pacific Ocean islands, Americas, Caribbean etc.; wild ancestor is *Sus scrofa* Eurasian Wild Boar

❏ *Sus heureni* **Flores Warty Pig**

Flores (Lesser Sundas)

❏ *Sus philippensis* **Philippine Warty Pig**

Philippines

❏ *Sus salvanius* **Pygmy Hog**

Riparian woods and grasslands of S Nepal, N India and Bhutan; Critically Endangered

❏ *Sus scrofa* **Eurasian Wild Boar**

Forests, scrub, grasslands and farmland of Europe and N Africa to China and Bali

❏ *Sus timoriensis* **Timor Wild Boar**

Timor (Lesser Sundas)

❏ *Sus verrucosus* **Javan Pig**

Java, Madura and Bawean; Endangered

❏ *Potamochoerus larvatus* **Bush Pig**

Forests, scrub and dense grasslands of S Sudan and Ethiopia, S to South Africa; Madagascar and Comoro Is.

❏ *Potamochoerus porcus* **Red River Hog**

Rainforests of Senegal to D.R. Congo

❏ *Hylochoerus meinertzhageni* **Giant Forest Hog**

Forests and dense scrub of Guinea to Ghana; E Nigeria to SW Ethiopia, Kenya and N Tanzania

❏ *Phacochoerus aethiopicus* **Desert Warthog**

Arid open areas of Somalia and N Kenya; formerly Cape Prov. (South Africa, extinct)

❏ *Phacochoerus africanus* **Warthog**

Savannas and open areas of subsaharan Africa

❏ *Babyrousa babyrussa* **Babirusa**

N & C Sulawesi incl. Togian Is., Sula Is. and Buru (WC Moluccas)

Family: TAYASSUIDAE (Peccaries—3)

❏ *Tayassu pecari* **White-lipped Peccary**

Undisturbed forests and scrub of S Mexico to Suriname, S to Paraguay, NE Argentina, S Brazil (and Uruguay?); feral Cuba

❏ *Dicotyles tajacu* **Collared Peccary (Javelina)**

Forests, scrub, grasslands and deserts of SW & SC USA to French Guiana, S to NW Peru, Paraguay, N Argentina and SE Brazil; feral Cuba

❏ *Catagonus wagneri* **Chacoan Peccary**

Arid Gran Chaco region of SC Bolivia, W Paraguay and N Argentina; Endangered

Family: HIPPOPOTAMIDAE (Hippopotamuses—2)

❏ *Hexaprotodon liberiensis* **Pygmy Hippopotamus**

Riparian habitats and swamps in rainforests of Sierra Leone to Ivory Coast; SC Nigeria

❏ *Hippopotamus amphibius* **Hippopotamus**

Rivers and lakes of subsaharan Africa, S to E South Africa

Family: CAMELIDAE (Camels—7)

❏ *Lama glama* **Llama**

Domesticated, mainly in S Peru, W Bolivia and NW Argentina; wild ancestor is *Lama guanicoe* Guanaco

❏ *Lama guanicoe* **Guanaco**

Grasslands and shrub of S Peru and extreme W Paraguay to S Chile and S Argentina

❏ *Vicugna pacos* Alpaca

Domesticated, mainly in S Peru and W Bolivia; wild ancestor is *Vicugna vicugna* Vicuña

❏ *Vicugna vicugna* Vicuña

Open grasslands of altiplano of Andes of S Peru, W Bolivia, N Chile and NW Argentina

❏ *Camelus bactrianus* Domestic Bactrian Camel

Domesticated in Iran, Afghanistan and Pakistan to Kazakhstan; wild ancestor is *Camelus ferus* Wild Bactrian Camel

❏ *Camelus dromedarius* Dromedary

Domesticated, mainly in N Africa, feral C Australia; wild ancestor unknown

❏ *Camelus ferus* Wild Bactrian Camel

NW & NC China and SW Mongolia; Critically Endangered

Family: TRAGULIDAE (Chevrotains—4)

Tragulus (Moschiola)

❏ *Tragulus meminna* Indian Spotted Chevrotain

Forests and grasslands of Nepal, India and Sri Lanka

Tragulus (Tragulus)

❏ *Tragulus javanicus* Lesser Mouse-Deer

Yunnan (SW China), Thailand, Indochina, Malaya, Sumatra, Java, Borneo, and near islands

❏ *Tragulus napu* Greater Mouse-Deer

Thailand, Indochina, Malaya, Sumatra, Borneo, Balabac (SW Philippines), and near islands

❏ *Hyemoschus aquaticus* Water Chevrotain

Riparian habitats and swamps in rainforests of Sierra Leone to Ghana; Nigeria to D.R. Congo and Uganda

Family: MOSCHIDAE (Musk Deer—4)

❏ *Moschus berezovskii* Chinese Forest Musk Deer

C & S China and N Vietnam

❏ *Moschus chrysogaster* Alpine Musk Deer

Montane forests of N Afghanistan, N Pakistan, N India, Nepal, Tibet and W & C China

❏ *Moschus fuscus* Dusky Musk Deer

Nepal, NE India, Bhutan, N Myanmar, SE Tibet and W Yunnan (SW China)

❏ *Moschus moschiferus* Siberian Musk Deer

Forests of E Siberia incl. Sakhalin I. (Russia), N Mongolia, N China and Korea

Family: ANTILOCAPRIDAE (Pronghorn—1)

❏ *Antilocapra americana* Pronghorn

SC Canada, interior W USA and locally N Mexico; feral Lanai (Hawaiian Is.)

Family: CERVIDAE (Deer—47)

❏ *Hydropotes inermis* Chinese Water Deer

Woods and marshes of S & E China and Korea; feral England and C France

❏ *Muntiacus atherodes* Bornean Yellow Muntjac

Borneo

❏ *Muntiacus crinifrons* Black Muntjac

E China

❏ *Muntiacus feae* Fea's Muntjac

SE Yunnan (SW China), S Myanmar, Thailand and Laos

❏ *Muntiacus gongshanensis* Gong Shan Muntjac

SE Tibet, N Myanmar and W Yunnan (SW China)

❏ *Muntiacus muntjak* Indian Muntjac

Dense forests of NE Pakistan and India to S China, Indochina, Malaya, Sumatra, Java, Bali, Lombok (Lesser Sundas) and Borneo

❏ *Muntiacus putaoensis* Leaf Muntjac

Highlands of NC Myanmar

❏ *Muntiacus reevesi* Reeves' Muntjac

Forests of C & S China and Taiwan; feral Great Britain

❏ *Muntiacus trungsonensis* Truong Son Muntjac

Truong Son Mtns (C Vietnam)

❏ *Muntiacus vuquangensis* Giant Muntjac

Trung Son Mtns (C Vietnam) and Laos

❏ *Elaphodus cephalophus* Tufted Deer

N Myanmar, C & S China

❏ *Cervus albirostris* Thorold's Deer

Tibet, W & C China

❏ *Cervus alfredi* Visayan Spotted Deer

C Philippines; Endangered

❏ *Cervus duvaucelii* Swamp Deer (Barasingha)

Swamp forests and marshes of SW Nepal and N & C India; formerly Pakistan (extinct)

❏ *Cervus elaphus* Red Deer (Elk)

Forests and moorlands of Europe and N Africa to Siberia, Mongolia and N China; W North America; feral USA, S South America and Australasia

❏ *Cervus eldii* Eld's Deer (Thamin)

Riparian forests, swamps and wet grasslands of NE India, Myanmar, Thailand, Laos, Vietnam, Cambodia and Hainan (S China)

❏ *Cervus mariannus* Philippine Brown Deer

Philippines; feral Bonin Is. (Japan, now extinct), Mariana Is. and Caroline Is. (Palau and Fed. States of Micronesia)

❏ *Cervus nippon* Sika Deer

Forests of E Siberia, Korea, E China, Japan incl. Tsushima Is., and Vietnam; feral Europe, Caucasus Mtns, New Zealand, USA and Pacific Ocean islands

❏ †*Cervus schomburgki* Schomburgk's Deer

Formerly Thailand; Extinct (last specimen killed in 1932)

❏ *Cervus timorensis* Rusa

Java, Bali, Sulawesi, Lesser Sundas and Moluccas; feral Indian Ocean islands, Borneo, New Guinea, Australasia, etc.

❏ *Cervus unicolor* Sambar

Forests and grasslands of India and Sri Lanka to S China, Hainan and Taiwan, S to Malaya, Sumatra and Borneo; feral S USA and Australasia

❏ *Axis axis* Chital (Axis Deer)

Riparian grasslands of Pakistan to Bhutan, Bangladesh, India and Sri Lanka; feral Croatia, Andaman Is. (India), Australasia, Hawaii, Texas (USA) and South America

❏ *Axis calamianensis* Calamian Deer

Calamian Is. (WC Philippines); Endangered

❏ *Axis kuhlii* Bawean Deer

Bawean I. (between Java and Borneo); Endangered

❏ *Axis porcinus* Hog Deer

Riparian forests and grasslands of Pakistan, N India and Sri Lanka to Yunnan (SW China) and Vietnam; feral S Australia

❏ *Elaphurus davidianus* Père David's Deer (Milu)

Formerly E China, but became extinct in the wild; reintrod. near Beijing and near Shanghai (EC China); Critically Endangered

❏ *Dama dama* Fallow Deer

Woods of S Turkey; feral Europe, South Africa, Australasia, Fiji, W Canada, USA, S South America and Lesser Antilles

❏ *Dama mesopotamica* Mesopotamian Fallow Deer

Formerly widely in Middle East, now confined to W Iran

❏ *Capreolus capreolus* Western Roe Deer

Woods and farmland of Europe to W Russia and Middle East; ?feral N America

❏ *Capreolus pygargus* Eastern Roe Deer

E Russia and Siberia, S to N Caucasus Mtns, Kazakhstan, Tien Shan, C China and Korea

❏ *Alces alces* Moose (Elk)

Boreal forests of N & E Europe to E Siberia, S to Ukraine and N China; Alaska, Canada and N USA; feral New Zealand

❏ *Odocoileus hemionus* Mule Deer

Forests, brush and deserts of S Alaska, SW & SC Canada, W & C USA and N Mexico

❏ *Odocoileus virginianus* White-tailed Deer

Woods, savannas and regrowth areas of S Canada and USA to French Guiana, S to W Bolivia and N Brazil; feral E Europe, New Zealand and West Indies

❏ *Ozotoceros bezoarticus* Pampas Deer

Open grasslands of S Bolivia, Paraguay, N Argentina, Uruguay and SC & E Brazil

❏ *Mazama americana* Red Brocket

Forests, plantations and gardens of NE Mexico to French Guiana, S to S Bolivia, Paraguay, N Argentina and SE Brazil; Trinidad & Tobago

❏ *Mazama bricenii* Merioa Brocket

Mountains of Cordillera Oriental (NE Colombia and W Venezuela)

❏ *Mazama chunyi* Dwarf Brocket

Montane forests of Andes of S Peru and NW Bolivia

❏ *Mazama gouazoupira* Grey Brocket

Forests, regrowth areas, cerrado and chaco of San Jose I. (Panama) and Colombia to French Guiana, S to Bolivia, Paraguay, N Argentina, Uruguay and S Brazil

❏ *Mazama nana* Pygmy Brocket (Bororó)

Dense montane bamboo thickets of SE Paraguay, Misiones Prov. (NE Argentina) and SE Brazil

❏ *Mazama pandora* Yucatán Brown Brocket

S Mexico

❏ *Mazama rufina* Little Red Brocket

Submontane forests of S Colombia and N Ecuador

❏ *Pudu mephistophiles* Northern Pudu

High montane forests of Cordillera Central (C Colombia), Ecuador and Peru

❏ *Pudu puda* Southern Pudu

Dense forests of S Chile and SW Argentina

❏ *Blastocerus dichotomus* Marsh Deer

Marshes of C Brazil to Paraguay and N & E Argentina

❏ *Hippocamelus antisensis* Peruvian Guemal

High-elevation grasslands and scrub of Andes from Ecuador to N Chile and NW Argentina

❏ *Hippocamelus bisulcus* Chilean Guemal

Steep rocky slopes of Andes of S Chile and S Argentina; Endangered

❏ *Rangifer tarandus* Caribou (Reindeer)

Boreal forests and tundra of N Europe, N Asia, N North America and Greenland; feral Iceland and subantarctic islands

Family: GIRAFFIDAE (Giraffes—2)

❏ *Okapia johnstoni* Okapi

Dense rainforests of N & E D.R. Congo, formerly Uganda

❏ *Giraffa camelopardalis* Giraffe

Locally in dry savannas of subsaharan Africa, S to NE South Africa

Family: BOVIDAE (Cattle, Antelope, Sheep and Goats—141)

❏ *Pelea capreolus* Rhebok

Open grasslands and sand-dunes of South Africa, Lesotho and Swaziland

❏ †*Gazella arabica* Arabian Gazelle

Formerly "Farasan Is." (Saudi Arabia) (type locality uncertain) and N Yemen; Extinct (reported as very common in 1951 but no further reports)

❏ *Gazella bennettii* Indian Gazelle

Arid scrub and semi-deserts of Iran to Pakistan and NW & C India

❏ *Gazella cuvieri* Cuvier's Gazelle

Upland woods and scrub of Morocco and N Algeria (formerly C Tunisia); Endangered

❏ *Gazella dama* Dama Gazelle

Formerly widely in semi-desert grasslands in countries bordering Sahara Desert, now confined to Mali, Burkina Faso, Niger, Chad and Sudan; Endangered

❏ *Gazella dorcas* Dorcas Gazelle

Arid stony deserts of Morocco to S Israel, S to Mauritania, NE Ethiopia and N Somalia

❏ *Gazella gazella* Mountain Gazelle

Lebanon, Syria, Israel and Arabian Peninsula

❏ *Gazella granti* Grant's Gazelle

Upland and seasonally flooded grasslands of SE Sudan and S Ethiopia to N Tanzania

❏ *Gazella leptoceros* Rhim Gazelle

Sand-dune deserts of Algeria to W Egypt; Mali, Niger, N Chad and Sudan; Endangered

❏ *Gazella rufifrons* Red-fronted Gazelle

Grasslands of Senegal to NE Ethiopia, S to N Togo, N Central African Republic and N Tanzania

❏ †*Gazella rufina* Red Gazelle

Formerly Algeria; Extinct (last seen before 1894)

❏ †*Gazella saudiya* Saudi Gazelle

Formerly S Iraq, Kuwait and Saudi Arabia; Extinct

❏ *Gazella soemmerringii* Sömmerring's Gazelle

Locally in highland scrub and shortgrass plains of EC Sudan, Ethiopia and N Somalia

❏ *Gazella spekei* Speke's Gazelle

Stony semi-deserts of E Ethiopia and Somalia

❏ *Gazella subgutturosa* Goitered Gazelle

Middle East and Arabian Peninsula to Mongolia and W China

❏ *Gazella thomsonii* Thomson's Gazelle

SE Sudan, SW Ethiopia, Kenya and N Tanzania

❏ *Gazella gutturosa* Mongolian Gazelle

Formerly E Siberia, Mongolia and N China, now confined to E Mongolia and Nei Mongol (N China)

❏ *Gazella picticaudata* Tibetan Gazelle

Tibet, NE India and C China

❏ *Gazella przewalskii* Przewalski's Gazelle

N China; Critically Endangered

❏ *Antilope cervicapra* Blackbuck

Grasslands and farmland of India; reintrod. E Pakistan; formerly Bangladesh; feral Nepal, Texas (USA) and Argentina

❏ *Antidorcas marsupialis* Springbok

Open arid grasslands and scrub of SW Angola, Namibia, Botswana and NW South Africa

❏ *Pantholops hodgsonii* Chiru

High-altitude desert grasslands of N Ladakh (India), Tibet and C China; Endangered

❏ *Litocranius walleri* Gerenuk

Savanna scrub of E Ethiopia, Somalia, Kenya and NE Tanzania

❏ *Saiga tatarica* Saiga

N Caucasus Mtns (Russia), Kazakhstan, Uzbekistan, Xinjiang (NW China) and SW Mongolia; Critically Endangered

❏ *Ammodorcas clarkei* Dibatag

Camel-brush thickets of E Ethiopia and N Somalia

❏ *Raphicerus campestris* Steenbok

Open arid areas and stony savannas of S Kenya and Tanzania; Angola and W Zambia S to South Africa, Swaziland and S Mozambique

❏ *Raphicerus melanotis* Cape Grysbok

Dense fynbos and scrub of S Cape Prov. (South Africa)

❏ *Raphicerus sharpei* Sharpe's Grysbok

Dense scrub of SE D.R. Congo and Tanzania, S to Swaziland and NE South Africa

❏ *Madoqua guentheri* Günther's Dik-dik

Arid desert and semi-desert scrub of SE Sudan, S Ethiopia, C & S Somalia, N Kenya and NE Uganda

❏ *Madoqua kirkii* Kirk's Dik-dik

Dense dry woods and scrub of S Somalia, Kenya, N & C Tanzania; SW Angola and Namibia

❏ *Madoqua piacentinii* Silver Dik-dik

Dense thickets on sandy littoral of E Somalia

❏ *Madoqua saltiana* Salt's Dik-dik

Dense scrub of NE Sudan, Djibouti, N Ethiopia and Somalia

❏ *Oreotragus oreotragus* Klipspringer

Dense grassy areas and scrub in rocky habitats of C Nigeria, N Central African Republic, E Sudan and Ethiopia, S to South Africa

❏ *Ourebia ourebi* Oribi

Open shortgrass habitats of Senegal to C Ethiopia and S Somalia, S to N Namibia, N Botswana and E South Africa

❏ *Neotragus batesi* Dwarf Antelope

Dense riparian scrub, regrowth areas and plantations of SE Nigeria, SE Cameroon, E D.R. Congo, W Uganda, NE Gabon and N Rep. Congo

❏ *Neotragus moschatus* Suni

Dense dry woods and scrub of SE Kenya and Tanzania incl. Zanzibar and Mafia Is., S to E South Africa

❏ *Neotragus pygmaeus* Royal Antelope

Forest edge, regrowth areas and farmland of Sierra Leone to Ghana

❏ *Dorcatragus megalotis* Beira

Highland rocky slopes of Djibouti, NE Ethiopia and N Somalia

❏ *Capra aegagrus* Wild Goat

Rocky slopes of Crete (Greece), Turkey and Caucasus Mtns to Pakistan

❏ *Capra caucasica* West Caucasian Tur

W Caucasus Mtns (Russia); Endangered

❏ *Capra cylindricornis* East Caucasian Tur

E Caucasus Mtns (Russia, E Georgia and Azerbaijan)

❏ *Capra falconeri* Markhor

Steep rocky montane slopes of S Uzbekistan, Tajikistan, Afghanistan, N & C Pakistan, Kashmir and N India; Endangered

❏ *Capra hircus* Domestic Goat

Domesticated worldwide; feral S Europe, North and South America, New Zealand, Pacific and Indian Ocean islands; wild ancestor is *Capra aegagrus* Wild Goat

❏ *Capra ibex* Alpine Ibex

Steep rocky montane slopes of Alps from France to Austria (reintrod.) and Gran Paradiso Massif (N Italy); feral Slovenia, Bulgaria and USA

❏ *Capra nubiana* Nubian Ibex

Steep rocky slopes of E Egypt and NE Sudan; Syria to Yemen and Oman; Endangered

❏ *Capra pyrenaica* Spanish Ibex

Montane cliffs and meadows of Spain

❏ *Capra sibirica* Siberian Ibex

Mountains of E Kazakhstan, C Siberia and W Mongolia, S to N Pakistan, N India and NW China

❏ *Capra walie* Walia Ibex

Montane scrub and grasslands of Simen Mtns (N Ethiopia); Critically Endangered

❏ *Ovis ammon* Argali

Mountains of E Kazakhstan, C Siberia and Mongolia, S to N Pakistan, N India and Nei Mongol (N China)

❏ *Ovis aries* Domestic Sheep

Domesticated worldwide; feral widely in Europe, New Zealand, USA, S South America and Kerguelen Is.; wild ancestor is *Ovis orientalis* Mouflon

❏ *Ovis canadensis* Bighorn Sheep

Montane cliffs and deserts of interior SW Canada, W USA and NW Mexico

❏ *Ovis dalli* Dall's Sheep

Montane cliffs of Alaska (USA) and NW Canada

❏ *Ovis nivicola* Snow Sheep

NC & NE Siberia incl. Kamchatka Peninsula

❏ *Ovis orientalis* Mouflon

Mountains and semi-deserts of S & E Turkey, Armenia, S Azerbaijan, N Iraq and W Iran

❏ *Ovis vignei* Urial

Uzbekistan and NE Iran to Pakistan and NW India; ?feral Oman

❏ *Hemitragus hylocrius* Nilgiri Tahr

Rocky slopes in hill forest edges and grasslands of SW & S India; Endangered

❏ *Hemitragus jayakari* Arabian Tahr

United Arab Emirates and Oman; Endangered

❏ *Hemitragus jemlahicus* Himalayan Tahr

Montane forests and moorlands of Kashmir, N & NE India, Nepal and S Tibet; feral SW Western Cape Prov. (South Africa) and New Zealand

❏ *Ammotragus lervia* Barbary Sheep (Aoudad)

Rocky areas in deserts of Western Sahara to W Egypt, S to Mali and Sudan; feral SE Spain, Czech Republic, USA and N Mexico

❏ *Rupicapra pyrenaica* Pyrenean Chamois

Montane forests and alpine meadows of N Spain, SW France and Abruzzo Mtns (C Italy)

❏ *Rupicapra rupicapra* Chamois

Montane cliffs and alpine meadows of C & SE Europe, Turkey and Caucasus Mtns; feral New Zealand and Argentina

❏ *Oreamnos americanus* Mountain Goat

High mountain cliffs and meadows of SE Alaska, W Canada and NW USA

❏ *Pseudois nayaur* Bharal

High-altitude montane grasslands of Tajikistan, N Pakistan, N India, Tibet and SW & C China

❏ *Pseudois schaeferi* Dwarf Bharal

Upper Yangtze R. gorge (W China); Endangered

❏ *Budorcas taxicolor* Takin

Mid-elevation and montane forests and bamboo thickets of NE India, Bhutan, SE Tibet, N Myanmar and SW & C China

❏ *Ovibos moschatus* Muskox

Tundra of NE Alaska, far N Canada and N Greenland; feral C Scandinavia, N Siberia and W Alaska (USA)

❏ *Naemorhedus baileyi* Red Goral

Assam (NE India), SE Tibet, N Myanmar and Yunnan (SW China)

❏ *Naemorhedus caudatus* Chinese Goral

E Myanmar and W Thailand; E Siberia and E China

❏ *Naemorhedus goral* **Common Goral**

Hill forests and grasslands of N Pakistan, N & NE India, Nepal and Bhutan

❏ *Capricornis crispus* **Japanese Serow**

Honshu, Shikoku and Kyushu (C & S Japan)

❏ *Capricornis sumatraensis* **Sumatran Serow**

Mid-elevation and montane forests of N & NE India, Nepal, Myanmar and C China, S to Indochina, Malay Peninsula and Sumatra

❏ *Capricornis swinhoei* **Taiwan Serow**

Taiwan

❏ *Pseudoryx nghetinhensis* **Saola (Vu Quang Ox)**

Highlands of NC Vietnam; Endangered

❏ *Boselaphus tragocamelus* **Nilgai**

Forests, forest edges and scrub of E Pakistan, SW Nepal and India; feral S Texas (USA)

❏ *Tetracerus quadricornis*
Four-horned Antelope (Chousingha)

Open forests, woods and grasslands of C Nepal and India

❏ *Syncerus caffer* **African Buffalo**

Savannas and grasslands of subsaharan Africa, S to NE South Africa

❏ *Bos frontalis* **Mithan**

Domesticated in hills of Assam (NE India); wild ancestor is *Bos gaurus* Gaur

❏ *Bos gaurus* **Gaur**

Very locally in forests and grasslands of India and Nepal to Yunnan (SW China), S to Malay Peninsula

❏ *Bos grunniens* **Domestic Yak**

Domesticated in C Asia; wild ancestor is *Bos mutus* Wild Yak

❏ *Bos indicus* **Zebu (Humped Cattle)**

Domesticated subsaharan Africa, Asia, N America etc.; wild ancestor is extinct *Bos namadicus* Indian Aurochs

❏ *Bos javanicus* **Banteng (Bali Cattle)**

Domesticated in SE Asia (Bali Cattle), feral Enggano (Sumatra), Bali, Sangihe Is. (N Moluccas) and Australia, wild in Java; Endangered

❏ *Bos mutus* **Wild Yak**

High-altitude grasslands of N India, Nepal, Bhutan, Tibet and W & C China

❏ *Bos sauveli* **Kouprey**

SE Thailand, Cambodia, S Laos and W Vietnam; Critically Endangered

❏ *Bos taurus* **European Domestic Cattle**

Domesticated worldwide, feral Europe, Australasia, North and South America, Pacific Ocean islands etc.; wild ancestor is extinct *Bos primigenius* Aurochs

❏ *Bubalus arnee* **Wild Water Buffalo**

Riparian forests and grasslands of N India, Nepal, S Bhutan and N Thailand; Endangered

❏ *Bubalus bubalis* **Domestic Water Buffalo**

Domesticated in S Europe, N Africa, Asia and South America, feral Sri Lanka, Borneo, Bismarck Archipelago, Australia etc.; wild ancestor is *Bubalus arnee* Wild Water Buffalo

❏ *Bison bison* **American Bison (Buffalo)**

Formerly prairies of NW Canada to N Mexico, now mainly reintrod. to scattered localities in Alaska, C Canada and W & SC USA

❏ *Bison bonasus* **European Bison (Wisent)**

Formerly widely in forests of Europe and Russia, now confined to Poland, W Russia and Caucasus Mtns; Endangered

❏ *Anoa depressicornis* **Lowland Anoa**

Sulawesi; Endangered

❏ *Anoa mindorensis* **Tamaraw**

Mindoro (WC Philippines); Critically Endangered

❏ *Anoa quarlesi* **Mountain Anoa**

Mountains of Sulawesi; Endangered

❏ *Tragelaphus angasii* **Nyala**

Savannas and riparian woods of S Malawi, Zimbabwe, Mozambique, E South Africa and Swaziland

❏ *Tragelaphus buxtoni* **Mountain Nyala**

Montane woods and scrub of Bale Mtns (EC Ethiopia), formerly more widely in Ethiopia; Endangered

❏ *Tragelaphus imberbis* **Lesser Kudu**

Deciduous woods and scrub of SE Sudan, SE Ethiopia, Somalia, Kenya, NE Uganda and E Tanzania; SW Saudi Arabia and Yemen

❑ *Tragelaphus scriptus* Bushbuck

Dense woods and scrub of subsaharan Africa

❑ *Tragelaphus spekii* Sitatunga

Forest swamps and reedbeds of Senegal to S Sudan, S to NE Namibia, N Botswana and NW Zimbabwe

❑ *Tragelaphus strepsiceros* Greater Kudu

Dense riparian and montane woods of S Chad and N Central African Republic to E Sudan, S to South Africa

❑ *Taurotragus derbianus* Giant Eland

Isoberlinia woodland remnants of Senegal to Guinea; Cameroon to S Sudan; formerly S to NW Uganda

❑ *Taurotragus oryx* Eland

Savannas and scrub-covered plains of SE Sudan and SW Ethiopia, S to South Africa and Lesotho

❑ *Boocercus eurycerus* Bongo

Locally in forests of Sierra Leone to S Sudan, S to N Gabon, D.R. Congo and S Kenya

❑ *Redunca arundinum* Southern Reedbuck

Reedbeds and dense grasslands of S Gabon, S D.R. Congo and Tanzania, S to E South Africa

❑ *Redunca fulvorufula* Mountain Reedbuck

Rocky slopes in highlands of E Nigeria to SE Sudan and C Ethiopia, S to C South Africa, Swaziland and Lesotho

❑ *Redunca redunca* Bohor Reedbuck

Riparian grasslands of Senegal to C Ethiopia, S to N D.R. Congo, Rwanda, Burundi and Tanzania

❑ *Kobus ellipsiprymnus* Waterbuck

Riparian woods and scrub of Senegal to Somalia, S to E South Africa

❑ *Kobus kob* Kob

Riparian grasslands of Senegal to Sudan and W Ethiopia, S to N D.R. Congo, Uganda and W Kenya; formerly NW Tanzania

❑ *Kobus leche* Lechwe

Seasonal wetlands of SE D.R. Congo, SE Angola, Zambia, extreme NE Namibia and N Botswana

❑ *Kobus megaceros* Nile Lechwe

Seasonally flooded riparian grasslands of Sudd region (upper White Nile R., S Sudan and SW Ethiopia)

❑ *Kobus vardonii* Puku

Riparian grasslands and marshes of S D.R. Congo and S Tanzania, S to NE Namibia and N Botswana; vagrant N Zimbabwe

❑ *Hippotragus equinus* Roan Antelope

Open and lightly wooded grasslands of Senegal to W Ethiopia, S to NE South Africa

❑ *Hippotragus niger* Sable Antelope

Miombo (*Brachystegia*) woods and grasslands of SE Kenya to Angola and NE South Africa

❑ *Oryx dammah* Scimitar-horned Oryx

Formerly semi-desert grasslands of Western Sahara and Mauritania to Egypt and Sudan; Extinct in the Wild

❑ *Oryx gazella* Gemsbok (Beisa Oryx)

Arid open grasslands and scrub of SE Sudan, NE Ethiopia and Somalia to N Tanzania; SW Angola, Botswana and W Zimbabwe to NW South Africa

❑ *Oryx leucoryx* Arabian Oryx

SE Arabian Peninsula (reintrod.) and formerly Iraq; Endangered

❑ *Addax nasomaculatus* Addax

Sand-dune deserts of Niger (and Chad?); formerly Tunisia and Mauritania to Egypt and Sudan; Critically Endangered

❑ *Alcelaphus buselaphus* Hartebeest

Savannas and grasslands of Senegal to Ethiopia, locally S to South Africa; formerly Morocco to Egypt

❑ *Connochaetes gnou* Black Wildebeest

Formerly widely in open grasslands of South Africa, now extinct in the wild but locally reintrod. in South Africa, Lesotho and Swaziland

❑ *Connochaetes taurinus* Blue Wildebeest

Open savannas and grasslands of S Kenya and Tanzania; S Angola and W Zambia to E South Africa

❑ *Damaliscus hunteri* Hirola (Hunter's Hartebeest)

Very locally in seasonally arid grassy plains of SE Kenya (formerly S Somalia); Critically Endangered

❑ *Damaliscus lunatus* Topi (Tsessebe)

Formerly widely in savannas of subsaharan Africa, now locally from Burkina Faso to Sudan, S to NE South Africa

❑ *Damaliscus pygargus* Bontebok (Blesbok)

Formerly widely in open grasslands and fynbos of South Africa, now locally reintrod. in South Africa, Lesotho and Swaziland

❏ *Sigmoceros lichtensteinii*
Lichtenstein's Hartebeest

Savannas of Angola, SE D.R. Congo and Tanzania to E Zimbabwe and C Mozambique; introd. NE South Africa

❏ *Aepyceros melampus* **Impala**

Open savannas of Angola, S D.R. Congo, Uganda and Kenya, S to NE South Africa

❏ *Cephalophus adersi* **Aders' Duiker**

Dense forests of Sokoke Forest (SE Kenya) and Zanzibar I. (Tanzania); Endangered

❏ *Cephalophus callipygus* **Peters' Duiker**

Dense forests and regrowth areas of S Cameroon, S Central African Republic, Gabon and Rep. Congo

❏ *Cephalophus dorsalis* **Bay Duiker**

Lowland primary rainforests of Sierra Leone to Central African Republic, S to N Angola and D.R. Congo

❏ *Cephalophus harveyi* **Harvey's Duiker**

Forests, woods and regrowth areas of E Ethiopia and S Somalia, S to E Zambia and N Malawi

❏ *Cephalophus jentinki* **Jentink's Duiker**

Forests, regrowth areas, plantations and farmland of Sierra Leone, Liberia and W Ivory Coast

❏ *Cephalophus leucogaster* **White-bellied Duiker**

Very locally in forests of Cameroon, Gabon, Rep. Congo and SW & E D.R. Congo

❏ *Cephalophus maxwellii* **Maxwell's Duiker**

Rainforests of Senegal and Gambia to SW Nigeria

❏ *Cephalophus monticola* **Blue Duiker**

Forests and dense scrub of E Nigeria and Bioko (Equatorial Guinea) to Kenya and Tanzania incl. Zanzibar and Pemba I., S to SE South Africa

❏ *Cephalophus natalensis* **Natal Duiker**

Forests and dense woods of S Tanzania, S Malawi, Mozambique, Swaziland and E South Africa

❏ *Cephalophus niger* **Black Duiker**

Forests and regrowth areas of Guinea to lower Niger R. (Nigeria)

❏ *Cephalophus nigrifrons* **Black-fronted Duiker**

Forest swamps and marshes of Cameroon to Uganda and Kenya, S to N Angola

❏ *Cephalophus ogilbyi* **Ogilby's Duiker**

Locally in forests of Sierra Leone to S Cameroon, S Gabon and Bioko (Equatorial Guinea)

❏ *Cephalophus rubidus* **Ruwenzori Duiker**

Montane woods and bamboo thickets of Ruwenzori Mtns (E D.R. Congo and W Uganda)

❏ *Cephalophus rufilatus* **Red-flanked Duiker**

Forests and riparian woods of Senegal to SW Sudan and NE Uganda, S to Cameroon and N D.R. Congo

❏ *Cephalophus silvicultor* **Yellow-backed Duiker**

Forests and woods of Senegal to SW Sudan, W Uganda and W Kenya, S to Angola and Zambia

❏ *Cephalophus spadix* **Abbott's Duiker**

Montane forests, scrub and swamps of E Tanzania

❏ *Cephalophus weynsi* **Weyns' Duiker**

Dense forests and regrowth areas of D.R. Congo, Uganda, W Kenya and Rwanda

❏ *Cephalophus zebra* **Zebra Duiker**

Forests, forest edge and regrowth areas of E Sierra Leone, Liberia and W Ivory Coast

❏ *Sylvicapra grimmia* **Bush Duiker**

Savannas, scrub and farmland of subsaharan Africa

Mirorder: ALTUNGULATA
Order: PERISSODACTYLA
Family: EQUIDAE (Horses—11)

❏ *Equus africanus* **Wild Ass**

Semi-desert grasslands and scrub of NE Sudan, NE Ethiopia and N Somalia; Critically Endangered

❏ *Equus asinus* **Domestic Donkey**

Domesticated worldwide, feral N Africa, Saudi Arabia, Sri Lanka, Australia, USA etc.; wild ancestor is *Equus africanus* Wild Ass

❏ *Equus burchellii* **Burchell's Zebra**

Grasslands and savannas of SE Sudan and SW Ethiopia, S to N Namibia, N & E Botswana and E South Africa

❏ *Equus caballus* **Domestic Horse**

Domesticated worldwide, feral S & SE Europe, Iran, Sri Lanka, Australasia, North and South America etc.; wild ancestor is *Equus ferus* Przewalski's Horse

❏ *Equus ferus* Przewalski's Horse

Mongolia (reintrod.), formerly N China (extinct)

❏ *Equus grevyi* Grevy's Zebra

Deserts and seasonally flooded grasslands of S & E Ethiopia, S Somalia and N Kenya; Endangered

❏ *Equus hemionus* Kulan

E Siberia (extinct), Mongolia and N China

❏ *Equus kiang* Kiang

High-altitude grasslands of NW & NE India, Nepal, Tibet and C China

❏ *Equus onager* Onager

C Iran, Turkmenistan and Rann of Kutch (India)

❏ †*Equus quagga* Quagga

Formerly South Africa; Extinct (last captive specimen died 1872)

❏ *Equus zebra* Mountain Zebra

Upland arid scrub of S Angola, Namibia and S Cape Prov. (South Africa); Endangered

Family: RHINOCEROTIDAE (Rhinoceroses—5)

❏ *Dicerorhinus sumatrensis* Sumatran Rhinoceros

Very locally in forests of Malay Peninsula, Sumatra and Borneo (formerly to NE India and Vietnam); Critically Endangered

❏ *Rhinoceros sondaicus* Javan Rhinoceros

Very locally in forests of W Java and Vietnam (formerly to Bangladesh, Malay Peninsula and Sumatra); Critically Endangered

❏ *Rhinoceros unicornis* Indian Rhinoceros

Locally in riparian woods and grasslands of Nepal and NE India; Endangered

❏ *Ceratotherium simum* White Rhinoceros

Savannas and shortgrass habitats of NE D.R. Congo; Namibia, Botswana, Zimbabwe, Mozambique, E South Africa and Swaziland (formerly to S Chad and S Sudan)

❏ *Diceros bicornis* Black Rhinoceros

Savannas and scrub of Kenya, Tanzania, Namibia, Zambia, Zimbabwe and E South Africa (formerly to Chad and S Sudan); Critically Endangered

Family: TAPIRIDAE (Tapirs—4)

❏ *Tapirus bairdii* Baird's Tapir

Undisturbed forests and wet grasslands of S Mexico to Panama and W of the Andes in Colombia and Ecuador; Endangered

❏ *Tapirus indicus* Malayan Tapir

S Myanmar, Thailand, Malaya and Sumatra; Endangered

❏ *Tapirus pinchaque* Mountain Tapir

High montane forests and páramo of Andes of Colombia and Ecuador; Endangered

❏ *Tapirus terrestris* Brazilian Tapir

Forests and wet grasslands of Colombia to Suriname, S to Paraguay, N Argentina and S Brazil

Order: URANOTHERIA
Family: PROCAVIIDAE (Hyraxes—6)

❏ *Procavia capensis* Rock Hyrax

Rocky habitats of N Africa; Senegal to Sudan, S to N Tanzania; Angola and S Malawi to South Africa; Middle East and Arabian Peninsula

❏ *Heterohyrax antineae* Hoggar Hyrax

Rocky habitats of Hoggar Mtns (SE Algeria)

❏ *Heterohyrax brucei* Yellow-spotted Hyrax

Rocky habitats of Egypt, S to Angola and N South Africa

❏ *Dendrohyrax arboreus* Southern Tree Hyrax

Forests and woods of Sudan, D.R. Congo, Kenya, Tanzania, Zambia, Malawi, Mozambique and SE South Africa

❏ *Dendrohyrax dorsalis* Western Tree Hyrax

Forests and woods of Gambia to NE D.R. Congo and N Uganda, S to N Angola; Bioko (Equatorial Guinea)

❏ *Dendrohyrax validus* Eastern Tree Hyrax

Forests and woods of S Kenya and E Tanzania incl. Pemba and Zanzibar Is.

Family: TRICHECHIDAE (Manatees—3)

❏ *Trichechus inunguis* Amazonian Manatee

Amazon R. system of extreme SE Colombia to Guyana, S to Ecuador, Peru and Brazil

❏ *Trichechus manatus* West Indian Manatee

Coastal tropical and subtropical waters and river systems of W
Atlantic Ocean and Caribbean Sea

❏ *Trichechus senegalensis* West African Manatee

Coastal tropical waters and river systems of W Africa from S
Mauritania to Angola

Family: DUGONGIDAE (Dugong—1)

❏ *Dugong dugon* Dugong

Discontinuously in coastal tropical waters of Indian and W
Pacific Oceans

Family: ELEPHANTIDAE (Elephants—3)

❏ *Loxodonta africana* African Savanna Elephant

Savannas and grasslands of S Mauritania to Sudan, S to South
Africa; Endangered

❏ *Loxodonta cyclotis* African Forest Elephant

Forests of W & C Africa; Endangered

❏ *Elephas maximus* Asian Elephant

Forests, bamboo thickets and grasslands of India, Nepal and
Sri Lanka to Myanmar, China, Thailand, Laos, Vietnam,
Cambodia, Malaya and Indonesia; Endangered

APPENDIX
Changes compared to Wilson & Reeder (1993)

This appendix details the major changes to the species classification in the present work, compared to the 2nd edition (1993) of Wilson & Reeder's catalogue of mammalian species. In addition to the changes listed below, note that many of the genus assignments differ since the higher classification we used is based on McKenna & Bell (1997).

EXTINCT SPECIES NOT INCLUDED

The catalogue of mammal species by Wilson & Reeder (1993) includes "existing or recently extinct species (possibly alive during the preceding 500 years)" (*ibid.*, p. 9), although a few species that certainly became extinct much earlier than that are also listed e.g. Short-horned Water Buffalo *Bubalus mephistopheles*. The present checklist is mainly intended for keeping a 'life list' of mammals seen in the wild, so we use a more recent cut-off date and exclude species which certainly became extinct before 1800.

The following 41 species of mammals listed by Wilson & Reeder (1993) are either known only from subfossil remains or certainly became extinct before 1800, and have therefore been excluded from the main checklist:

OCHOTONIDAE (Pikas)

Prolagus sardus Sardinian Pika – Corsica and Sardinia, extinct by 18th Century

MURIDAE (Mice, Rats, Voles and Gerbils)

Juscelinomys talpinus Mole-like Burrowing Mouse – Minas Gerais (Brazil), subfossil

Bibimys labiosus Large-lipped Crimson-nosed Rat – Minas Gerais (Brazil), subfossil

Canariomys tamarani Canary Mouse – Canary Is., subfossil

Malpaisomys insularis Lava Mouse – Canary Is., subfossil

Rattus sanila New Ireland Rat – New Ireland (Bismarck Archipelago), subfossil

Papagomys theodorverhoeveni Verhoeven's Giant Tree Rat – Flores (Lesser Sundas), subfossil

Solomys spriggsarum Buka Naked-tailed Rat – Buka I. (Solomon Is.), subfossil

Spelaeomys florensis Flores Cave Rat – Flores (Lesser Sundas), subfossil

Coryphomys buehleri Buhler's Rat – Timor (Lesser Sundas), subfossil

ECHIMYIDAE (Spiny Rats)

Boromys offella Oriente Cave Rat – Cuba and Isla de Juventud, subfossil

Boromys torrei Torre's Cave Rat – Cuba and Isla de Juventud, subfossil

Heteropsomys antillensis Antillean Cave Rat – Puerto Rico, subfossil

Heteropsomys insulans Insular Cave Rat – Puerto Rico, subfossil

Brotomys contractus Haitian Edible Rat – Hispaniola, subfossil

Brotomys voratus Hispaniolan Edible Rat – Hispaniola and La Gonave I., subfossil. Wilson & Reeder (1993) state that *Brotomys voratus* became extinct "in the last 50 years", but McFarlane (2000) claims that this is not supported by the published radiometric data.

Puertoricomys corozalus Corozal Rat – Puerto Rico, subfossil

CAPROMYIDAE (Hutias)

Hexolobodon phenax Imposter Hutia – Hispaniola and La Gonave I., subfossil

Quemisia gravis Twisted-toothed Hutia – Hispaniola, subfossil

Isolobodon montanus Montane Hutia – Hispaniola and La Gonave I., subfossil

Plagiodontia araeum San Rafael Hutia – Hispaniola, subfossil

Plagiodontia ipnaeum Samana Hutia – Hispaniola, subfossil

Rhizoplagiodontia lemkei Lemke's Hutia – SW Haiti, subfossil

HEPTAXODONTIDAE (Giant Hutias)

Elasmodontomys obliquus Plate-toothed Giant Hutia – Puerto Rico, subfossil

Clidomys osborni (including *C. parvus* as a synonym, per D. McFarlane pers. comm.) Jamaican Giant Hutia – Jamaica, subfossil

Amblyrhiza inundata Lesser Antillean Giant Hutia – Anguilla and St Martin, subfossil

NESOPHONTIDAE (Nesophontes)

Nesophontes edithiae Puerto Rican Nesophontes – Puerto Rico, subfossil

Nesophontes hypomicrus Atalaye Nesophontes – Haiti and La Gonave I., subfossil

Nesophontes longirostris Slender Cuban Nesophontes – Cuba, subfossil

Nesophontes major Greater Cuban Nesophontes – Cuba, subfossil

Nesophontes micrus Western Cuban Nesophontes – Cuba, Haiti and Isla de Juventud, subfossil

Nesophontes paramicrus St Michel Nesophontes – Haiti, subfossil

Nesophontes submicrus Lesser Cuban Nesophontes – Cuba, subfossil

Nesophontes zamicrus Haitian Nesophontes – Haiti, subfossil

SOLENODONTIDAE (Solenodons)

Solenodon marcanoi Marcano's Solenodon – Dominican Republic, subfossil

HIPPOPOTAMIDAE (Hippopotamuses)

Hexaprotodon madagascariensis Madagascar Pygmy Hippopotamus – Madagascar, extinct c.1500

Hippopotamus lemerlei Madagascar Dwarf Hippopotamus – Madagascar, extinct c.1500

BOVIDAE (Cattle, Antelope, Sheep and Goats)

Hippotragus leucophaeus Bluebuck – S Cape Prov. (South Africa), extinct by 1800

Bubalus mephistopheles Short-horned Water Buffalo – NE China, extinct by 12th Century BC

DUGONGIDAE (Dugongs)

Hydrodamalis gigas Steller's Sea Cow – Bering I. and Copper I. (Alaska, USA), extinct by 1768

NEW SPECIES

We have attempted to bring the species taxonomy up to date by including all species newly described since Wilson & Reeder (1993) was published. The following new species have come to our attention.

TACHYGLOSSIDAE (Echidnas)

Zaglossus attenboroughi Attenborough's Echidna – Flannery & Groves, 1998

Zaglossus bartoni Barton's Echidna split from *Zaglossus bruijni* Long-beaked Echidna – Flannery & Groves, 1998

DASYURIDAE (Dasyurids)

Antechinus (Antechinus) adustus Rusty Antechinus split from *A. (Antechinus) stuartii* Brown Antechinus – Van Dyck, 1982; Van Dyck & Crowther, 2000; Menkhorst & Knight, 2001

Antechinus (Antechinus) agilis Agile Antechinus – Dickman, Parnaby, Crowther & King, 1998

Antechinus (Antechinus) habbema Habbema Antechinus split from *A. (Antechinus) naso* Long-nosed Antechinus – Flannery, 1995a

Antechinus (Antechinus) subtropicus Subtropical Antechinus – Van Dyck & Crowther, 2000

Antechinus (Pseudantechinus) mimulus Carpentarian Pseudantechinus split from *A. (Pseudantechinus) macdonnellensis* Fat-tailed Pseudantechinus – Menkhorst & Knight, 2001

Antechinus (Pseudantechinus) roryi Tan Pseudantechinus – Cooper, Aplin & Adams, 2000

Dasycercus hillieri Ampurta split from *D. cristicauda* Mulgara – Menkhorst & Knight, 2001

Sminthopsis bindi Kakadu Dunnart – Van Dyck, Woinarski & Press, 1994

PHALANGERIDAE (Cuscuses and Brushtail Possums)

Trichosurus cunninghamii Mountain Brushtail Possum split from *T. caninus* Short-eared Brushtail Possum – Lindenmayer, Dubach & Viggers, 2002

Spilocuscus kraemeri Admiralty Cuscus split from *S. maculatus* Common Spotted Cuscus -Flannery, 1995b

Spilocuscus papuensis Waigeo Cuscus split from *S. maculatus* Common Spotted Cuscus -Flannery, Archer & Maynes, 1987; Flannery, 1995a

Phalanger alexandrae Gebe Cuscus – Flannery & Boeadi, 1995

Phalanger intercastellanus Southern Common Cuscus split from *P. orientalis* Northern Common Cuscus – Flannery 1995a

Phalanger mimicus Cryptic Cuscus split from *P. orientalis* Northern Common Cuscus – Norris & Musser, 2001

MACROPODIDAE (Kangaroos and Wallabies)

Bettongia tropica Northern Bettong split from *B. penicillata* Brush-tailed Bettong – Menkhorst & Knight, 2001

Potorous gilbertii Gilbert's Potoroo split from *P. tridactylus* Long-nosed Potoroo – Menkhorst & Knight, 2001

Thylogale browni New Guinea Pademelon split from *T. brunii* Dusky Pademelon – Flannery, 1995a

Thylogale calabyi Calaby's Pademelon – Flannery, 1992

Petrogale coenensis Cape York Rock Wallaby split from *P. godmani* Godman's Rock Wallaby – Eldridge & Close, 1992; Menkhorst & Knight, 2001

Petrogale herberti Herbert's Rock Wallaby split from *P. penicillata* Brush-tailed Rock Wallaby – Eldridge & Close, 1997; Menkhorst & Knight, 2001

Petrogale mareeba Mareeba Rock Wallaby split from *P. godmani* Godman's Rock Wallaby – Eldridge & Close, 1992; Menkhorst & Knight, 2001

Petrogale purpureicollis Purple-necked Rock Wallaby split from *P. lateralis* Black-footed Rock Wallaby – Eldridge *et al.*, 2001; Menkhorst & Knight, 2001

Petrogale sharmani Sharman's Rock Wallaby split from *P. assimilis* Allied Rock Wallaby – Eldridge & Close, 1992; Menkhorst & Knight, 2001

Dendrolagus mbaiso Dingiso Tree Kangaroo – Flannery, Boeadi & Szalay, 1995

PETAURIDAE (Ringtail Possums and Gliders)

Pseudocheirus cinereus Daintree River Ringtail Possum split from *P. herbertensis* Herbert River Ringtail Possum – Menkhorst & Knight, 2001

Pseudocheirus occidentalis Western Ringtail Possum split from *P. peregrinus* Common Ringtail Possum – Kerle, 2001; Menkhorst & Knight, 2001

Pseudochirops coronatus Reclusive Ringtail Possum split from *P. albertisii* D'Albertis' Ringtail Possum – Flannery, 1995a

Petaurus biacensis Biak Glider split from *P. breviceps* Sugar Glider – Flannery, 1995a

DIDELPHIDAE (American Opossums)

Marmosa (Marmosops) bishopi Bishop's Slender Mouse-Opossum split from *M. (Marmosops) parvidens* Delicate Slender Mouse-Opossum – Voss, Lunde & Simmons, 2001

Marmosa (Marmosops) juninensis Junín Slender Mouse-Opossum split from *M. (Marmosops) parvidens* Delicate Slender Mouse-Opossum – Voss, Lunde & Simmons, 2001

Marmosa (Marmosops) neblina Cerro Neblina Slender Mouse-Opossum split from *M. (Marmosops) impavidus* Andean Slender Mouse-Opossum – Emmons & Feer, 1997

Marmosa (Marmosops) paulensis São Paulo Slender Mouse-Opossum split from *M. (Marmosops) incana* Grey Slender Mouse-Opossum – Mustrangi & Patton, 1997

Marmosa (Marmosops) pinheiroi Pinheiro's Slender Mouse-Opossum split from *M. (Marmosops) parvidens* Delicate Slender Mouse-Opossum – Voss, Lunde & Simmons, 2001

Gracilinanus ignitus Red-bellied Gracile Mouse-Opossum – Díaz, Flores & Barquez, 2002

Gracilinanus kalinowskii Kalinowski's Gracile Mouse-Opossum – Hershkovitz, 1992

Gracilinanus longicaudus Long-tailed Gracile Mouse-Opossum – Hershkovitz, 1992

Gracilinanus perijae Sierra de Perijá Gracile Mouse-Opossum – Hershkovitz, 1992

Monodelphis orinoci Orinoco Short-tailed Opossum split from *M. brevicaudata* Red-legged Short-tailed Opossum – Perez-Hernandez, Soriano & Lew, 1993; Ventura, Perez-Hernandez & Lopez-Fuster, 1998

Thylamys cinderella Cinderella Fat-tailed Opossum split from *T. elegans* Elegant Fat-tailed Opossum – Flores *et al.*, 2000

Thylamys sponsorius Sponsorial Fat-tailed Opossum split from *T. elegans* Elegant Fat-tailed Opossum – Flores *et al.*, 2000

Thylamys venustus Pretty Fat-tailed Opossum split from *T. elegans* Elegant Fat-tailed Opossum – Palma & Yates, 1998

Micoureus phaea Little Woolly Mouse-Opossum split from *M. regina* Short-furred Woolly Mouse-Opossum – Emmons & Feer, 1997

Philander frenata Bridled Four-eyed Opossum split from *P. opossum* Grey Four-eyed Opossum – Patton & da Silva, 1997

Philander mcilhennyi McIlhenny's Four-eyed Opossum split from *P. andersoni* Anderson's Four-eyed Opossum – Patton & da Silva, 1997

CAENOLESTIDAE (Shrew-Opossums)

Caenolestes condorensis Condor Shrew-Opossum – Albuja & Patterson, 1996

DASYPODIDAE (Armadillos)

Dasypus yepesi Yepes' Long-nosed Armadillo – Vizcaíno, 1995

BRADYPODIDAE (Three-toed Sloths)

Bradypus pygmaeus Pygmy Three-toed Sloth – Anderson & Handley, 2001

OCHOTONIDAE (Pikas)

Ochotona nigritia Pianma Black Pika – Gong, Wang, Li & Li, 2000

LEPORIDAE (Rabbits and Hares)

Sylvilagus (Sylvilagus) obscurus Appalachian Cottontail split from *S. (Sylvilagus) floridanus* Eastern Cottontail – Chapman, Kramer, Dippenaar & Robinson, 1992

Sylvilagus (Tapeti) varynaensis Barinas Rabbit – Durant & Guevara, 2000

Nesolagus timminsi Annamite Striped Rabbit – Averianov, Abramov & Tikhonov, 2000

SCIURIDAE (Squirrels)

Sciurus (Guerlinguetus) argentinius South Yungas Red Squirrel split from *S. (Guerlinguetus) ignitus* Bolivian Squirrel – Emmons & Feer, 1997

Prosciurillus rosenbergii Rosenberg's Dwarf Squirrel split from *P. murinus* Sulawesi Dwarf Squirrel – Flannery, 1995b [as *Callosciurus rosenbergi*]

Olisthomys morrisi Morris' Flying Squirrel split from *Petinomys setosus* Temminck's Flying Squirrel – McKenna & Bell, 1997

MURIDAE (Mice, Rats, Voles and Gerbils)

Neotoma (Neotoma) magister Allegheny Woodrat split from *N. (Neotoma) floridana* Eastern Woodrat – Hays & Harrison, 1992; Hays & Richmond, 1993

Peromyscus keeni Northwestern Deer Mouse split from *P. maniculatus* North American Deer Mouse – Hogan, Hedin, Hung Sun Koh, Davis & Greenbaum, 1993; Kays & Wilson, 2002

Oryzomys caracolus Caracol Rice Rat split from *O. albigularis* Tomes' Rice Rat – Aguilera, Perez-Zapata & Martino, 1995

Oryzomys emmonsae Emmons' Rice Rat – Musser, Carleton, Brothers & Gardner, 1998

Oryzomys seuanezi Seuánez's Rice Rat – Weksler, Geise & Cerqueira, 1999

Oryzomys tatei Tate's Rice Rat – Musser, Carleton, Brothers & Gardner, 1998

Amphinectomys savamis Amphibious Rat – Malygin in Malygin et al., 1994

Oligoryzomys stramineus Straw-coloured Pygmy Rice Rat – Bonvicino & Weskler, 1998

Neacomys dubosti Dubost's Bristly Mouse – Voss, Lunde & Simmons, 2001

Neacomys minutus Small Bristly Mouse – Patton, da Silva & Malcolm, 2000

Neacomys musseri Musser's Bristly Mouse – Patton, da Silva & Malcolm, 2000

Neacomys paracou Paracou Bristly Mouse – Voss, Lunde & Simmons, 2001

Microakodontomys transitorius Intermediate Lesser Grass Mouse – Hershkovitz, 1993

Oecomys auyantepui North Amazonian Arboreal Rice Rat split from *O. paricola* South Amazonian Arboreal Rice Rat – Voss, Lunde & Simmons, 2001

Scolomys juruaense Rio Juruá Spiny Mouse – Patton & da Silva, 1995

Thomasomys apeco Apeco Oldfield Mouse – Leo & Gardner, 1993

Thomasomys macrotis Large-eared Oldfield Mouse – Gardner & Romo, 1993

Thomasomys onkiro Onkiro Oldfield Mouse – Luna & Pacheco, 2002

Juliomys rimofrons Long-haired Julio Mouse – de Oliveira & Bonvicino, 2002

Aepeomys reigi Reig's Montane Mouse – Ochoa, Aguilera, Pacheco & Soriano, 2001

Akodon (Akodon) aliquantulus Tucumán Grass Mouse – Díaz, Barquez, Braun & Mares, 1999

Akodon (Akodon) mystax Caparao Grass Mouse – Hershkovitz, 1998

Akodon (Akodon) oenos Wine Grass Mouse – Braun, Mares & Ojeda, 2000

Akodon (Akodon) paranaensis Paraná Grass Mouse – Christoff, Fagundes, Sbalqueiro, Mattevi & Yonenaga-Yassuda, 2000

Akodon (Akodon) pervalens Robust Grass Mouse split from *A. (Akodon) sylvanus* Forest Grass Mouse – Díaz, Braun, Mares & Barquez, 2000

Brucepattersonius albinasus White-nosed Brucie – Hershkovitz, 1998

Brucepattersonius griserufescens Grey-bellied Brucie – Hershkovitz, 1998

Brucepattersonius guarani Guaraní Brucie – Mares & Braun, 2000

Brucepattersonius igniventris Red-bellied Brucie – Hershkovitz, 1998

Brucepattersonius misionensis Misiones Brucie – Mares & Braun, 2000

Brucepattersonius paradisus Beautiful Brucie – Mares & Braun, 2000

Brucepattersonius soricinus Soricine Brucie – Hershkovitz, 1998

Oxymycterus amazonicus Amazon Hocicudo – Hershkovitz, 1994

Oxymycterus caparaoe Caparao Hocicudo – Hershkovitz, 1998

Oxymycterus josei José's Hocicudo – Hoffmann, Lessa & Smith, 2002

Juscelinomys guaporensis Rio Guaporé Burrowing Mouse – Emmons, 1999b

Juscelinomys huanchacae Huanchaca Burrowing Mouse – Emmons, 1999b

Pearsonomys annectens Pearson's Long-clawed Mouse – Patterson, 1992

Calomys expulsus Rejected Vesper Mouse split from *C. callosus* Large Vesper Mouse – Bonvicino & Almeida, 2000

Calomys venustus Pretty Vesper Mouse split from *C. callosus* Large Vesper Mouse – Díaz, Braun, Mares & Barquez, 2000

Graomys (Andalgalomys) roigi Roig's Chaco Mouse – Mares & Braun, 1996

Tapecomys primus Tapecua Rat – Anderson & Yates, 2000

Salinomys delicatus Delicate Salt-flat Mouse – Braun & Mares, 1995

Phyllotis limatus Narrow-toothed Leaf-eared Mouse split from *P. xanthopygus* Yellow-rumped Leaf-eared Mouse – Steppan, 1995a, 1995b

Punomys kofordi Koford's Puna Mouse – Pacheco & Patton, 1995

Chibchanomys orcesi Orces' Chibchan Water Mouse – Jenkins & Barnett, 1997

Microtus (Microtus) dogramacii Dogramaci's Vole – Kefelioglu & Krystufek, 1999

Microtus (Mynomes) mogollonensis Mogollon Vole split from *M. (Mynomes) mexicanus* Mexican Vole – Frey & LaRue, 1993; Kays & Wilson, 2002

Lemmus trimucronatus North American Brown Lemming split from *L. sibiricus* Siberian Brown Lemming – Jarrell & Fredga, 1993; Kays & Wilson, 2002

Monticolomys koopmani Malagasy Mountain Mouse – Carleton & Goodman, 1996

Voalavo gymnocaudus Naked-tailed Voalavo – Carleton & Goodman, 1998

Eliurus antsingy Antsingy Tufted-tailed Rat – Carleton, Goodman & Rakotondravony, 2001

Eliurus ellermani Ellerman's Tufted-tailed Rat – Carleton, 1994

Eliurus petteri Petter's Tufted-tailed Rat – Carleton, 1994

Lophuromys (Lophuromys) angolensis Angolan Brush-furred Rat – Verheyen, Dierckx & Hulselmans, 2000

Lophuromys (Lophuromys) aquilus Blackish Brush-furred Rat split from *L. (Lophuromys) flavopunctatus* Yellow-spotted Brush-furred Rat – Verheyen, Hulselmans, Dierckx & Verheyen, 2002

Lophuromys (Lophuromys) brevicaudus Short-tailed Brush-furred Rat split from *L. (Lophuromys) flavopunctatus* Yellow-spotted Brush-furred Rat – Lavrenchenko, Verheyen & Hulselmans, 1998

Lophuromys (Lophuromys) brunneus Brown Brush-furred Rat split from *L. (Lophuromys) flavopunctatus* Yellow-spotted Brush-furred Rat – Verheyen, Hulselmans, Dierckx & Verheyen, 2002

Lophuromys (Lophuromys) chrysopus Gold-footed Brush-furred Rat split from *L. (Lophuromys) flavopunctatus* Yellow-spotted Brush-furred Rat – Lavrenchenko, Verheyen & Hulselmans, 1998

Lophuromys (Lophuromys) dieterleni Dieterlen's Brush-furred Rat – Verheyen, Hulselmans, Colyn & Hutterer, 1997

Lophuromys (Lophuromys) dudui Dudu's Brush-furred Rat split from *L. (Lophuromys) flavopunctatus* Yellow-spotted Brush-furred Rat – Verheyen, Hulselmans, Dierckx & Verheyen, 2002

Lophuromys (Lophuromys) huttereri Hutterer's Brush-furred Rat – Verheyen, Colyn & Hulselmans, 1996

Lophuromys (Lophuromys) roseveari Rosevear's Brush-furred Rat – Verheyen, Hulselmans, Colyn & Hutterer, 1997

Lophuromys (Lophuromys) verhageni Verhagen's Brush-furred Rat split from *L. (Lophuromys) flavopunctatus* Yellow-spotted Brush-furred Rat – Verheyen, Hulselmans, Dierckx & Verheyen, 2002

Lophuromys (Lophuromys) zena Zena Brush-furred Rat split from *L. (Lophuromys) flavopunctatus* Yellow-spotted Brush-furred Rat – Verheyen, Hulselmans, Dierckx & Verheyen, 2002

Gerbillus rupicola Rock-loving Gerbil – Granjon, Aniskin, Volobouev & Sicard, 2002

Mus (Mus) domesticus Western House Mouse split from *M. (Mus) musculus* Eastern House Mouse – Mitchell-Jones et al., 1999

Nilopegamys plumbeus Ethiopian Water Mouse split from *Colomys goslingi* African Water Rat – McKenna & Bell, 1997

Praomys degraafi De Graaf's Soft-furred Mouse – Van der Straeten & Kerbis Peterhans, 1999

Mastomys awashensis Awash Multimammate Mouse – Lavrenchenko, Linkhnova, Baskevic & Bekele, 1998

Bandicota maxima Giant Bandicoot-Rat split from *B. indica* Greater Bandicoot-Rat – Pradhan, Mondal, Bhagwat & Agrawal, 1993

Bullimus gamay Camiguin Forest Rat – Rickart, Heaney & Tabaranza, 2002

Stenomys omichlodes Arianus' Rat split from *S. richardsoni* Glacier Rat – Flannery, 1995a

Crateromys heaneyi Panay Bushy-tailed Cloud Rat – Gonzales & Kennedy, 1996

Batomys russatus Dinagat Hairy-tailed Rat – Musser, Heaney & Tabaranza, 1998

Apomys gracilirostris Large Mindoro Forest Mouse – Ruedas, 1995

Crunomys suncoides Mount Katanglad Shrew-Mouse – Rickart, Heaney, Tabaranza & Balete 1998

Sommeromys macrorhinos Long-nosed Shrew-Mouse – Musser & Durden, 2002

Archboldomys musseri Palanan Shrew-Mouse – Rickart, Heaney, Tabaranza & Balete 1998

Pseudomys calabyi Calaby's Pebble-mound Mouse split from *P. laborifex* Kimberley Mouse – Menkhorst & Knight, 2001

Melomys bannisteri Bannister's Melomys – Kitchener & Maryanto, 1993b

Melomys caurinus Short-tailed Talaud Melomys split from *M. leucogaster* White-bellied Melomys – Flannery, 1995b

Melomys cooperae Cooper's Melomys – Kitchener & Marayanto 1995

Melomys fulgens Orange Melomys split from *M. leucogaster* White-bellied Melomys – Flannery, 1995b

Melomys gressitti Gressitt's Melomys – Menzies, 1996 [as *Paramelomys gressitti*]

Melomys howi How's Melomys – Kitchener & Suyanto, 1996a

Melomys lutillus Grassland Melomys split from *M. burtoni* Burton's Melomys – Flannery, 1995a

Melomys matambuai Manus Melomys – Flannery, Colgan & Trimble 1994

Melomys talaudium Long-tailed Talaud Melomys split from *M. leucogaster* White-bellied Melomys – Flannery, 1995b

Uromys (Uromys) boeadii Biak Giant Rat – Groves & Flannery, 1994

Uromys (Uromys) emmae Emma's Giant Rat – Groves & Flannery, 1994

GLIRIDAE (Dormice)

Dryomys niethammeri Niethammer's Dormouse – Holden, 1996

GEOMYIDAE (Pocket Gophers)

Geomys attwateri Attwater's Pocket Gopher split from *G. bursarius* Plains Pocket Gopher – Kays & Wilson, 2002

Geomys breviceps Baird's Pocket Gopher split from *G. bursarius* Plains Pocket Gopher – Kays & Wilson, 2002

Geomys knoxjonesi Knox Jones' Pocket Gopher split from *G. bursarius* Plains Pocket Gopher – Kays & Wilson, 2002

Geomys texensis Central Texas Pocket Gopher split from *G. bursarius* Plains Pocket Gopher – Kays & Wilson, 2002

Heteromys (Heteromys) teleus Southern Chocó Spiny Pocket Mouse – Anderson & Jarrín-V, 2002

Chaetodipus eremicus Chihuahuan Desert Pocket Mouse split from *C. penicillatus* Sonoran Desert Pocket Mouse – Lee, Riddle & Lee, 1996

Dipodomys simulans Dulzura Kangaroo-Rat split from *D. agilis* Agile Kangaroo-Rat – Williams, Genoways & Braun, 1993

ERETHIZONTIDAE (New World Porcupines)

Coendou (Sphiggurus) ichillus Long-tailed Hairy Dwarf Porcupine – Voss & Da Silva, 2001

Coendou (Sphiggurus) melanurus Black-tailed Hairy Dwarf Porcupine split from *C. (Sphiggurus) insidiosus* Bahia Hairy Dwarf Porcupine – Emmons & Feer, 1997; Voss & Angermann, 1997; Bonvicino, Penna-Firme & Braggio, 2002

Coendou (Sphiggurus) paragayensis Paraguay Hairy Dwarf Porcupine split from *C. (Sphiggurus) spinosus* Orange-spined Hairy Dwarf Porcupine – Emmons & Feer, 1997

Coendou (Sphiggurus) pruinosus Frosted Hairy Dwarf Porcupine split from *C. (Sphiggurus) vestitus* Brown Hairy Dwarf Porcupine – Emmons & Feer, 1997

Coendou (Sphiggurus?) quichua Quichua Hairy Dwarf Porcupine split from *C. (Coendou) bicolor* Bicolour-spined Porcupine – Emmons & Feer, 1997

Coendou (Sphiggurus) roosmalenorum Van Roosmalens' Hairy Dwarf Porcupine – Voss & Da Silva, 2001

BATHYERGIDAE (African Mole-Rats)

Cryptomys anselli Ansell's Mole-Rat – Burda, Zima, Scharff, Macholán & Kawalika, 1999

Cryptomys kafuensis Kafue Mole-Rat – Burda, Zima, Scharff, Macholán & Kawalika, 1999

AGOUTIDAE (Agoutis and Pacas)

Dasyprocta variegata Brown Agouti split from *D. punctata* Central American Agouti – Emmons & Feer, 1997; Eisenberg & Redford, 1999

CAVIIDAE (Cavies and Guinea-Pigs)

Cavia intermedia Intermediate Guinea-Pig – Cherem, Olimpio & Ximenez, 1999

Kerodon acrobata Acrobat Cavy – Moojen, Locks & Langguth, 1997

OCTODONTIDAE (Degus and Tuco-tucos)

Octodon pacificus Isla Mocha Degu – Hutterer, 1994

Pipanacoctomys aureus Golden Viscacha-Rat – Mares, Braun, Barquez & Díaz, 2000

Salinoctomys loschalchalerosorum Chalchalero Viscacha-Rat – Mares, Braun, Barquez & Díaz, 2000

Ctenomys coyhaiquensis Coyhaique Tuco-tuco – Kelt & Gallardo, 1994

Ctenomys osvaldoreigi Osvaldo Reig's Tuco-tuco – Contreras, 1995

Ctenomys roigi Roig's Tuco-tuco – Contreras, 1989

ECHIMYIDAE (Spiny Rats)

Proechimys (Proechimys) barinas Barinas Spiny Rat – Aguilera & Corti, 1994

Proechimys (Proechimys) echinothrix Hedgehog Spiny Rat – da Silva, 1998

Proechimys (Proechimys) gardneri Gardner's Spiny Rat – da Silva, 1998

Proechimys (Proechimys) kulinae Kulinas' Spiny Rat – da Silva, 1998

Proechimys (Proechimys) pattoni Patton's Spiny Rat – da Silva, 1998

Proechimys (Trinomys) mirapatanga Mirapatanga Spiny Rat – Lara, Patton & Hingst-Zaher, 2002

Proechimys (Trinomys) yonenagae Yonenaga-Yassuda's Spiny Rat – da Rocha, 1996

Mesomys occultus Furtive Spiny Tree Rat – Patton, da Silva & Malcolm, 2000

Echimys pattoni Rusty-sided Atlantic Tree Rat – Emmons, Leite, Kock & Costa, 2002 [as *Phyllomys pattoni*; also described as *Nelomys* sp. in Emmons & Feer, 1997]

Isothrix sinnamarensis Sinnamary Brush-tailed Rat – Vié, Volobouev, Patton & Granjon, 1996

Makalata occasius Bare-tailed Tree Rat split from *M. didelphoides* Armored Tree Rat – Emmons & Feer, 1997

ABROCOMIDAE (Chinchilla-Rats)

Cuscomys ashaninka White-fronted Cusco Rat – Emmons, 1999a

Abrocoma budini Budin's Chinchilla-Rat split from *A. cinerea* Ashy Chinchilla-Rat – Braun & Mares, 2002

Abrocoma famatina Famatina Chinchilla-Rat split from *A. cinerea* Ashy Chinchilla-Rat – Braun & Mares, 2002

Abrocoma schistacea Tontal Chinchilla-Rat split from *A. cinerea* Ashy Chinchilla-Rat – Braun & Mares, 2002

Abrocoma uspallata Uspallata Chinchilla-Rat – Braun & Mares, 2002

Abrocoma vaccarum Vacas Chinchilla-Rat split from *A. cinerea* Ashy Chinchilla-Rat – Braun & Mares, 2002

VIVERRIDAE (Genets and Civets)

Viverra tainguensis Tainguen Civet – Sokolov, Rozhnov & Anh, 1997

Genetta bourloni Bourlon's Genet – Gaubert, 2003

Genetta pardina Forest Genet split from *G. maculata* Rusty-spotted Genet – Gaubert, 2003

Genetta poensis King Genet split from *G. maculata* Rusty-spotted Genet – Gaubert, 2003

FELIDAE (Cats)

Felis (Felis) catus Domestic Cat split from *F. (Felis) silvestris* Wild Cat – Clutton-Brock, 1999

Felis (Leopardus) braccatus Pantanal Cat split from *F. (Leopardus) colocolo* Colocolo – Garcia-Perea, 1994

Felis (Leopardus) pajeros Pampas Cat split from *F. (Leopardus) colocolo* Colocolo – Garcia-Perea, 1994

HERPESTIDAE (Mongooses)

Herpestes (Herpestes) auropunctatus Small Indian Mongoose split from *H. (Herpestes) javanicus* Javan Mongoose – Gurung & Singh, 1996; Mitchell-Jones *et al.*, 1999

CANIDAE (Dogs and Foxes)

Vulpes (Vulpes) macrotis Kit Fox split from *V. (Vulpes) velox* Swift Fox – Mercure, Ralls, Koepfli & Wayne, 1993; Kays & Wilson, 2002

Canis familiaris Domestic Dog (Dingo) split from *C. lupus* Grey Wolf – Clutton-Brock, 1999

MUSTELIDAE (Otters, Weasels and Badgers)

Spilogale gracilis Western Spotted Skunk split from *S. putorius* Eastern Spotted Skunk – Kays & Wilson, 2002

Mustela formosana Taiwan Mountain Weasel – Lin & Harada, 1998

Mustela furo Ferret split from European Polecat *M. putorius* – Davison *et al.*, 1999

CHRYSOCHLORIDAE (Golden-Moles)

Amblysomus corriae Western Cape Golden-Mole split from *A. iris* Zulu Golden-Mole – Bronner, 1996

Amblysomus marleyi Marley's Golden-Mole split from *A. hottentotus* Hottentot Golden-Mole – Bronner, 1996

Amblysomus robustus Robust Golden-Mole – Bronner, 2000

Amblysomus septentrionalis Highveld Golden-Mole split from *A. iris* Zulu Golden-Mole – Bronner, 1996

SORICIDAE (Shrews)

Sorex arunchi Italian Shrew – Lapini & Testone, 1998

Sorex maritimensis Maritime Shrew split from *S. arcticus* Arctic Shrew – Stewart, Perry & Fumagalli, 2002

Sorex yukonicus Alaska Tiny Shrew – Dokuchaev, 1997

Notiosorex cockrumi Cockrum's Desert Shrew split from *N. crawfordi* Desert Shrew – Baker, O'Neill & McAliley, 2003

Notiosorex evotis Large-eared Grey Shrew split from *N. crawfordi* Desert Shrew – Carraway & Timm, 2000

Notiosorex villai Villa's Grey Shrew – Carraway & Timm, 2000

Cryptotis colombiana Colombian Small-eared Shrew – Woodman & Timm, 1993

Cryptotis mayensis Maya Small-eared Shrew split from *C. nigrescens* Blackish Small-eared Shrew – Woodman & Timm, 1993

Cryptotis mera Darién Small-eared Shrew split from *C. nigrescens* Blackish Small-eared Shrew – Woodman & Timm, 1993

Cryptotis merriami Merriam's Small-eared Shrew split from *C. nigrescens* Blackish Small-eared Shrew – Woodman & Timm, 1993

Cryptotis peruviensis Peruvian Small-eared Shrew – Vivar, Pacheco & Valqui, 1997

Cryptotis tamensis Páramo de Tamá Small-eared Shrew – Woodman, 2002

Crocidura brunnea Thick-tailed Shrew split from *C. fuliginosa* Southeast Asian Shrew – Ruedi, 1995

Crocidura foetida Lowland Bornean Shrew split from *C. fuliginosa* Southeast Asian Shrew – Ruedi, 1995

Crocidura gmelini Gmelin's Shrew split from *Sorex minutus* Eurasian Pygmy Shrew – Jiang & Hoffmann, 2001

Crocidura hilliana Hill's Shrew – Jenkins & Smith, 1995

Crocidura hutanis Hutan Shrew – Ruedi, 1995

Crocidura lepidura Sumatran Giant Shrew split from *C. fuliginosa* Southeast Asian Shrew – Ruedi, 1995

Crocidura musseri Mossy Forest Shrew – Ruedi, 1995

Crocidura orientalis Oriental Shrew split from *C. fuliginosa* Southeast Asian Shrew – Ruedi, 1995

Crocidura ramona Negev Shrew – Ivanitskaya, Shenbrot & Nevo, 1996

Crocidura rapax Rapacious Shrew split from *C. russula* Greater White-toothed Shrew – Jiang & Hoffmann, 2001

Crocidura shantungensis Shantung Shrew split from *C. suaveolens* Lesser White-toothed Shrew – Jiang & Hoffmann, 2001

Crocidura vorax Voracious Shrew split from *C. russula* Greater White-toothed Shrew – Jiang & Hoffmann, 2001

Crocidura vosmaeri Banka Shrew split from *C. fuliginosa* Southeast Asian Shrew – Ruedi, 1995

Myosorex (Myosorex) kihaulei Udzungwa Mouse-Shrew – Stanley & Hutterer, 2000

Sylvisorex konganensis Central African Shrew – Ray & Hutterer, 1996

Sylvisorex pluvialis Rainforest Shrew – Hutterer & Schlitter, 1996

TENRECIDAE (Tenrecs)

Hemicentetes nigriceps Highland Streaked Tenrec split from *H. semispinosus* Lowland Streaked Tenrec – Garbutt, 1999

Microgale drouhardi Striped Shrew-Tenrec split from *M. cowani* Cowan's Shrew-Tenrec – Garbutt, 1999

Microgale fotsifotsy Pale-footed Shrew-Tenrec – Jenkins, Raxworthy & Nussbaum, 1997

Microgale gymnorhyncha Naked-nosed Shrew-Tenrec – Jenkins, Goodman & Raxworthy, 1996

Microgale monticola Montane Shrew-Tenrec – Goodman & Jenkins, 1998

Microgale nasoloi Nasolo's Shrew-Tenrec – Jenkins & Goodman, 1999

Microgale soricoides Soricine Shrew-Tenrec – Jenkins, 1993

Microgale taiva Taiva Shrew-Tenrec split from *M. cowani* Cowan's Shrew-Tenrec – Garbutt, 1999

PTEROPODIDAE (Old World Fruit Bats)

Pteropus banakrisi Torresian Flying Fox – Richards & Hall, 2002

Pteropus capistratus Bismarck Flying Fox split from *P. temmincki* Temminck's Flying Fox – Flannery, 1995b

Pteropus cognatus Makira Flying Fox split from *P. rayneri* Solomons Flying Fox – Flannery, 1995b

Pteropus pelewensis Palau Flying Fox split from *P. mariannus* Marianas Flying Fox – Flannery, 1995b

Pteropus rennelli Rennell Flying Fox split from *P. rayneri* Solomons Flying Fox – Flannery, 1995b

Pteropus ualanus Kosrae Flying Fox split from *P. mariannus* Marianas Flying Fox – Flannery, 1995b

Pteropus yapensis Yap Flying Fox split from *P. mariannus* Marianas Flying Fox – Flannery, 1995b

Pteralopex taki New Georgia Monkey-faced Bat – Parnaby, 2002b

Dobsonia anderseni Andersen's Bare-backed Fruit Bat split from *D. moluccensis* Moluccan Bare-backed Fruit Bat – Flannery, 1995b

Dobsonia crenulata Halmahera Bare-backed Fruit Bat split from *D. viridis* Green Bare-backed Fruit Bat – Flannery, 1995b

Dobsonia magna Great Bare-backed Fruit Bat split from *D. moluccensis* Moluccan Bare-backed Fruit Bat – Flannery, 1995b

Cynopterus luzoniensis Sulawesi Short-nosed Fruit Bat split from *C. brachyotis* Lesser Short-nosed Fruit Bat – Kitchener & Maharadatunkamsi, 1991

Cynopterus minutus Small Short-nosed Fruit Bat split from *C. brachyotis* Lesser Short-nosed Fruit Bat – Kitchener & Maharadatunkamsi, 1991

Paranyctimene tenax Greater Unstriped Tube-nosed Bat split from *P. raptor* Unstriped Tube-nosed Bat – Bergmans, 2001

Nyctimene bougainville Solomons Tube-nosed Fruit Bat split from *N. vizcaccia* Umboi Tube-nosed Fruit Bat – Flannery, 1995b

Melonycteris (Nesonycteris) fardoulisi Fardoulis' Fruit Bat – Flannery, 1993

Notopteris neocaledonica New Caledonia Blossom Bat split from *N. macdonaldi* Fijian Blossom Bat – Flannery, 1995b

EMBALLONURIDAE (Sheath-tailed Bats)

Taphozous (Taphozous) achates Brown-bearded Tomb Bat split from *T. (Taphozous) melanopogon* Black-bearded Tomb Bat – Kitchener & Suyanto, 1996b

Taphozous (Taphozous) troughtoni Troughton's Tomb Bat split from *T. (Taphozous) georgianus* Sharp-nosed Tomb Bat – Menkhorst & Knight, 2001

Emballonura serii Seri's Sheath-tailed Bat – Flannery, 1995c

Peropteryx (Peropteryx) trinitatis Trinidad Dog-like Bat split from *P. (Peropteryx) macrotis* Lesser Dog-like Bat – Simmons & Voss, 1998

NYCTERIDAE (Slit-faced Bats)

Nycteris madagascariensis Madagascar Slit-faced Bat split from *N. macrotis* Large-eared Slit-faced Bat – Garbutt, 1999

RHINOLOPHIDAE (Horseshoe Bats)

Rhinolophus beddomei Indian Woolly Horseshoe Bat split from *R. luctus* Woolly Horseshoe Bat – Bates & Harrison, 1997

Rhinolophus convexus Convex Horseshoe Bat – Csorba, 1997

Rhinolophus hillorum Hills' Horseshoe Bat split from *H. clivosus* Geoffroy's Horseshoe Bat – Cotterill, 2002

Rhinolophus maendeleo Maendeleo Horseshoe Bat – Kock, Csorba & Howell 2000

Rhinolophus sakejiensis Sakeji Horseshoe Bat – Cotterill, 2002

Rhinolophus ziama Ziama Horseshoe Bat – Fahr, Vierhaus, Hutterer & Kock, 2002

Hipposideros demissus Makira Roundleaf Bat split from *H. diadema* Diadem Roundleaf Bat – Flannery, 1995b

Hipposideros edwardshilli John Hill's Roundleaf Bat – Flannery & Colgan, 1993

Hipposideros hypophyllus Kolar Roundleaf Bat – Kock & Bhat, 1994

Hipposideros madurae Maduran Roundleaf Bat – Kitchener & Maryanto, 1993a

Hipposideros orbiculus Orb-faced Roundleaf Bat – Francis, Kock & Habersetzer, 1999

Hipposideros rotalis Laotian Roundleaf Bat – Francis, Kock & Habersetzer, 1999

Hipposideros sorenseni Pangandaran Roundleaf Bat – Kitchener & Maryanto, 1993a

Hipposideros sumbae Sumba Roundleaf Bat split from *H. larvatus* Intermediate Roundleaf Bat – Kitchener & Maryanto, 1993a

Triaenops rufus Rufous Trident Bat split from *T. persicus* Persian Trident Bat – Garbutt, 1999

PHYLLOSTOMIDAE (American Leaf-nosed Bats)

Micronycteris (Micronycteris) brosseti Brosset's Big-eared Bat – Simmons & Voss, 1998

Micronycteris (Micronycteris) homezi Homez's Big-eared Bat – Pirlot, 1967; Simmons & Voss, 1998

Micronycteris (Micronycteris) matses Matses' Big-eared Bat – Simmons, Voss & Fleck, 2002

Micronycteris (Micronycteris) microtis Common Big-eared Bat split from *M. (Micronycteris) megalotis* Little Big-eared Bat – Reid, 1997; Simmons & Voss, 1998

Micronycteris (Micronycteris) sanborni Sanborn's Big-eared Bat – Simmons, 1996

Lonchorhina inusitata Strange Sword-nosed Bat – Handley & Ochoa, 1996

Tonatia saurophila Stripe-headed Round-eared Bat split from *T. bidens* Greater Round-eared Bat – Williams, Willig & Reid, 1995; Reid, 1997

Mimon (Mimon) cozumelae Cozumel Golden Bat split from *M. (Mimon) bennettii* Golden Bat – Simmons & Voss, 1998

Leptonycteris yerbabuenae North American Long-nosed Bat split from *L. curasoae* Southern Long-nosed Bat – Kays & Wilson, 2002

Anoura luismanueli Luis Manuel's Tailless Bat – Molinari, 1994

Carollia sowelli Sowell's Short-tailed Bat – Baker, Solari & Hoffmann, 2002

Sturnira (Sturnira) mistratensis Mistrató Yellow-shouldered Bat – Contreras Vega & Cadena, 2000

Artibeus (Artibeus) intermedius Intermediate Fruit-eating Bat split from *A. (Artibeus) lituratus* Great Fruit-eating Bat – Reid, 1997

Artibeus (Artibeus) watsoni Thomas' Fruit-eating Bat split from *A. (Artibeus) glaucus* Silver Fruit-eating Bat – Reid, 1997

Artibeus (Dermanura) incomitatus Solitary Fruit-eating Bat – Kalko & Handley, 1994

THYROPTERIDAE (Disk-winged Bats)

Thyroptera lavali LaVal's Disk-winged Bat – Pine, 1993 (= *T. robusta* Czaplewski, 1996, Czaplewski, 1996)

MOLOSSIDAE (Free-tailed Bats)

Mormopterus (Mormopterus) loriae Little Northern Free-tailed Bat split from *M. (Mormopterus) planiceps* Southern Free-tailed Bat – Flannery, 1995b; Menkhorst & Knight, 2001

Molossops (Cynomops) mexicanus Mexican Dog-faced Bat split from *M. (Cynomops) greenhalli* Greenhall's Dog-faced Bat – Peters, Lim & Engstrom, 2002

Molossops (Cynomops) paranus Brown-bellied Dog-faced Bat split from *M. (Cynomops) planirostris* Southern Dog-faced Bat – Simmons & Voss, 1998

Cheiromeles parvidens Lesser Hairless Bat split from *C. torquatus* Hairless Bat – Ingle & Heaney, 1992; Flannery, 1995b

Tadarida kuboriensis New Guinea Mastiff Bat split from *T. australis* White-striped Free-tailed Bat – Flannery, 1995a

Tadarida latouchei Latouche's Free-tailed Bat split from *T. teniotis* European Free-tailed Bat – Kock, 1999; Funakoshi & Kunisaki, 2000

Chaerephon bregullae Fijian Mastiff Bat split from *C. jobensis* Northern Mastiff Bat – Flannery, 1995b

Chaerephon leucogaster Pale-bellied Free-tailed Bat split from *C. pumila* Little Free-tailed Bat – Garbutt, 1999

Chaerephon solomonis Solomons Mastiff Bat split from *C. jobensis* Northern Mastiff Bat – Flannery, 1995b

Chaerephon tomensis São Tomé Free-tailed Bat – Juste & Ibanez, 1993

Mops (Mops) leucostigma Pale-marked Free-tailed Bat split from *M. (Mops) condylurus* Angolan Free-tailed Bat – Garbutt, 1999

Otomops johnstonei Alor Mastiff Bat – Kitchener, How & Maryanto, 1992

Eumops patagonicus Argentine Dwarf Bonneted Bat split from *E. bonariensis* Dwarf Bonneted Bat – Díaz, Braun, Mares & Barquez, 2000

Molossus barnesi Barnes' Mastiff Bat split from *M. molossus* Pallas' Mastiff Bat – Simmons & Voss, 1998

VESPERTILIONIDAE (Vesper Bats)

Myotis (Leuconoe) annamiticus Annamite Myotis – Kruskop & Tsytsulina, 2001

Myotis (Leuconoe) csorbai Csorba's Myotis – Topal, 1997

Myotis (Myotis) septentrionalis Northern Myotis split from *M. (Myotis) keenii* Keen's Myotis – Kays & Wilson, 2002

Myotis (Selysius) alcathoe Alcathoe's Myotis – Von Helversen, Heller, Mayer, Nemeth, Volleth & Gombkötö, 2001

Myotis (Selysius) ater Small Black Myotis split from *M. (Selysius) muricola* Wall-roosting Myotis – Francis & Hill, 1998; Payne & Francis, 1998

Myotis (Selysius) aurascens Golden Myotis split from *M. (Selysius) mystacinus* European Whiskered Bat – Benda & Tsytsulina, 2000

Myotis (Selysius) ciliolabrum Western Small-footed Myotis split from *M. (Selysius) leibii* Eastern Small-footed Myotis – Kays & Wilson, 2002

Myotis (Selysius) gomantongensis Gomantong Myotis – Francis & Hill, 1998

Myotis (Selysius) hajastanicus Armenian Myotis split from *M. (Selysius) mystacinus* European Whiskered Bat – Benda & Tsytsulina, 2000

Myotis (Selysius) nipalensis Asian Whiskered Bat split from *M. (Selysius) mystacinus* European Whiskered Bat – Benda & Tsytsulina, 2000

Myotis (Selysius) yanbarensis Okinawa Myotis – Maeda & Matsumara, 1998

Plecotus (Plecotus) alpinus Alpine Long-eared Bat – Kiefer & Veith, 2001 (= *P. microdontus* Spitzenberger, Haring & Tvrtkovic, 2002)

Plecotus (Plecotus) balensis Bale Long-eared Bat – Kruskop & Lavrenchenko, 2000

Plecotus (Plecotus) kolombatovici Kolombatovic's Long-eared Bat split from *P. (Plecotus) austriacus* Grey Long-eared Bat – Spitzenberger, Pialek & Haring, 2001

Plecotus (Plecotus) sardus Sardinian Long-eared Bat – Mucedda, Kiefer, Pidinchedda & Veith, 2002

Pipistrellus (Arielulus) aureocollaris Gold-collared Gilded Pipistrelle – Kock & Storch, 1996, described as *Thainycteris aureocollaris* Kock & Storch, 1996, type species of new genus *Thainycteris* which was later synonymized with *Pipistrellus* subgenus *Arielulus* by Csorba & Lee, 1999

Pipistrellus (Arielulus) torquatus Taiwan Gilded Pipistrelle – Csorba & Lee, 1999

Pipistrellus (Falsistrellus) mackenziei Western False Pipistrelle split from *P. (Falsistrellus) tasmaniensis* Eastern False Pipistrelle – Menkhorst & Knight, 2001

Pipistrellus (Neoromicia) matroka Madagascar Serotine split from *P. (Neoromicia) capensis* Cape Serotine – Garbutt, 1999

Pipistrellus (Pipistrellus) adamsi Cape York Pipistrelle split from *P. (Pipistrellus) tenuis* Least Pipistrelle – Menkhorst & Knight, 2001

Pipistrellus (Pipistrellus) angulatus New Guinea Pipistrelle split from *P. (Pipistrellus) tenuis* Least Pipistrelle – Flannery, 1995a

Pipistrellus (Pipistrellus) collinus Mountain Pipistrelle split from *P. (Pipistrellus) tenuis* Least Pipistrelle – Flannery, 1995a

Pipistrellus (Pipistrellus) papuanus Papuan Pipistrelle split from *P. (Pipistrellus) tenuis* Least Pipistrelle – Flannery, 1995a

Pipistrellus (Pipistrellus) pygmaeus Soprano Pipistrelle split from *P. (Pipistrellus) pipistrellus* Common Pipistrelle – Jones & Barratt, 1999

Pipistrellus (Pipistrellus) vordermanni White-winged Pipistrelle split from *P. (Pipistrellus) macrotis* Big-eared Pipistrelle – Corbet & Hill, 1992

Pipistrellus (Pipistrellus) wattsi Watts' Pipistrelle split from *P. (Pipistrellus) tenuis* Least Pipistrelle – Flannery, 1995a

Pipistrellus (Pipistrellus) westralis Mangrove Pipistrelle split from *P. (Pipistrellus) tenuis* Least Pipistrelle – Menkhorst & Knight, 2001

Pipistrellus (Vespadelus) caurinus Northern Cave Bat split from *P. (Vespadelus) pumilus* Eastern Forest Bat – Menkhorst & Knight, 2001

Pipistrellus (Vespadelus) finlaysoni Inland Cave Bat split from *P. (Vespadelus) pumilus* Eastern Forest Bat – Menkhorst & Knight, 2001

Pipistrellus (Vespadelus) troughtoni Eastern Cave Bat split from *P. (Vespadelus) pumilus* Eastern Forest Bat – Menkhorst & Knight, 2001

Histiotus humboldti Humboldt's Big-eared Brown Bat – Handley, 1996

Chalinolobus (Chalinolobus) neocaledonicus New Caledonia Wattled Bat split from *C. (Chalinolobus) gouldii* Gould's Wattled Bat – Flannery, 1995b

Chalinolobus (Glauconycteris) curryi Curry's Bat – Eger & Schlitter, 2001

Nycticeius (Scotorepens) orion Eastern Broad-nosed Bat split from *N. (Scotorepens) balstoni* Inland Broad-nosed Bat – Menkhorst & Knight, 2001

Rhogeessa (Rhogeessa) hussoni Husson's Yellow Bat – Genoways & Baker, 1996

Scotophilus collinus Small Asiatic Yellow Bat split from *S. kuhlii* Lesser Asiatic Yellow Bat – Kitchener, Packer & Maryanto, 1997

Lasiurus (Dasypterus) xanthinus Western Yellow Bat split from *L. (Dasypterus) ega* Southern Yellow Bat – Baker, Patton, Genoways & Bickham, 1988; Kays & Wilson, 2002

Lasiurus (Lasiurus) atratus Mourning Bat – Handley, 1996

Lasiurus (Lasiurus) blossevillii Western Red Bat split from *L. (Lasiurus) borealis* Eastern Red Bat – Reid, 1997

Lasiurus (Lasiurus) ebenus Ebony Bat – Fazzolari-Corrêa, 1994

Nyctophilus bifax Northern Nyctophilus split from *N. gouldi* Gould's Nyctophilus – Menkhorst & Knight, 2001

Nyctophilus nebulosus New Caledonia Nyctophilus – Parnaby, 2002a [as *N. nebulosa*]

Miniopterus gleni Glen's Long-fingered Bat – Peterson, Eger & Mitchell, 1995

Miniopterus macrocneme Small Melanesian Bent-winged Bat split from *M. pusillus* Small Bent-winged Bat – Flannery, 1995a, 1995b

Miniopterus majori Major's Long-fingered Bat split from *M. schreibersi* Common Bent-winged Bat – Garbutt, 1999

Miniopterus manavi Malagasy Least Long-fingered Bat split from *M. schreibersi* Common Bent-winged Bat – Garbutt, 1999

Miniopterus medius Medium Bent-winged Bat split from *M. fuscus* Ryukyu Bent-winged Bat – Corbet & Hill, 1992; Flannery, 1995a

Miniopterus propritristis Large Melanesian Bent-winged Bat split from *M. tristis* Great Bent-winged Bat – Flannery, 1995a

Murina (Murina) ryukyuana Ryukyu Tube-nosed Bat – Maeda & Matsumura, 1998

LEMURIDAE (Lemurs)

Hapalemur alaotrensis Alaotran Bamboo Lemur split from *H. griseus* Grey Bamboo Lemur – Groves, 2001

Hapalemur occidentalis Sambirano Bamboo Lemur split from *H. griseus* Grey Bamboo Lemur – Groves, 2001

Varecia rubra Red Ruffed Lemur split from *V. variegata* Black-and-white Ruffed Lemur – Groves, 2001

Eulemur albifrons White-fronted Lemur split from *E. fulvus* Brown Lemur – Groves, 2001

Eulemur albocollaris White-collared Lemur split from *E. fulvus* Brown Lemur – Groves, 2001

Eulemur cinereiceps Grey-headed Lemur split from *E. fulvus* Brown Lemur – Groves, 2001

Eulemur collaris Red-collared Lemur split from *E. fulvus* Brown Lemur – Groves, 2001

Eulemur rufus Red-fronted Lemur split from *E. fulvus* Brown Lemur – Groves, 2001

Eulemur sanfordi Sanford's Lemur split from *E. fulvus* Brown Lemur – Groves, 2001

LORIDAE (Galagos and Lorises)

Pseudopotto martini Martin's False Potto – Schwartz, 1996

Galago (Galago) cameronensis Cross River Squirrel Galago split from *G. (Galago) alleni* Allen's Squirrel Galago – Groves, 2001

Galago (Galago) gabonensis Gabon Squirrel Galago split from *G. (Galago) alleni* Allen's Squirrel Galago – Groves, 2001

Galago (Galagoides) granti Mozambique Galago split from *G. (Galagoides) zanzibaricus* Zanzibar Galago – Kingdon, 1997; Groves, 2001

Galago (Galagoides) nyasae Malawi Galago split from *G. (Galago) moholi* South African Galago – Groves, 2001

Galago (Galagoides) orinus Uluguru Galago split from *G. (Galagoides) demidoff* Demidoff's Galago – Groves, 2001

Galago (Galagoides) rondoensis Rondo Galago – Honess 1996

Galago (Galagoides) thomasi Thomas' Galago split from *G. (Galagoides) demidoff* Demidoff's Galago – Kingdon, 1997; Groves, 2001

Galago (Galagoides) udzungwensis Matundu Galago – Honess 1996

Otolemur monteiri Silvery Greater Galago split from *O. crassicaudatus* Brown Greater Galago – Kingdon, 1997; Groves, 2001

Nycticebus bengalensis Bengal Slow Loris split from *N. coucang* Sunda Slow Loris – Groves, 2001

Loris lydekkerianus Grey Slender Loris split from *L. tardigradus* Red Slender Loris – Groves, 2001

CHEIROGALEIDAE (Dwarf Lemurs and Mouse-Lemurs)

Microcebus berthae Berthe's Mouse-Lemur – Rasoloarison, Goodman & Ganzhorn, 2000

Microcebus myoxinus Pygmy Mouse-Lemur split from *M. murinus* Grey Mouse-Lemur – Groves, 2001

Microcebus ravelobensis Golden Mouse-Lemur – Zimmerman, Cepok, Rakotoarison, Zietemann & Radespiel, 1998

Microcebus sambiranensis Sambirano Mouse-Lemur – Rasoloarison, Goodman & Ganzhorn, 2000

Microcebus tavaratra Northern Rufous Mouse-Lemur – Rasoloarison, Goodman & Ganzhorn, 2000

Cheirogaleus adipicaudatus Southern Fat-tailed Dwarf Lemur split from *C. medius* Western Fat-tailed Dwarf Lemur – Groves, 2001

Cheirogaleus crossleyi Furry-eared Dwarf Lemur split from *C. major* Greater Dwarf Lemur – Groves, 2001

Cheirogaleus minusculus Lesser Iron-grey Dwarf Lemur – Groves, 2000

Cheirogaleus ravus Greater Iron-grey Dwarf Lemur – Groves, 2000

Cheirogaleus sibreei Sibree's Dwarf Lemur split from *C. major* Greater Dwarf Lemur – Groves, 2001

Phaner electromontis Amber Mountain Fork-crowned Lemur split from *P. furcifer* Masoala Fork-crowned Lemur – Groves, 2001

Phaner pallescens Western Fork-crowned Lemur split from *P. furcifer* Masoala Fork-crowned Lemur – Groves, 2001

Phaner parienti Sambirano Fork-crowned Lemur split from *P. furcifer* Masoala Fork-crowned Lemur – Groves, 2001

INDRIIDAE (Indris and Sifakas)

Propithecus coquereli Coquerel's Sifaka split from *P. verreauxi* Verreaux's Sifaka – Groves, 2001

Propithecus deckenii Van der Decken's Sifaka split from *P. verreauxi* Verreaux's Sifaka – Groves, 2001

Propithecus edwardsi Milne-Edwards' Sifaka split from *P. diadema* Diademed Sifaka – Groves, 2001

Propithecus perrieri Perrier's Sifaka split from *P. diadema* Diademed Sifaka – Groves, 2001

Avahi occidentalis Western Avahi split from *A. laniger* Eastern Avahi – Groves, 2001

Avahi unicolor Unicoloured Avahi – Thalmann & Geissmann, 2000

TARSIIDAE (Tarsiers)

Tarsius pelengensis Peleng Tarsier split from *T. spectrum* Spectral Tarsier – Groves, 2001

Tarsius sangirensis Sangihe Tarsier split from *T. spectrum* Spectral Tarsier – Groves, 2001

CERCOPITHECIDAE (Old World Monkeys)

Colobus vellerosus Geoffroy's Pied Colobus split from *C. polykomos* Western Pied Colobus – Kingdon, 1997; Groves, 2001

Piliocolobus foai Central African Red Colobus split from *P. pennantii* Pennant's Red Colobus – Groves, 2001

Piliocolobus gordonorum Udzungwa Red Colobus split from *P. pennantii* Pennant's Red Colobus – Kingdon, 1997; Groves, 2001

Piliocolobus kirkii Zanzibar Red Colobus split from *P. pennantii* Pennant's Red Colobus – Kingdon, 1997; Groves, 2001

Piliocolobus tephrosceles Ugandan Red Colobus split from *P. pennantii* Pennant's Red Colobus – Groves, 2001

Piliocolobus tholloni Thollon's Red Colobus split from *P. pennantii* Pennant's Red Colobus – Kingdon, 1997; Groves, 2001

Presbytis chrysomelas Sarawak Leaf Monkey split from *P. femoralis* Banded Leaf Monkey – Groves, 2001

Presbytis natunae Natuna Islands Leaf Monkey split from *P. femoralis* Banded Leaf Monkey – Groves, 2001

Presbytis siamensis White-thighed Leaf Monkey split from *P. femoralis* Banded Leaf Monkey – Groves, 2001

Semnopithecus ajax Kashmir Grey Langur split from *S. entellus* Northern Plains Grey Langur – Groves, 2001

Semnopithecus dussumieri Southern Plains Grey Langur split from *S. entellus* Northern Plains Grey Langur – Groves, 2001

Semnopithecus hector Tarai Grey Langur split from *S. entellus* Northern Plains Grey Langur – Groves, 2001

Semnopithecus hypoleucos Black-footed Grey Langur split from *S. entellus* Northern Plains Grey Langur – Groves, 2001

Semnopithecus priam Tufted Grey Langur split from *S. entellus* Northern Plains Grey Langur – Groves, 2001

Semnopithecus schistaceus Nepal Grey Langur split from *S. entellus* Northern Plains Grey Langur – Groves, 2001

Trachypithecus barbei Tenasserim Leaf Monkey split from *T. cristatus* Silvered Leaf Monkey – Groves, 2001

Trachypithecus delacouri Delacour's Leaf Monkey split from *T. francoisi* François' Leaf Monkey – Groves, 2001

Trachypithecus ebenus Black Leaf Monkey – Brandon-Jones, 1995 [as *Semnopithecus auratus ebenus*] split from *T. auratus* Javan Langur – Groves, 2001

Trachypithecus germaini Indochinese Leaf Monkey split from *T. cristatus* Silvered Leaf Monkey – Groves, 2001

Trachypithecus hatinhensis Hatinh Leaf Monkey split from *T. francoisi* François' Leaf Monkey – Groves, 2001

Trachypithecus laotum Laotian Leaf Monkey split from *T. francoisi* François' Leaf Monkey – Groves, 2001

Trachypithecus poliocephalus White-headed Leaf Monkey split from *T. francoisi* François' Leaf Monkey – Groves, 2001

Trachypithecus shortridgei Shortridge's Leaf Monkey split from *T. pileatus* Capped Leaf Monkey – Groves, 2001

Pygathrix cinerea Grey-shanked Douc Langur – Nadler, 1997 [as *Pygathrix nemaeus cinereus*] split from *P. nemaeus* Red-shanked Douc Langur – Groves, 2001

Pygathrix nigripes Black-shanked Douc Langur split from *P. nemaeus* Red-shanked Douc Langur – Groves, 2001

Macaca hecki Heck's Macaque split from *M. tonkeana* Tonkean Macaque – Groves, 2001

Macaca leonina Northern Pig-tailed Macaque split from *M. nemestrina* Sunda Pig-tailed Macaque – Groves, 2001

Macaca nigrescens Gorontalo Macaque split from *M. nigra* Sulawesi Crested Macaque – Groves, 2001

Macaca pagensis Mentawai Macaque split from *M. nemestrina* Sunda Pig-tailed Macaque – Groves, 2001

Macaca siberu Siberut Macaque split from *M. pagensis* Mentawai Macaque – Fuentes & Olson, 1995; Kitchener & Groves, 2002

Papio anubis Olive Baboon split from *P. hamadryas* Hamadryas Baboon – Kingdon, 1997; Groves, 2001

Papio cynocephalus Yellow Baboon split from *P. hamadryas* Hamadryas Baboon – Kingdon, 1997; Groves, 2001

Papio papio Guinea Baboon split from *P. hamadryas* Hamadryas Baboon – Kingdon, 1997; Groves, 2001

Papio ursinus Chacma Baboon split from *P. hamadryas* Hamadryas Baboon – Kingdon, 1997; Groves, 2001

Cercocebus atys Sooty Mangabey split from *C. torquatus* Red-capped Mangabey – Kingdon, 1997; Groves, 2001

Cercocebus chrysogaster Golden-bellied Mangabey split from *C. agilis* Agile Mangabey – Kingdon, 1997; Groves, 2001

Cercocebus sanjei Sanje Mangabey split from *C. agilis* Agile Mangabey – Kingdon, 1997; Groves, 2001

Lophocebus aterrimus Black Mangabey split from *L. albigena* Grey-cheeked Mangabey – Kingdon, 1997; Groves, 2001

Lophocebus opdenboschi Opdenbosch's Mangabey split from *L. albigena* Grey-cheeked Mangabey – Groves, 2001

Cercopithecus albogularis Sykes' Monkey split from *C. mitis* Blue Monkey – Groves, 2001

Cercopithecus denti Dent's Monkey split from *C. wolfi* Wolf's Monkey – Kingdon, 1997; Groves, 2001

Cercopithecus doggetti Silver Monkey split from *C. mitis* Blue Monkey – Groves, 2001

Cercopithecus kandti Golden Monkey split from *C. mitis* Blue Monkey – Groves, 2001

Cercopithecus lowei Lowe's Monkey split from *C. campbelli* Campbell's Monkey – Kingdon, 1997; Groves, 2001

Cercopithecus roloway Roloway Monkey split from *C. diana* Diana Monkey – Groves, 2001

Chlorocebus cynosuros Malbrouck Monkey split from *C. aethiops* Grivet Monkey – Groves, 2001

Chlorocebus djamdjamensis Bale Monkey split from *C. aethiops* Grivet Monkey – Kingdon, 1997; Groves, 2001

Chlorocebus pygerythrus Vervet Monkey split from *C. aethiops* Grivet Monkey – Kingdon, 1997; Groves, 2001

Chlorocebus sabaeus Green Monkey split from *C. aethiops* Grivet Monkey – Kingdon, 1997; Groves, 2001

Chlorocebus tantalus Tantalus Monkey split from *C. aethiops* Grivet Monkey – Kingdon, 1997; Groves, 2001

Miopithecus ogouensis Northern Talapoin split from *M. talapoin* Southern Talapoin – Kingdon, 1997

HOMINIDAE (Apes)

Pongo abelii Sumatran Orangutan split from *P. pygmaeus* Bornean Orangutan – Groves, 2001

Gorilla beringei Eastern Gorilla split from *G. gorilla* Western Gorilla – Groves, 2001

Hylobates (Hylobates) albibarbis White-bearded Bornean Gibbon split from *H. (Hylobates) agilis* Agile Gibbon – Groves, 2001

Hylobates (Nomascus) hainanus Hainan Gibbon split from *H. (Nomascus) concolor* Black Crested Gibbon – Groves, 2001

Hylobates (Nomascus) siki Southern White-cheeked Gibbon split from *H. (Nomascus) leucogenys* Northern White-cheeked Gibbon – Groves, 2001

CALLITRICHIDAE (Tamarins and Marmosets)

Saguinus graellsi Graells' Black-mantled Tamarin split from *S. nigricollis* Black-mantled Tamarin – Groves, 2001

Saguinus martinsi Martins' Bare-faced Tamarin split from *S. bicolor* Pied Bare-faced Tamarin – Groves, 2001

Saguinus melanoleucus White-mantled Tamarin split from *S. fuscicollis* Saddle-backed Tamarin – Groves, 2001

Saguinus niger Black Tamarin split from *S. midas* Red-handed Tamarin – Groves, 2001

Saguinus pileatus Red-capped Moustached Tamarin split from *S. mystax* Black-chested Moustached Tamarin – Groves, 2001

Callithrix acariensis Rio Acarí Marmoset – Van Roosmalen, Van Roosmalen, Mittermeier & Rylands, 2000

Callithrix chrysoleuca Gold-and-white Marmoset split from *C. humeralifera* Santarém Marmoset – Groves, 2001

Callithrix emiliae Emilia's Marmoset split from *C. argentata* Silvery Marmoset – Emmons & Feer, 1997; Groves, 2001

Callithrix humilis Dwarf Marmoset – Van Roosmalen, Van Roosmalen, Mittermeier & da Fonseca, 1998

Callithrix intermedia Hershkovitz's Marmoset split from *C. argentata* Silvery Marmoset – Groves, 2001

Callithrix leucippe White Marmoset split from *C. argentata* Silvery Marmoset – Groves, 2001

Callithrix manicorensis Rio Manicoré Marmoset – Van Roosmalen, Van Roosmalen, Mittermeier & Rylands, 2000

Callithrix marcai Marca's Marmoset – Alperin, 1993; Groves, 2001

Callithrix mauesi Rio Maués Marmoset – Mittermeier, Ayres & Schwarz, 1992

Callithrix melanura Black-tailed Marmoset split from *C. argentata* Silvery Marmoset – Groves, 2001

Callithrix nigriceps Black-headed Marmoset – Ferrari & Lopes, 1992

Callithrix saterei Saterê-Maués' Marmoset – de Sousa e Silva & de Noronha, 1998

ATELIDAE (New World Monkeys)

Cebus kaapori Ka'apor Capuchin – Queiroz, 1992

Cebus libidinosus Black-striped Tufted Capuchin split from *C. apella* Brown Tufted Capuchin – Groves, 2001

Cebus nigritus Black Tufted Capuchin split from *C. apella* Brown Tufted Capuchin – Groves, 2001

Cebus xanthosternos Golden-bellied Tufted Capuchin split from *C. apella* Brown Tufted Capuchin – Groves, 2001

Callicebus baptista Baptista Lake Titi split from *C. hoffmannsi* Hoffmanns' Titi – Groves, 2001; Van Roosmalen, Van Roosmalen & Mittermeier, 2002

Callicebus barbarabrownae Blond Titi split from *C. personatus* Masked Titi – Van Roosmalen, Van Roosmalen & Mittermeier, 2002

Callicebus bernhardi Prince Bernhard's Titi – Van Roosmalen, Van Roosmalen & Mittermeier, 2002

Callicebus coimbrai Coimbra-Filho's Titi – Kobayashi & Langguth 1999

Callicebus discolor Double-browed Titi split from *C. cupreus* Coppery Titi – Van Roosmalen, Van Roosmalen & Mittermeier, 2002

Callicebus lucifer Lucifer Titi split from *C. torquatus* Collared Titi – Van Roosmalen, Van Roosmalen & Mittermeier, 2002

Callicebus lugens Mourning Titi split from *C. torquatus* Collared Titi – Van Roosmalen, Van Roosmalen & Mittermeier, 2002

Callicebus medemi Medem's Titi split from *C. torquatus* Collared Titi – Groves, 2001; Van Roosmalen, Van Roosmalen & Mittermeier, 2002

Callicebus melanochir Black-handed Titi split from *C. personatus* Masked Titi – Van Roosmalen, Van Roosmalen & Mittermeier, 2002

Callicebus nigrifrons Black-fronted Titi split from *C. personatus* Masked Titi – Van Roosmalen, Van Roosmalen & Mittermeier, 2002

Callicebus ornatus Ornate Titi split from *C. cupreus* Coppery Titi – Groves, 2001; Van Roosmalen, Van Roosmalen & Mittermeier, 2002

Callicebus pallescens Pallid Titi split from *C. donacophilus* White-eared Titi – Groves, 2001; Van Roosmalen, Van Roosmalen & Mittermeier, 2002

Callicebus purinus Red-crowned Titi split from *C. torquatus* Collared Titi – Van Roosmalen, Van Roosmalen & Mittermeier, 2002

Callicebus regulus Kinglet Titi split from *C. torquatus* Collared Titi – Van Roosmalen, Van Roosmalen & Mittermeier, 2002

Callicebus stephennashi Stephen Nash's Titi – Van Roosmalen, Van Roosmalen & Mittermeier, 2002

Ateles hybridus Brown Spider Monkey split from *A. belzebuth* White-bellied Spider Monkey – Groves, 2001

Brachyteles hypoxanthus Northern Muriqui split from *B. arachnoides* Southern Muriqui – Groves, 2001

Lagothrix cana Grey Woolly Monkey split from *L. lagotricha* Brown Woolly Monkey – Groves, 2001

Lagothrix lugens Colombian Woolly Monkey split from *L. lagotricha* Brown Woolly Monkey – Groves, 2001

Lagothrix poeppigii Silvery Woolly Monkey split from *L. lagotricha* Brown Woolly Monkey – Groves, 2001

Alouatta macconnelli Guyanan Red Howler Monkey split from *A. seniculus* Venezuelan Red Howler Monkey – Groves, 2001

Alouatta nigerrima Amazon Black Howler Monkey split from *A. belzebul* Red-handed Howler Monkey – Groves, 2001

BALAENOPTERIDAE (Rorquals)

Balaenoptera bonaerensis Antarctic Minke Whale split from *B. acutorostrata* Northern Minke Whale – Rice, 1998; Reeves *et al.*, 2002

BALAENIDAE (Right Whales)

Balaena (Eubalaena) japonica North Pacific Right Whale split from *B. (Eubalaena) glacialis* North Atlantic Right Whale – Reeves *et al.*, 2002

HYPEROODONTIDAE (Beaked Whales)

Mesoplodon traversii Spade-toothed Whale split from *M. layardii* Strap-toothed Whale – = *M. bahamondi* Reyes, van Waerebeek, Cárdenas & Yáñez, 1995; van Helden, Baker, Dalebout, Reyes, van Waerebeek & Baker, 2002

Mesoplodon perrini Perrin's Beaked Whale – Dalebout, Mead, Baker, Baker & van Helden, 2002

DELPHINIDAE (Marine Dolphins)

Delphinus capensis Long-beaked Common Dolphin split from *D. delphis* Short-beaked Common Dolphin – Heyning & Perrin, 1994

Delphinus tropicalis Arabian Common Dolphin split from *D. delphis* Short-beaked Common Dolphin – Rice, 1998

Tursiops aduncus Indo-Pacific Bottle-nosed Dolphin split from *T. truncatus* Common Bottle-nosed Dolphin – Rice, 1998

SUIDAE (Pigs)

Sus domesticus Domestic Pig split from *S. scrofa* Eurasian Wild Boar – Clutton-Brock, 1999

CAMELIDAE (Camels)

Camelus ferus Wild Bactrian Camel split from *C. bactrianus* Domestic Bactrian Camel – Clutton-Brock, 1999

CERVIDAE (Deer)

Muntiacus putaoensis Leaf Muntjac – Amato, Egan & Rabinowitz, 1999

Muntiacus trungsonensis Truong Son Muntjac – Giao, Tuoc, Dung, Wikramanayake, Amato, Arctander & MacKinnon, 1998

Muntiacus vuquangensis Giant Muntjac – Tuoc, Dung, Dawson, Arctander & Mackinnon, 1994 [as *Megamuntiacus vuquangensis*, type species of new genus *Megamuntiacus* but synonymized with *Muntiacus* by McKenna & Bell, 1997]

Mazama pandora Yucatán Brown Brocket split from *M. americana* Red Brocket – Medellín, Gardner & Aranda, 1998

BOVIDAE (Cattle, Antelope, Sheep and Goats)

Capra aegagrus Wild Goat split from *C. hircus* Domestic Goat – Clutton-Brock, 1999

Ovis orientalis Mouflon split from *O. aries* Domestic Sheep – Clutton-Brock, 1999

Pseudoryx nghetinhensis Saola (Vu Quang Ox) – Dung, Giao, Chinh, Tuoc, Arctander & MacKinnon, 1993

Bos gaurus Gaur split from *B. frontalis* Mithan – Clutton-Brock, 1999

Bos indicus Zebu (Humped Cattle) split from *B. taurus* European Domestic Cattle – Clutton-Brock, 1999

Bos mutus Wild Yak split from *B. grunniens* Domestic Yak – Clutton-Brock, 1999

Bubalus arnee Wild Water Buffalo split from *B. bubalis* Domestic Water Buffalo – Clutton-Brock, 1999

EQUIDAE (Horses)

Equus africanus Wild Ass split from *E. asinus* Domestic Donkey – Clutton-Brock, 1999

Equus ferus Przewalski's Horse split from *E. caballus* Domestic Horse – Clutton-Brock, 1999

ELEPHANTIDAE (Elephants)

Loxodonta cyclotis African Forest Elephant split from *L. africana* African Savanna Elephant – Grubb, Groves, Dudley & Shoshani, 2000

SPECIES DEMOTED INTO SYNONYMY

PHALANGERIDAE (Cuscuses and Brushtail Possums)

Trichosurus arnhemensis Northern Brushtail Possum becomes a synonym of *T. vulpecula* Common Brushtail Possum (Kerle, 2001; Menkhorst & Knight, 2001)

DIDELPHIDAE (American Opossums)

Gracilinanus longicaudus Long-tailed Gracile Mouse-Opossum becomes a synonym of *G. emiliae* Emilia's Gracile Mouse-Opossum (Voss, Lunde & Simmons, 2001)

MURIDAE (Mice, Rats, Voles and Gerbils)

Peromyscus caniceps Burt's Deer Mouse becomes a synonym of subspecies *P. eremicus fraterculus* Cactus Deer Mouse (Hafner, Riddle & Alvarez-Castañeda, 2001)

Peromyscus dickeyi Dickey's Deer Mouse becomes a subspecies of *P. merriami* Merriam's Deer Mouse (Hafner, Riddle & Alvarez-Castañeda, 2001)

Peromyscus interparietalis San Lorenzo Deer Mouse becomes a synonym of subspecies *P. eremicus eremicus* Cactus Deer Mouse (Hafner, Riddle & Alvarez-Castañeda, 2001)

Peromyscus oreas Columbian Mouse and *Peromyscus sitkensis* Sitka Mouse become synonyms of *P. keeni* Northwestern Deer Mouse (new status, Hogan, Hedin, Hung Sun Koh, Davis & Greenbaum, 1993)

Peromyscus stephani San Esteban Island Deer Mouse becomes a subspecies of *P. boylii* Brush Deer Mouse (Hafner, Riddle & Alvarez-Castañeda, 2001)

Dicrostonyx exsul St Lawrence Island Collared Lemming, *D. kilangmiutak* Victoria Collared Lemming, *D. nelsoni* Nelson's Collared Lemming, *D. nunatakensis* Ogilvie Mountain Collared Lemming, *D. rubricatus* Bering Collared Lemming and *D. unalascensis* Unalaska Collared Lemming become synonyms of *D. groenlandicus* Northern Collared Lemming (Jarrell & Fredga, 1993; Kays & Wilson, 2002)

Otomys saundersiae Saunders' Vlei Rat becomes a synonym of *O. irroratus* Common Vlei Rat (Taylor, Meester & Kearney, 1993)

Crunomys rabori Leyte Shrew-Mouse becomes a synonym of *C. melanius* Southern Philippine Shrew-Mouse (Musser & Heaney, 1992)

GEOMYIDAE (Pocket Gophers)

Perognathus xanthonotus Yellow-eared Pocket Mouse becomes a synonym of *P. parvus* Great Basin Pocket Mouse (Williams, Genoways & Braun, 1993)

Dipodomys elephantinus Big-eared Kangaroo-Rat becomes a subspecies of *D. venustus* Narrow-faced Kangaroo-Rat (Best, Chesser, McCullough & Baumgardner, 1996)

ERETHIZONTIDAE (New World Porcupines)

Coendou (Sphiggurus) pallidus Pallid Hairy Dwarf Porcupine becomes a synonym of *C. (Sphiggurus) insidiosus* Bahia Hairy Dwarf Porcupine (Voss & Angermann, 1997)

Coendou (Sphiggurus) villosus Atlantic Forest Hairy Dwarf Porcupine becomes a synonym of *C. (Sphiggurus) spinosus* Orange-spined Hairy Dwarf Porcupine (Emmons & Feer, 1997)

ECHIMYIDAE (Spiny Rats)

Mesomys didelphoides Brazilian Spiny Tree Rat becomes a senior synonym of *Makalata armata* Armored Tree Rat (hence *Makalata didelphoides*, Emmons, 1993)

CANIDAE (Dogs and Foxes)

Pseudalopex griseus Argentine Grey Fox becomes a synonym of *P. gymnocercus* Pampas Fox (Zunino *et al.*, 1995)

CHRYSOCHLORIDAE (Golden-Moles)

Amblysomus iris Zulu Golden-Mole becomes a subspecies of *A. hottentotus* Hottentot Golden-Mole (Bronner, 1996)

SORICIDAE (Shrews)

Sorex jacksoni St Lawrence Island Shrew becomes a subspecies of *S. cinereus* Cinereous Shrew (Rausch & Rausch, 1995)

Cryptotis avia Andean Small-eared Shrew becomes a synonym of *C. thomasi* Thomas' Small-eared Shrew (Woodman, 1996)

Crocidura maxi Max's Shrew and *Crocidura minuta* Minute Shrew become synonyms of *C. monticola* Sunda Shrew (Ruedi, 1995)

Suncus madagascariensis Madagascar Shrew becomes a subspecies of *S. etruscus* Pygmy White-toothed Shrew (Garbutt, 1999)

TENRECIDAE (Tenrecs)

Microgale pulla Dark Shrew-Tenrec becomes a synonym of *M. parvula* Pygmy Shrew-Tenrec (Jenkins, Goodman & Raxworthy, 1996)

Oryzorictes talpoides Mole-like Rice Tenrec becomes a synonym of *O. hova* Hova Rice Tenrec (Goodman, Jenkins & Pidgeon, 1999)

PTEROPODIDAE (Old World Fruit Bats)

Pteropus mearnsi Mearns' Flying Fox becomes a synonym of *P. speciosus* Philippine Flying Fox (Flannery, 1995b)

Nyctimene celaeno Dark Tube-nosed Fruit Bat becomes a synonym of *N. aello* Greater Tube-nosed Fruit Bat (Flannery, 1995a)

Nyctimene malaitensis Malaita Tube-nosed Fruit Bat becomes a subspecies of *N. bougainville* Solomons Tube-nosed Fruit Bat (Flannery, 1995b)

Nyctimene masalai Demonic Tube-nosed Fruit Bat becomes a synonym of *N. albiventer* Common Tube-nosed Fruit Bat (Flannery, 1995b)

Nyctimene vizcaccia Umboi Tube-nosed Fruit Bat becomes a synonym of *N. albiventer* Common Tube-nosed Fruit Bat (Flannery, 1995b)

Melonycteris (Nesonycteris) aurantius Orange Blossom Bat becomes a subspecies of *M. (Nesonycteris) woodfordi* Woodford's Blossom Bat (Flannery, 1993)

PHYLLOSTOMIDAE (American Leaf-nosed Bats)

Choeroniscus intermedius Intermediate Whiskered Long-nosed Bat becomes a synonym of *C. minor* Lesser Whiskered Long-nosed Bat (Simmons & Voss, 1998)

MOLOSSIDAE (Free-tailed Bats)

Tadarida espiritosantensis Espírito Santo Free-tailed Bat becomes a synonym of *Nyctinomops laticaudatus* Broad-eared Bat (Eisenberg & Redford, 1999)

ATELIDAE (New World Monkeys)

Aotus brumbacki Brumback's Night Monkey becomes a subspecies of *A. lemurinus* Lemurine Night Monkey (Groves, 2001)

Aotus infulatus Feline Night Monkey becomes a subspecies of *A. azarai* Azara's Night Monkey (Groves, 2001)

BOVIDAE (Cattle, Antelope, Sheep and Goats)

Gazella bilkis Queen of Sheba's Gazelle becomes a subspecies of *G. arabica* Arabian Gazelle (Groves, 1997)

NOMENCLATURAL CHANGES

PHALANGERIDAE (Cuscuses and Brushtail Possums)

Strigocuscus gymnotis Ground Cuscus becomes *Phalanger gymnotis* (Flannery, 1995b)

MURIDAE (Mice, Rats, Voles and Gerbils)

Oryzomys capito Large-headed Rice Rat becomes *O. megacephalus* (Musser, Brothers, Carleton & Gardner, 1998)

Oryzomys legatus Big-headed Rice Rat becomes *O. russatus* (Díaz, Braun, Mares & Barquez, 2000)

Nectomys parvipes Small-footed Water Rat becomes *N. melanius* (Voss, Lunde & Simmons, 2001)

Holochilus magnus Greater Marsh Rat becomes *Lundomys molitor* (McKenna & Bell, 1997; Eisenberg & Redford, 1999)

Akodon azarae Azara's Grass Mouse becomes *Akodon azarai*, a justified emendation

Hydromys neobrittanicus New Britain Water Rat becomes *N. neobritannicus,* a justified emendation (Flannery, 1995b)

Oxymycterus iheringi Ihering's Hocicudo becomes *Brucepattersonius iheringi* Ihering's Brucie (Hershkovitz, 1998; Mares & Braun, 2000)

Uromys neobritanicus Bismarck Giant Rat becomes *U. neobritannicus*, a justified emendation (Flannery, 1995b)

ERETHIZONTIDAE (New World Porcupines)

Coendou koopmani Koopman's Porcupine becomes *C. nycthemera* (Voss & Angermann, 1997)

AGOUTIDAE (Agoutis and Pacas)

Dasyprocta azarae Azara's Agouti becomes *D. azarai*, a justified emendation

OCTODONTIDAE (Degus and Tuco-tucos)

Ctenomys azarae Azara's Tuco-tuco becomes *C. azarai*, a justified emendation

ECHIMYIDAE (Spiny Rats)

Proechimys (Proechimys) cayennensis Cayenne Spiny Rat becomes *P. (Proechimys) guyannensis* (Patton, 1987; Eisenberg & Redford, 1999)

Proechimys (Proechimys) oris Pará Spiny Rat becomes *P. (Proechimys) roberti* (Weksler, Bonvicino, Otazu & Silva, 2001)

Makalata armata Armoured Tree Rat becomes *Makalata didelphoides* (Emmons, 1993; Emmons & Feer, 1997)

OTARIIDAE (Eared Seals)

Otaria byronia South American Sea-Lion becomes *O. flavescens* (Reeves *et al.*, 2002)

SORICIDAE (Shrews)

Sorex hydrodromus Pribilof Island Shrew becomes *S. pribilofensis* (Rausch & Rausch, 1997)

PTEROPODIDAE (Old World Fruit Bats)

Pteropus gilliardi New Britain Flying Fox becomes *P. gilliardorum*, a justified emendation (Flannery, 1995b)

PHYLLOSTOMIDAE (American Leaf-nosed Bats)

Sturnira bogotensis Bogotá Yellow-shouldered Bat becomes *S. oporaphilum* (Díaz, Braun, Mares & Barquez, 2000)

MOLOSSIDAE (Free-tailed Bats)

Molossops mattogrossensis Mato Grosso Dog-faced Bat becomes *M. matogrossensis*, a justified emendation (Emmons & Feer, 1997)

Molossus bondae Bonda Mastiff Bat becomes *M. currentium* (López-González & Presley, 2001)

VESPERTILIONIDAE (Vesper Bats)

Eptesicus sagittula Large Forest Bat becomes *E. darlingtoni* (Menkhorst & Knight, 2001)

Vespertilio superans Asian Particoloured Bat becomes *V. sinensis* (Horácek, 1997)

ATELIDAE (New World Monkeys)

Alouatta fusca Brown Howler Monkey becomes *A. guariba* (Groves, 2001)

BALAENOPTERIDAE (Rorquals)

Balaenoptera edeni Bryde's Whale becomes *B. brydei* (Dizon *et al.*, 1995)

KOGIIDAE (Pygmy Sperm Whales)

Kogia simus Dwarf Sperm Whale becomes *K. sima* (Jones *et al.*, 1997)

CAMELIDAE (Camels)

Lama pacos Alpaca becomes *Vicugna pacos* (Clutton-Brock, 1999)

BIBLIOGRAPHY

AGUILERA, M. & CORTI, M. 1994. Craniometric differentiation and chromosomal speciation of the genus *Proechimys*. *Mamm. Biol. (Z. Säugert.)* 59: 366-377.

AGUILERA, M., PEREZ-ZAPATA, A. & MARTINO, A. 1995. Cytogenetics and karyosystematics of *Oryzomys albigularis* from Venezuela. *Cytogenet. Cell Genet.* 69: 44-49.

ALBUJA, V.L. & PATTERSON, B.D. 1996. A new species of northern shrew-opossum (Paucituberculata: Caenolestidae) from the Cordillera del Condor, Ecuador. *J. Mammalogy* 77(1): 41-53.

ALPERIN, R. 1993. *Callithrix argentata* (Linnaeus, 1771): considerações taxonômicas e descriçao de subespécie nova. *Boletim del Museo Paraense Emilio Goeldi Série Zoologia* 9: 317-328.

AMATO, G., EGAN, M.G. & RABINOWITZ, A. 1999. A new species of muntjac, *Muntiacus putaoensis* (Artiodactyla: Cervidae) from northern Myanmar. *Animal Conservation* 2: 1-7.

ANDERSON, R.P. & HANDLEY, C.O., Jr. 2001. A new species of three-toed sloth (Mammalia: Xenarthra) from Panamá, with a review of the genus *Bradypus*. *Proc. Biol. Soc. Washington* 114: 1-33.

ANDERSON, R.P. & JARRÍN-V., P. 2002. A new species of spiny pocket mouse (Heteromyidae: *Heteromys*) endemic to western Ecuador. *Amer. Mus. Novitates* 3382.

ANDERSON, S. & YATES, T.L. 2000. A new genus and species of phyllotine rodent from Bolivia. *J. Mammalogy* 81(1): 18-36.

AVERIANOV, A.O., ABRAMOV, A.V. & TIKHONOV, A.N. 2000. A new species of *Nesolagus* (Lagomorpha, Leoporidae) from Vietnam with osteological description. *Contributions from the Zoological Institute, St. Petersburg* 3: 1-22.

BAKER, R.J., O'NEILL, M.B. & McALILEY, L.R. 2003. A new species of desert shrew, *Notiosorex*, based on nuclear and mitochondrial sequence data. *Occasional Papers, Museum of Texas Tech University* Number 222.

BAKER, R.J., PATTON, J.C., GENOWAYS, H.H. & BICKHAM, J.W. 1988. Genic studies of *Lasiurus* (Chiroptera: Vespertilionidae). *Occasional Papers, Museum of Texas Tech University* Number 117.

BAKER, R.J., SOLARI, S. & HOFFMANN, F.G. 2002. A new Central American species from the *Carollia brevicauda* complex. *Occasional Papers, Museum of Texas Tech University* Number 217.

BATES, P.J.J. & HARRISON, D.L. 1997. *Bats of the Indian Subcontinent*. Harrison Zoological Museum.

BEARDER, S. 1997-1998. *Pseudopotto*: When is a potto not a potto? *African Primates* 3(1-2): 43-44.

BENDA, P. & TSYTSULINA, K.A. 2000. Taxonomic revision of *Myotis mystacinus* group (Mammalia: Chiroptera) in the western Palearctic. *Actas Societatis Zoologicae Bohemicae* 64(4): 331-398.

BERGMANS, W. 2001. Notes on distribution and taxonomy of Australian bats, Pteropodinae and Nyctimeninae (Mammalia), Megachiroptera, Pteropodidae. *Beaufortia* 51(8): 119-152.

BEST, T.L., CHESSER, R.K., McCULLOUGH, D.A. & BAUMGARDNER, G.D. 1996. Genic and morphometric variation in kangaroo rats, genus *Dipodomys*, from coastal California. *J. Mammalogy* 77: 785-800.

BONVICINO, C. & ALMEIDA, F.C. 2000. Karyotype, morphology and taxonomic status of *Calomys expulsus* (Rodentia: Sigmodontinae). *Mammalia* 64(3): 339-351.

BONVICINO, C.R., PENNA-FIRME, V. & BRAGGIO, E. 2002. Molecular and karyologic evidence of the taxonomic status of *Coendou* and *Sphiggurus* (Rodentia: Hystricognathi). *J. Mammalogy* 83(4): 1071-1076.

BONVICINO, C.R. & WEKSLER, M. 1998. A new species of *Oligoryzomys* (Rodentia, Sigmodontinae) from northeastern and central Brazil. *Mamm. Biol. (Z. Säugert.)* 63(2): 90-103.

BRANDON-JONES, C. 1995. A revision of the Asian pied leaf-monkeys (Mammalia: Cercopithecidae: superspecies *Semnopithecus auratus*), with a description of a new subspecies. *Raffles Bulletin of Zoology* 43: 3-43.

BRAUN, J.K. & MARES, M.A. 1995. A new genus and species of phyllotine rodent (Rodentia: Muridae: Sigmodontinae: Phyllotini) from South America. *J. Mammalogy* 76: 504-521.

BRAUN, J.K. & MARES, M.A. 2002. Systematics of the *Abrocoma cinerea* species complex (Rodentia: Abrocomidae), with a description of a new species of *Abrocoma*. *J. Mammalogy* 83(1): 1-19.

BRAUN, J.K., MARES, M.A. & OJEDA, R.A. 2000. A new species of grass mouse, genus *Akodon* (Muridae: Sigmodontinae), from Mendoza Province, Argentina. *Mamm. Biol. (Z. Säugert.)* 65(4): 216-225.

BRONNER, G.N. 1996. Geographic patterns of morphometric variation in the Hottentot golden mole, *Amblysomus hottentotus* (Insectivora: Chrysochloridae). A multivariate analysis. *Mammalia* 60(4): 729-751.

BRONNER, G.N. 2000. New species and subspecies of Golden Mole (Chrysochloridae: *Amblysomus*) from Mpumalanga, South Africa. *Mammalia* 64(1): 41-54.

BURDA, H., ZIMA, J., SCHARFF, A., MACHOLÁN, M. & KAWALIKA, M. 1999. The karyotypes of *Cryptomys anselli* sp. nova and *Cryptomys kafuensis* sp. nova: new species of the common mole-rat from Zambia (Rodentia, Bathyergidae). *Mamm. Biol. (Z. Säugert.)* 64(1): 36-50.

CARLETON, M.D. 1994. Systematic studies of Madagascar's endemic rodents (Muroidea: Nesomyinae): revision of the genus *Eliurus*. *Amer. Mus. Novitates* 3087.

CARLETON, M.D. & GOODMAN, S.M. 1996. Systematic study of Madagascar's endemic rodent (Muroidea: Nesomyinae): a new genus and species from the central highlands. *Fieldiana: Zoology* (n.s.) 85: 231-256.

CARLETON, M.D. & GOODMAN, S.M. 1998. New taxa of nesomyine rodents (Muroidea: Muridae) from Madagascar's northern highlands, with taxonomic comments on previously described forms. *Fieldiana: Zoology* (n.s.) 139: 161-198.

CARLETON, M.D., GOODMAN, S.M. & RAKOTONDRAVONY, D. 2001. A new species of tufted-tailed rat, genus *Eliurus* (Muridae: Nesomyinae), from western Madagascar, with notes on the distribution of *E. myoxinus*. *Proc. Biol. Soc. Washington* 14: 972-987.

CARRAWAY, L.N. & TIMM, R.M. 2000. Revision of the extant taxa of the genus *Notiosorex* (Mammalia: Insectivora: Soricidae). *Proc. Biol. Soc. Washington* 113(1): 302-318.

CARWARDINE, M. & CAMM, M. 1995. *Whales, Dolphins and Porpoises.* London: Dorling Kindersley.

CHAPMAN, J.A., KRAMER, K.L., DIPPENAAR, N.J. & ROBINSON, T.J. 1992. Systematics and biogeography of the New England cottontail, *Sylvilagus transitionalis* (Bangs, 1895), with the description of a new species from the Appalachian mountains. *Proc. Biol. Soc. Washington* 105(4): 841-866.

CHEREM, J.J., OLIMPIO, J. & XIMENEZ, A. 1999. Descricao de uma nova especie do genero *Cavia* Pallas, 1776 (Mammalis – Cavidae) das Ilhas dos Moleques do Sul, Santa Catarina, Sul do Brasil. *Biotemas* 12(1): 95-117.

CHRISTOFF, A.U., FAGUNDES, V., SBALQUEIRO, I.J., MATTEVI, M.S. & YONENAGA-YASSUDA, Y. 2000. Description of a new species of *Akodon* (Rodentia: Sigmodontinae) from southern Brazil. *J. Mammalogy* 81(3): 838-851.

CLUTTON-BROCK, J. 1999. *A Natural History of Domesticated Mammals.* 2nd edition. Cambridge: Cambridge University Press.

CONTRERAS, J.R. 1989. *Ctenomys roigi*, una nueva especie de 'anguya tutu' de la Provincia de Corrientes, Argentina (Rodentia: Ctenomyidae). *Boletin de Istituto de Estudios Almerienses* 1988 (extra)(1989): 51-67.

CONTRERAS, J.R. 1995. *Ctenomys osvaldoreigi*, una nueva especie de tuco-tuco procedente de las sierras de Cordoba, Republica Argentina (Rodentia: Ctenomyidae). *Notulas Faunisticas* 84: 1-4.

CONTRERAS VEGA, M. & CADENA, A. 2000. Una nueva especie del género *Sturnira* (Chiroptera: Phyllostomidae) de los Andes colombianos. *Revista de la Academia Colombiana de Ciencias Exactas, Fisicas y Naturales* 24(91): 285-287.

COOPER, S.J.B., APLIN, K.P. & ADAMS, M. 2000. A new species of false antechinus (Marsupialia: Dasyuromorphia: Dasyuridae) from the Pilbara region, western Australia. *Records of the Western Australian Museum* 20: 115-136.

CORBET, G.C. & HILL, J.E. 1992. *The Mammals of the Indomalayan Region: a systematic review.* Oxford: Oxford University Press.

COTTERILL, F. P. D. 2002. A new species of horseshoe bat (Microchiroptera: Rhinolophidae) from south-central Africa: with comments on its affinities and evolution, and the characterization of rhinolophid species. *J. Zool., Lond.* 256: 165-179.

CSORBA, G. 1997. Description of a new species of *Rhinolophus* (Chiroptera: Rhinolophidae) from Malaysia. *J. Mammalogy* 78(2): 342-347.

CSORBA, G. & LEE, L.-L. 1999. A new species of vespertilionid bat from Taiwan and a revision of the taxonomic status of *Arielulus* and *Thainycteris* (Chiroptera: Vespertilionidae). *J. Zool., Lond.* 248: 361-367.

CZAPLEWSKI, N.J. 1996. *Thyroptera robusta* Czaplewski, 1996, is a junior synonym of *Thyroptera lavali* Pine, 1993 (Mammalia: Chiroptera). *Mammalia* 60(1): 153-155.

DALEBOUT, M.L., MEAD, J.G., BAKER, C.S., BAKER, A.N. & VAN HELDEN, A.L. 2002. A new species of beaked whale *Mesoplodon perrini* sp. n. (Cetacea: Ziphiidae) discovered through phylogenetic analyses of mitochondrial DNA sequences. *Marine Mammal Science* 18: 577-608.

DA ROCHA, P.L.B. 1996. *Proechimys yonenagae*, a new species of spiny rat (Rodentia: Echimyidae) from fossil sand dunes in the Brazilian Caatinga. *Mammalia* 59(4) [1995]: 537-549.

DA SILVA, M.N.F. 1998. Four new species of spiny rats of the genus *Proechimys* (Rodentia: Echimyidae) from the western Amazon of Brazil. *Proc. Biol. Soc. Washington* 111(2): 436-471.

DAVISON, A., BIRKS, J.D.S., GRIFFITHS, H.I., KITCHENER, A.C., BIGGINS, D. & BUTLIN, R.K. 1999. Hybridization and the phylogenetic relationship between polecats and domestic ferrets in Britain. *Biological Conservation* 87: 155-161.

DE OLIVEIRA, J.A. & BONVICINO, C.R. 2002. A new species of sigmodontine rodent from the Atlantic forest of eastern Brazil. *Acta Theriologica* 47(3): 307-322.

DE SOUSA E SILVA, J., Jr. & DE NORONHA, M. 1998. On a new species of bare-eared marmoset, genus *Callithrix* Erxleben, 1777, from central Amazonia, Brazil (Primates: Callitrichidae). *Goeldiana Zoologia* (21): 1-28.

DÍAZ, M.M., BARQUEZ, R.M., BRAUN, J.K. & MARES, M.A. 1999. A new species of *Akodon* (Muridae: Sigmodontinae) from northwestern Argentina. *J. Mammalogy* 80: 786-798.

DÍAZ, M.M., BRAUN, J.K., MARES, M.A. & BARQUEZ, R.M. 2000. An update of the taxonomy, systematics, and distribution of the mammals of Salta province, Argentina. *Occasional Papers, Sam Noble Oklahoma Museum of Natural History* 10: 1-52.

DÍAZ, M.M., FLORES, D.A. & BARQUEZ, R.M. 2002. A new species of gracile mouse opossum, genus *Gracilinanus* (Didelphimorphia: Didelphidae), from Argentina. *J. Mammalogy* 83(3): 824-833.

DICKMAN, C.R., PARNABY, H.E., CROWTHER, M.S. & KING, H.H. 1998. *Antechinus agilis* (Marsupialia: Dasyuridae), a new species from the *A. stuartii* complex in south-eastern Australia. *Australian J. Zool.* 46: 1-26.

DIZON, A., LUX, C., COSTA, S., LEDUC, R. & BROWNELL, R., Jr. 1995. Phylogenetic relationships of the closely related Sei and Bryde's Whales: a possible third species? *In: Abstracts: Eleventh biennial conference of the biology of marine mammals; 1995 14-18 December*. Orlando: The Society for Marine Mammalogy.

DOKUCHAEV, N.E. 1997. A new species of shrew (Soricidae, Insectivora) from Alaska. *J. Mammalogy* 78: 811-817.

DUNG, V.V., GIAO, P.M., CHINH, N.N., TUOC, D., ARCTANDER, P. & MACKINNON, J. 1993. A new species of living bovid from Vietnam. *Nature* 363: 443-445.

DURANT, P. & GUEVARA, M.A. 2000. A new rabbit species (*Sylvilagus*, Mammalia: Leporidae) from the lowlands of Venezuela. *Revista de Biologia Tropical* 49(1): 369-381.

EGER, J.L. & SCHLITTER, D.A. 2001. A new species of *Glauconycteris* from West Africa (Chiroptera: Vespertilionidae). *Acta Chiropterologica* 3(1): 1-10.

EISENBERG, J.F. 1989. *Mammals of the Neotropics. Volume 1. The Northern Neotropics: Panama, Colombia, Venezuela, Guyana, Suriname, French Guiana*. Chicago and London: University of Chicago Press.

EISENBERG, J.F. & REDFORD, K.H. 1999. *Mammals of the Neotropics. Volume 3. The Central Neotropics: Ecuador, Peru, Bolivia, Brazil*. Chicago and London: University of Chicago Press.

ELDRIDGE, M.D.B. & CLOSE, R.L. 1992. Taxonomy of rock wallabies, *Petrogale* (Marsupialia: Macropodidae). 1. A revision of the eastern *Petrogale* with the description of three new species. *Australian J. Zool.* 40(6): 605-625.

ELDRIDGE, M.D.B. & CLOSE, R.L. 1997. Chromosomes and evolution in rock-wallabies, *Petrogale* (Marsupialia). *Australian Mammalogy* 19: 123-135.

ELDRIDGE, M.D.B., WILSON, A.C.C., METCALFE, C.J., DOLLIN, A.E., BELL, J.N., JOHNSON, P.M., JOHNSTON, P.G. & CLOSE, R.L. 2001. Taxonomy of rock-wallabies, *Petrogale* (Marsupialia : Macropodidae). III. Molecular data confirms the species status of the purple-necked rock-wallaby (*Petrogale purpureicollis* Le Souef). *Australian J. Zool.* 49(4): 323-343.

EMMONS, L.H. 1993. On the identity of *Echimys didelphoides* Desmarest, 1817 (Mammalia: Rodentia: Echimyidae). *Proc. Biol. Soc. Washington* 106: 1-4.

EMMONS, L.H. 1999a. A new genus and species of abrocomid rodent from Peru (Rodentia: Abrocomidae). *Amer. Mus. Novitates* 3279.

EMMONS, L.H. 1999b. Two new species of *Juscelinomys* (Rodentia: Muridae) from Bolivia. *Amer. Mus. Novitates* 3280.

EMMONS, L.H., LEITE, Y.L.R., KOCK, D. & COSTA, L.P. 2002. A review of the named forms of *Phyllomys* (Rodentia: Echimyidae) with the description of a new species from coastal Brazil. *Amer. Mus. Novitates* 3380.

EMMONS, L.H. & FEER, F. 1997. *Neotropical Rainforest Mammals: a field guide*. 2nd edition. Chicago and London: The University of Chicago Press.

EMMONS, L.H. & VUCETICH, M.G. 1998. The identity of Winge's *Lasiuromys villosus* and the description of a new genus of echimyid rodent (Rodentia: Echimyidae). *Amer. Mus. Novitates* 3223.

FAHR, J., VIERHAUS, H., HUTTERER, R. & KOCK, D., 2002. A revision of the *Rhinolophus maclaudi* species group with the description of a new species from West Africa (Chiroptera: Rhinolophidae). *Myotis* 40: 95-126.

FAZZOLARI-CORRÊA, S. 1994. *Lasiurus ebenus*, a new vespertilionid bat from southeastern Brasil. *Mammalia* 58(1): 119-123.

FERRARI, S.F. & LOPES, M.A. 1992. A new species of marmoset, genus *Callithrix* Erxleben 1777 (Callitrichidae, Primates) from western Brazilian Amazonia. *Goeldiana Zoologia* (12): 1-3.

FLANNERY, T.F. 1992. Taxonomic revision of the *Thylogale brunii* complex (Macropodidae: Marsupialia) in Melanesia, with description of a new species. *Australian Mammalogy* 15: 7-23.

FLANNERY, T.F. 1993. Revision of the fruit bats of the genus *Melonycteris* (Pteropodidae: Mammalia). *Records of the Australian Museum* 45(1): 59-80.

FLANNERY, T.F. 1995a. *Mammals of New Guinea*. Revised and updated edn. Sydney: Australian Museum / Reed Books.

FLANNERY, T.F. 1995b. *Mammals of the South-west Pacific and Moluccan Islands*. Sydney: Australian Museum / Reed Books.

FLANNERY, T.F. 1995c. Systematic revision of *Emballonura furax* Thomas, 1911 and *E. dianae* Hill, 1956 (Chiroptera: Emballonuridae), with description of new species and subspecies. *Mammalia* 58(4)[1994]: 601-612.

FLANNERY, T.F., ARCHER, M. & MAYNES, G. 1987. The phylogenetic relationships of living phalangerids (Phalangeroidea: Marsupialia) with a suggested new taxonomy. *In:* M. Archer (ed.) 1987. *Possums and opossums*. Sydney: Surrey Beatty & Sons. pp. 477-506.

FLANNERY, T.F. & BOEADI, 1995. Systematic revision within the *Phalanger ornatus* complex, with description of a new species. *Australian Mammalogy* 18: 35-44.

FLANNERY, T.F., BOEADI & SZALAY, A.L. 1995. A new tree-kangaroo (*Dendrolagus*: Marsupialia) from Irian Jaya, Indonesia, with notes on ethnography and the evolution of tree-kangaroos. *Mammalia* 59(1): 65-84.

FLANNERY, T.F. & COLGAN, D.J. 1993. A new species and two new subspecies of *Hipposideros* (Chiroptera) from western Papua New Guinea. *Records of the Australian Museum* 45(1): 43-57.

FLANNERY, T., COLGAN, D. & TRIMBLE, J. 1994. A new species of *Melomys* from Manus Island, Papua New Guinea, with notes on the systematics of the *M. rufescens* complex (Muridae: Rodentia). *Proc. Linnaean Soc. N.S.W.* 114: 29-43.

FLANNERY, T.F. & GROVES, C.P. 1998. A revision of the genus *Zaglossus*, with description of a new species and subspecies. *Mammalia* 62(3): 367-396.

FLANNERY, T.F. & SCHOUTEN, P. 1994. *Possums of the World: a monograph of the Phalangeroidea*. Sydney: Geo Productions.

FLORES, D.A., DÍAZ, M.M. & BARQUEZ, R.M. 2000. Mouse opossums (Didelphimorphia, Didelphidae) of northwestern Argentina: systematics and distribution. *Mamm. Biol. (Z. Säugert.)* 65: 1-19.

FRANCIS, C.M. & HILL, J.E. 1998. New records and a new species of *Myotis* (Chiroptera, Vespertilionidae) from Malaysia. *Mammalia* 62(2): 241-252.

FRANCIS, C.M., KOCK, D. & HABERSETZER, J. 1999. Sibling species of *Hipposideros ridleyi* (Mammalia, Chiroptera, Hipposideridae). *Senckenbergiana biologica* 79(2): 255-270.

FREY, J.K. & LARUE, C.T. 1993. Notes on the distribution of the Mogollon vole (*Microtus mogollonensis*) in Arizona and New Mexico. *The Southwestern Naturalist* 38: 176-178.

FUENTES, A. & OLSON, M. 1995. Preliminary observations and status of the Pagai macaque (*Macaca pagensis*). *Asian Primates* 4(4): 1-4.

FUNAKOSHI, K. & KUNISAKI, T. 2000. On the validity of *Tadarida latouchei*, with reference to morphological divergence among *T. latouchei*, *T. insignis* and *T. teniotis* (Chiroptera, Molossidae). *Mammal Study* 25(2): 115-123.

GARBUTT, N. 1999. *Mammals of Madagascar*. Sussex: Pica Press.

GARCIA-PEREA, R. 1994. The pampas cat group (genus *Lynchailurus* Severtzov, 1858) (Carnivora: Felidae), a systematic and biogeographic review. *Amer. Mus. Novitates* 3096.

GARDNER, A.L. & ROMO, R.M. 1993. A new *Thomasomys* (Mammalia: Rodentia) from the Peruvian Andes. *Proc. Biol. Soc. Washington* 106: 762-774.

GAUBERT, P. 2003. Description of a new species of genet (Carnivora; Viverridae; genus *Genetta*) and taxonomic revision of forest forms related to the Large-spotted Genet complex. *Mammalia* 67(1): 85-108.

GENOWAYS, H.H. & BAKER, R.J. 1996. A new species of the genus *Rhogeessa*, with comments on geographic distribution and speciation in the genus. *In:* H.H. Genoways & R. Baker (eds.) *Contributions in Mammalogy: A Memorial Volume Honoring Dr. J. Knox Jones, Jr.* Museum of Texas Tech University, pp. 83-87.

GIAO, P., TUOC, D., DUNG, V., WIKRAMANAYAKE, E., AMATO, G., ARCTANDER, P. & MACKINNON, J. 1998. Description of *Muntiacus truongsonensis*, a new species of muntjac (Artiodactyla: Muntiacidae) from central Vietnam, and implications for conservation. *Animal Conservation* 1: 61-68.

GONG, Z.-D., WANG, Y.-X., LI, Z.-H. & LI, S.-Q. 2000. A new species of pika: Pianma blacked pika *Ochotona nigritia* (Lagomorpha: Ochotonidae) from Yunnan, China. *Zoological Research* 21(3): 204-209.

GONZALES, P.C. & KENNEDY, R.S. 1996. A new species of *Crateromys* (Rodentia: Muridae) from Panay, Philippines. *J. Mammalogy* 77(1): 25-40.

GONZÁLEZ, E.M. 2000. Un nuevo genero do roedor sigmodontino de Argentina y Brazil (Mammalia: Rodentia: Sigmodontinae). *Comunicaciones Zoologicas del Museo de Historia Natural de Montevideo* 12: 1-12.

GOODMAN, S.M. & JENKINS, P.D. 1998. The insectivores of the Réserve Spéciale d'Anjanaharibe-Sud, Madagascar. *Fieldiana: Zoology* (n.s.) 90: 139-161.

GOODMAN, S.M., JENKINS, P.D. & PIDGEON, M. 1999. Lipotyphla (Tenrecidae and Soricidae) of the Réserve Naturelle Intégrale d'Andohahela, Madagascar. *Fieldiana: Zoology* (n.s.) 94: 187-216.

GRANJON, L., ANISKIN, V.M., VOLOBOUEV, V. & SICARD, B. 2002. Sand-dwellers in rocky habitats: a new species of *Gerbillus* (Mammalia: Rodentia) from Mali. *J. Zool., Lond.* 256: 181-190.

GROVES, C.P. 1997. The taxonomy of Arabian gazelles. *In:* K.Habibi, A.H.Abuzinada & I.A.Nader (eds.) *The Gazelles of Arabia*. Riyadh: National Commission for Wildlife Conservation and Development, Publication No.29, English Series, pp. 4-51.

GROVES, C.P. 1997-1998. *Pseudopotto martini*: A new potto? *African Primates* 3(1-2): 42-43.

GROVES, C.P. 2000. The genus *Cheirogaleus*: unrecognized biodiversity in dwarf lemurs. *Int. J. Primatology* 21(6): 943-962.

GROVES, C.P. 2001. *Primate Taxonomy*. Washington, DC: Smithsonian Institution Press.

GROVES, C.P. & FLANNERY, T. 1994. A revision of the genus *Uromys* Peters, 1867 (Muridae, Mammalia) with description of two new species. *Records of the Australian Museum* 46: 145-170.

GRUBB, P., GROVES, C. P., DUDLEY, J.P. & SHOSHANI, J. 2000. Living African elephants belong to two species: *Loxodonta africana* (Blumenbach, 1797) and *Loxodonta cyclotis* (Matschie, 1900). *Elephant* 2(4):1-4.

GURUNG, K.K. & SINGH, R. 1996. *Field Guide to the Mammals of the Indian Subcontinent: where to watch mammals in India, Nepal, Bhutan, Bangladesh, Sri Lanka and Pakistan*. London: Academic Press.

HAFNER, D.J., RIDDLE, B.R. & ALVAREZ-CASTAÑEDA, S.T. 2001. Evolutionary relationships of white-footed mice (*Peromyscus*) on islands in the Sea of Cortéz, Mexico. *J. Mammalogy* 82(3): 775-790.

HANDLEY, C.O., Jr. 1996. New species of mammals from northern South America: bats of the genera *Histiotus* Gervais and *Lasiurus* Gray (Chiroptera: Vespertilionidae). *Proc. Biol. Soc. Washington* 109(1): 1-9.

HANDLEY, C.O., Jr. & OCHOA, J. 1996. New species of mammals from northern South America: a sword-nosed bat, genus *Lonchorhina* Tomes (Chiroptera: Phyllostomidae). *Memoria de la Sociedad de Ciencias Naturales La Salle* 57(148): 71-82.

HAYS, J.P. & HARRISON, R.G. 1992. Variation in mitochondrial DNA and the biogeographic history of woodrats (*Neotoma*) in the eastern United States. *Systematic Biology* 41: 331-334.

HAYS, J.P. & RICHMOND, M.E. 1993. Clinal variation and morphology of woodrats (*Neotoma*) of the eastern United States. *J. Mammalogy* 74: 204-216.

HERSHKOVITZ, P. 1992. The South American gracile mouse opossums, genus *Gracilinanus* Gardner and Creighton, 1989 (Marmosidae, Marsupialia): a taxonomic review with notes on general morphology and relationships. *Fieldiana: Zoology* (n.s.) 70: 1-56.

HERSHKOVITZ, P. 1993. A new central Brazilian genus and species of sigmodontine rodent (Sigmodontinae) transitional between akodonts and oryzomyines with discussion of muroid molar morphology and evolution. *Fieldiana: Zoology* (n.s.) 75: 1-18.

HERSHKOVITZ, P. 1994. The description of a new species of South American hocicudo, or long-nose mouse, genus *Oxymycterus* (Sigmodontinae, Muroidea), with a critical review of the generic content. *Fieldiana: Zoology* (n.s.) 79: 1-43.

HERSHKOVITZ, P. 1998. Report on some sigmodontine rodents collected in southeastern Brazil with descriptions of a new genus and six new species. *Bonn. Zool. Beitr.*, 47(3-4): 193-256.

HEYNING, J.E. & PERRIN, W.F. 1994. Evidence for two species of common dolphins (genus *Delphinus*) from the eastern North Pacific. *Contrib. Sci., Nat. Hist. Mus., Los Angeles County*, 442: 1-35.

HOFFMANN, F.G., LESSA, E.P. & SMITH, M.F. 2002. Systematics of *Oxymycterus* with description of a new species from Uruguay. *J. Mammalogy* 83(2): 408-420.

HOGAN, K.M., HEDIN, M.C., HUNG SUN KOH, DAVIS, S.K. & GREENBAUM, I.F. 1993. Systematic and taxonomic implications of karyotypic, electrophoretic, and mitochondrial-DNA variation in *Peromyscus* from the Pacific Northwest. *J. Mammalogy* 74: 819-831.

HOLDEN M.E. 1996. Description of a new species of *Dryomys* (Rodentia, Myoxidae) from Balochistan, Pakistan, including morphological comparisons with *Dryomys laniger* Felten & Storch, 1968 and *D. nitedula* (Pallas, 1778). *Bonn. Zool. Beitr.*, 46(1-4) [1995-1996]: 111-131.

HONESS, P.E. 1996. Speciation among galagos (Primates: Galagidae) in Tanzanian forests. Ph.D. thesis, Oxford Brookes University, Oxford, UK. [See also Honess, P.E. & Bearder, S. 1996. Descriptions of the dwarf galago species of Tanzania. *African Primates* 2(2): 75-79.]

HORÁCEK, I. 1997. Status of *Vesperus sinensis* Peters, 1880 and remarks on the genus *Vespertilio*. *Vespertilio* 2: 59-72.

HUTTERER, R. 1994. Island rodents: a new species of *Octodon* from Isla Mocha, Chile (Mammalia: Octodontidae). *Mamm. Biol. (Z. Säugert.)* 59: 27-41.

HUTTERER, R. & SCHLITTER, D.A. 1996. Shrews of Korup National Park, Cameroon, with the description of a new *Sylvisorex* (Mammalia: Soricidae). *In:* H.H. Genoways & R. Baker (eds.) *Contributions in Mammalogy: A Memorial Volume Honoring Dr. J. Knox Jones, Jr.* Museum of Texas Tech University, pp. 57-66.

INGLE, N.R. & HEANEY, L.R. 1992. A key to the bats of the Philippine Islands. *Fieldiana: Zoology* (n.s.) 69: 1-44.

IVANITSKAYA, E., SHENBROT, G., & NEVO, E. 1996. *Crocidura ramona* sp. nov. (Insectivora, Soricidae): a new species of shrew from the central Negev Desert, Israel. *Mamm. Biol. (Z. Säugert.)* 61(2): 93-103.

JARRELL, G.H. & FREDGA, K. 1993. How many kinds of lemmings? A taxonomic overview. pp. 45-57. *In:* N.C. Stenseth & R.A. Ims (eds.) *The Biology of Lemmings*. Linnaean Soc. Symposium Series No. 15. London: Academic Press.

JENKINS, P.D. 1993. A new species of *Microgale* (Insectivora: Tenrecidae) from Eastern Madagascar with an unusual dentition. *Amer. Mus. Novitates* 3067.

JENKINS, P.D. & BARNETT, A.A. 1997. A new species of water mouse, of the genus *Chibchanomys* (Rodentia: Muridae: Sigmodontinae) from Ecuador. *Bull. Nat. Hist. Mus. Lond. (Zool.)* 63(2): 123-128.

JENKINS, P.D. & GOODMAN, S.M. 1999. A new species of *Microgale* (Lipotyphla, Tenrecidae) from isolated forest in southwestern Madagascar. *Bull. Nat. Hist. Mus. Lond. (Zool.)* 65(2): 155-164.

JENKINS, P.D., GOODMAN, S.M. & RAXWORTHY, C.J. 1996. The shrew tenrecs (*Microgale*) (Insectivora: Tenrecidae) of the Réserve Naturelle Intégrale d'Andringitra, Madagascar. *In:* S.M. Goodman, A floral and faunal inventory of the eastern side of the Réserve Naturelle Intégrale d'Andringitra, Madagascar: with reference to elevational variation. *Fieldiana: Zoology* (n.s.) 85: 191-217.

JENKINS, P.D., RAXWORTHY, C.J. & NUSSBAUM, R.A. 1997. A new species of *Microgale* (Insectivora, Tenrecidae), with comments on the status of four other taxa of shrew tenrecs. *Bull. Nat. Hist. Mus. Lond. (Zool.)* 63(1): 1-12.

JENKINS, P.D. & SMITH, A.L. 1995. A new species of *Crocidura* (Insectivora: Soricidae) recovered from owl pellets in Thailand. *Bull. Nat. Hist. Mus. Lond. (Zool.)* 61(2): 103-109.

JIANG, X.-L. & HOFFMANN, R.S. 2001. A revision of the white-toothed shrews (*Crocidura*) of southern China. *J. Mammalogy* 82(4): 1059-1079.

JONES, C., HOFFMANN, R.S., RICE, D.W., BAKER, R.J., ENGSTROM, M.D., BRADLEY, R.D., SCHMIDLY, D.J. & JONES, C.A. 1997. Revised checklist of North American mammals north of Mexico, 1997. *Occasional Papers, Museum of Texas Tech University*, Number 173.

JONES, G. & BARRATT, E.M. 1999. *Vespertilio pipistrellus* Schreber, 1774 and *V. pygmaeus* Leach, 1825 (currently *Pipistrellus pipistrellus* and *P. pygmaeus*; Mammalia, Chiroptera): proposed designation of neotypes. *Bull. Zool. Nomenclature* 56: 182-186.

JUSTE, J. & IBANEZ, C. 1993. A new *Tadarida* of the subgenus *Chaerephon* (Chiroptera: Molossidae) from Sao Tome island, Gulf of Guinea (West Africa). *J. Mammalogy* 74(4): 901-907.

KALKO, E.K.V. & HANDLEY, C.O., Jr. 1994. Evolution, bio-geography, and description of a new species of fruit-eating bat, genus *Artibeus* Leach (1821), from Panamá. *Mamm. Biol. (Z. Säugert.)* 59: 257-273.

KAYS, R.W. & WILSON, D.E. 2002. *Mammals of North America.* Princeton & Oxford: Princeton University Press.

KEFELIOGLU, H. & KRYSTUFEK, B. 1999. The taxonomy of *Microtus socialis* group (Rodentia: Microtinae) in Turkey, with the description of a new species. *J. Natural History* 33(2): 289-303.

KELT, D.A. & GALLARDO, M.H. 1994. A new species of tuco-tuco, genus *Ctenomys* (Rodentia: Ctenomyidae) from Patagonian Chile. *J. Mammalogy* 75: 338-348.

KERLE, A. 2001. *Possums.* Sydney: University of New South Wales Press.

KIEFER, A. & VEITH, M. 2001. A new species of long-eared bat from Europe (Chiroptera: Vespertilionidae). *Myotis* 39: 5-16.

KINGDON, J. 1997. *The Kingdon Field Guide to African Mammals.* London: Academic Press.

KITCHENER, A.C. & GROVES, C. 2002. New insights into the taxonomy of the Mentawai Island macaques. *Mammalia* 66(4): 533-542.

KITCHENER, D.J., HOW, R.A. & MARYANTO, I. 1992. A new species of *Otomops* (Chiroptera: Molossidae) from Alor I., Nusa Tenggara, Indonesia. *Records of the Western Australian Museum* 15(4): 729-738.

KITCHENER, D.J. & MAHARADATUNKAMSI 1991. Description of a new species of *Cynopterus* (Chiroptera: Pteropodidae) from Nusa Tenggara, Indonesia. *Records of the Western Australian Museum* 15: 307-363.

KITCHENER, D.J. & MARYANTO, I. 1993a. Taxonomic reappraisal of the *Hipposideros larvatus* species complex (Chiroptera: Hipposideridae) in the Greater and Lesser Sunda Islands, Indonesia. *Records of the Western Australian Museum* 16: 199-173.

KITCHENER, D.J. & MARYANTO, I. 1993b. A new species of *Melomys* (Rodentia: Muridae) from Kai Baser Island. Maluku Tengah, Indonesia. *Records of the Western Australian Museum* 16: 427-436.

KITCHENER, D.J. & MARYANTO, I. 1995. A new speies of *Melomys* (Rodentia, Muridae) from Yamdana Island, Tanimbar group, Eastern Indonesia. *Records of the Western Australian Museum* 17: 43-50.

KITCHENER, D.J., PACKER, W.C. & MARYANTO, I. 1997. Morphological variation among island populations of *Scotophilus kuhlii (sensu lato)* Leach, 1821 (Chiroptera: Vespertilionidae) from the Greater and Lesser Sunda islands, Indonesia. *Tropical Biodiversity* 4: 53-81.

KITCHENER, D.J. & SUYANTO, A. 1996a. A new species of *Melomys* (Rodentia: Muridae) from Riama Island, Tanimbar Group. Maluku Tenggara, Indonesia. *Records of the Western Australian Museum* 18: 113-119.

KITCHENER, D.J. & SUYANTO, A. 1996b. Intraspecific morphological variation among island populations of small mammals in southern Indonesia. *In:* D.J. Kitchener & A. Suyanto (eds.) *Proceedings of the First International Conference on Eastern Indonesian – Australian Vertebrate Fauna.*

KOBAYASHI, S. & LANGGUTH, A. 1999. A new species of titi monkey, *Callicebus* Thomas, from north-eastern Brazil (Primates, Cebidae). *Revta. Bras. Zool.* 16(2): 531-551.

KOCK, D. 1999. *Tadarida (Tadarida) latouchei,* a separate species recorded from Thailand with remarks on related Asian taxa (Mammalia, Chiroptera, Molossidae). *Senckenbergiana biologica* 78(1/2): 237-240.

KOCK, D. & BHAT, H.R. 1994. *Hipposideros hypophyllus* n. sp. of the *H. bicolor*-group from peninsular India (Mammalia: Chiroptera: Hipposideridae). *Senckenbergiana biologica* 73(1/2): 25–31.

KOCK, D., CSORBA, G., & HOWELL, K.M. 2000. *Rhinolophus maendeleo* n. sp. from Tanzania, a horseshoe bat noteworthy for its systematics and biogeography (Mammalia, Chiroptera, Rhinolophidae). *Senckenbergiana biologica* 80(1/2): 233–239.

KOCK, D. & STORCH, G. 1996. *Thainycteris aureocollaris,* a remarkable new genus and species of vespertilionine bats from SE-Asia (Mammalia: Chiroptera: Vespertilionidae). *Senckenbergiana biologica* 76(1/2): 1-6.

KRUSKOP, S.V. & LAVRENCHENKO, L.A. 2000. A new species of long-eared bat (Vespertilionidae, Chiroptera) from Ethiopia. *Myotis* 38: 5-17.

KRUSKOP, S.V. & TSYTSULINA, K.A. 2001. A new big-footed mouse-eared bat *Myotis annamiticus* sp. nov. (Vespertilionidae, Chiroptera) from Vietnam. *Mammalia* 65(1): 63-72.

LAPINI, L. & TESTONE, R. 1998. A new *Sorex* from north-eastern Italy (Mammalia: Insectivora: Soricidae). *Gortania Atti del Museo Friulano di Storia Naturale* 20: 233-252.

LARA, M.C., PATTON, J.L. & HINGST-ZAHER, E. 2002. *Trinomys mirapitanga,* a new species of spiny rat (Rodentia: Echimyidae) from the Brazilian Atlantic Forest. *Mamm. Biol. (Z. Säugert.)* 67(4): 233-242.

LAVRENCHENKO, L.A., LINKHNOVA, O.P., BASKEVIC, M.I. & BEKELE, A. 1998. Systematics and distribution of *Mastomys* (Muridae, Rodentia) from Ethiopia with the description of a new species. *Mamm. Biol. (Z. Säugert.)* 63 (1): 37-51.

LAVRENCHENKO, L.A., VERHEYEN, W.N. & HULSELMANS, J. 1998. Systematic and distributional notes on the *Lophuromys flavopunctatus* Thomas 1888 species-complex in Ethiopia (Muridae, Rodentia). *Bulletin de l'Institut Royal des Sciences Naturelles de Belgique: Biologie* 68: 199-214.

LEE, T.E., Jr., RIDDLE, B.R. & LEE, P.L. 1996. Speciation in the desert pocket mouse (*Chaetodipus penicillatus* Woodhouse). *J. Mammalogy* 77: 58-68.

LEO, L.M. & GARDNER, A.L. 1993. A new species of giant *Thomasomys* (Mammalia: Muridae: Sigmodontinae) from the Andes of north central Peru. *Proc. Biol. Soc. Washington* 106: 417-428.

LIN, L.K. & HARADA, M. 1998. A new species of *Mustela* from Taiwan. *Proceedings Euro-American Mammal Congress, Spain.*

LINDENMAYER, D.B., DUBACH, J. & VIGGERS, K.L. 2002. Geographic dimorphism in the Mountain Brushtail Possum (*Trichosurus caninus*) – the case for a new species. *Australian J. Zool.* 50(4): 369-393.

LÓPEZ-GONZÁLEZ, C. & PRESLEY, S.J. 2001. Taxonomic status of *Molossus bondae* J.A. Allen, 1904 (Chiroptera: Molossidae), with description of a new subspecies. *J. Mammalogy* 82(3): 760-774.

LUNA, L. & PACHECO, V. 2002. A new species of *Thomasomys* (Muridae: Sigmodontinae) from the Andes of southeastern Peru. *J. Mammalogy* 83(3): 834-842.

MACDONALD, D. & BARRETT, P. 1993. *Collins Field Guide Mammals of Britain & Europe.* London: HarperCollins.

McFARLANE, D. 2000 *Extinct Mammals of the West Indies* website [http://www.jsd.claremont.edu/bio/extinct/extinctmammals/]

McKENNA, M.C. & BELL, S.K. 1997. *Classification of Mammals Above the Species Level.* New York: Columbia University Press.

McKENNA, M.C. & BELL, S.K. 31 October 2002. Mammal Classification. ftp://ftp.amnh.org/pub/people/mckenna/Mammalia.txn.sit [for Macintosh users] or Mammalia.zip [for PC users]. Read, displayed, and searchable with Unitaxon Browser. CD-ROM. 2001. Boulder, CO: Mathemaesthetics, Inc.

MAEDA, A. & MATSUMURA, S. 1998. Two new species of vespertilionid bats, *Myotis* and *Murina* (Vespertilionidae: Chiroptera) from Yanbaru, Okinawa Island, Okinawa Prefecture, Japan. *Zoological Science (Tokyo)* 15(2): 301-307.

MALYGIN, V.M., ANISKIN, V.M., ISAEV, S.I. & MILISHNIKOV, A.N. 1994. *Amphinectomys savamis* Malygin gen. et sp. n., a new genus and a new species of water rat (Cricetidae, Rodentia) from Peruvian Amazonia. *Zoologichaskii Zhurnal* 73: 195-208.

MARES, M.A. & BRAUN, J.K. 1996. A new species of phyllotine rodent, genus *Andalgalomys* (Muridae: Sigmodontinae), from Argentina. *J. Mammalogy* 77(4): 928-941.

MARES, M.A. & BRAUN, J.K. 2000. Three new species of *Brucepattersonius* (Rodentia: Muridae: Sigmodontinae) from Misiones Province, Argentina. *Occasional Papers, Sam Noble Museum of Natural History* 9: 1-13.

MARES, M.A., BRAUN, J.K., BARQUEZ, R.M. & DÍAZ, M.M. 2000. Two new genera and species of halophytic desert mammals from isolated salt flats in Argentina. *Occasional Papers, Museum of Texas Tech University* 203.

MEDELLÍN, R.A., GARDNER, A.L. & ARANDA, J.M. 1998. The taxonomic status of the Yucatán brown brocket *Mazama*

pandora (Mammalia: Cervidae). *Proc. Biol. Soc. Washington* 111(1): 1-14.

MENKHORST, P.W. & KNIGHT, F. 2001. *A Field Guide to the Mammals of Australia.* Oxford University Press.

MENZIES, J.I. 1996. A systematic revision of *Melomys* (Rodentia: Muridae) of New Guinea. *Australian J. Zoology* 44(4): 367-426.

MERCURE, A., RALLS, K., KOEPFLI, K.P. & WAYNE, R.K. 1993. Genetic subdivisions among small canids: mitochondrial DNA differentiation of swift, kit, and arctic foxes. *Evolution* 47: 1313-1328.

MITCHELL-JONES, A.J., AMORI, G., BOGDANOWICZ, W., KRY TUFEK, REIJNDERS, P.J.H., SPITZENBERGER, F., STUBBE, M., THISSEN, J.B.M., VOHRALÍK, V. & ZIMA, J. 1999. *The Atlas of European Mammals.* London: T. & A.D. Poyser Ltd / Academic Press.

MITTERMEIER, R.A., SCHWARZ, M. & AYRES, J.M. 1992. A new species of marmoset, genus *Callithrix* Erxleben 1777 (Callitrichidae, Primates), from the Rio Maues region, state of Amazonas, Central Brazilian Amazonia. *Goeldiana Zoologia* (14): 1-17.

MOLINARI, J. 1994. A new species of *Anoura* (Mammalia Chiroptera Phyllostomidae) from the Andes of northern South America. *Tropical Zoology* 7(1): 73-86.

MOOJEN, J., LOCKS, M. & LANGGUTH, A. 1997. A new species of *Kerodon* Cuvier, 1825 from the state of Goias, Brazil (Mammalia, Rodentia, Caviidae). *Boletim Museu Nacional Rio de Janeiro Zoologia* 377.

MUCEDDA, M., KIEFER, A., E. PIDINCHEDDA, E. & VEITH, M. 2002. A new species of long-eared bat (Chiroptera, Vespertilionidae) from Sardinia (Italy). *Acta Chiropterologica* 4(2): 121-135.

MUSSER, G.G., CARLETON, M.D., BROTHERS, E.M. & GARDNER A.L. 1998. Systematic studies of the oryzomyine rodents (Muridae, Sigmodontinae): diagnosis and distribution of species formerly assigned to *Oryzomys* "capito". *Bull. Amer. Mus. Nat. Hist.* 236: 1-376.

MUSSER, G.G. & DURDEN, L.A. 2002. Sulawesi rodents: description of a new genus and species of Murinae (Muridae, Rodentia) and its parasitic new species of sucking louse (Insecta, Anoplura). *Amer. Mus. Novitates* 3368.

MUSSER, G.G. & HEANEY, L.R. 1992. Philippine rodents: definitions of *Tarsomys* and *Limnomys* plus a preliminary assessment of phylogenetic patterns among native Philippine murines (Murinae, Muridae). *Bull. Amer. Mus. Nat. Hist.* 211: 1-138.

MUSSER, G.G., HEANEY, L.R. & TABARANZA, B.R., Jr. 1998. Philippine rodents: redefinitions of known species of *Batomys* (Muridae, Murinae) and description of a new species from Dinagat Island. *Amer. Mus. Novitates* 3237.

MUSTRANGI, M.A. & PATTON, J.L. 1997. Phylogeography and systematics of the slender mouse opossum *Marmosops* (Marsupialia, Didelphidae). *Univ. California Publ. Zool.* 130.

NADLER, T. 1997. A new subspecies of douc langur, *Pygathrix nemaeus cinereus* ssp. nov. *Zoologische Garten* 67: 165-176.

NORRIS, C.A. & MUSSER, G.G. 2001. Systematic revision within the *Phalanger orientalis* complex (Diprotodontia, Phalangeridae): a third species of lowland gray cuscus from New Guinea and Australia. *Amer. Mus. Novitates* 3356.

OCHOA, G.J., AGUILERA, M., PACHECO, V. & SORIANO, P.J. 2001. A new species of *Aepeomys* Thomas, 1898 (Rodentia: Muridae) from the Andes of Venezuela. *Mamm. Biol. (Z. Säugert.)* 66(4): 228-237.

PACHECO, V. & PATTON, J.L. 1995. A new species of puna mouse, *Punomys* Osgood, 1943 (Muridae, Sigmodontinae) from the Southeastern Andes of Peru. *Mamm. Biol. (Z. Säugert.)* 60: 85-96.

PALMA, R.E. & YATES, T.L. 1998. Phylogeny of southern South American mouse opossums (*Thylamys*, Didelphidae) based on allozyme and chromosomal data. *Mamm. Biol. (Z. Säugert.)* 63: 1-15.

PARNABY, H.E. 2002a. A new species of long-eared bat (*Nyctophilus*: Vespertilionidae) from New Caledonia. *Australian Mammalogy* 23: 115-124.

PARNABY, H.E. 2002b. A taxonomic review of the genus *Pteralopex* (Chiroptera: Pteropodidae), the monkey-faced bats of the south-western Pacific. *Australian Mammalogy* 23: 145-162.

PATTERSON, B.D. 1992. A new genus and species of long-clawed mouse (Rodentia: Muridae) from temperate rainforests of Chile. *Zool. J. Linn. Soc.* 106: 127-145.

PATTON, J.L. 1987. Species groups of spiny rats, genus *Proechimys* (Rodentia: Echimyidae). *Fieldiana: Zoology.* (n.s.) 39: 305-345.

PATTON, J.L. & DA SILVA, M.N.F. 1995. A review of the spiny mouse genus *Scolomys* (Rodentia: Muridae: Sigmodontinae) with the description of a new species from western Amazon of Brazil. *Proc. Biol. Soc. Washington* 108(2): 319-337.

PATTON, J.L. & DA SILVA, M.N.F. 1997. Definition of species of pouched four-eyed opossums (Didelphidae, *Philander*). *J. Mammalogy* 78(1): 90-102.

PATTON, J.L., DA SILVA, M.N.F. & MALCOLM, J.R. 2000. Mammals of the Rio Jurua and the evolutionary and ecological diversification of Amazonia. *Bull. Amer. Mus. Nat. Hist.* 244: 1-306.

PAYNE, J. & FRANCIS, C.M. 1998. *A Field Guide to the Mammals of Borneo.* 3rd reprint. Kota Kinabalu: Sabah Society.

PEREZ-HERNANDEZ, R., SORIANO, P. & LEW, D. 1993. *Marsupiales de Venezuela.* Cuadernos Lagoven: Lagoven, SA.

PETERS, S.L., LIM, B.. & ENGSTROM, M.D. 2002. Systematics of dog-faced bats (*Cynomops*) based on molecular and morphometric data. *J. Mammalogy* 83(4): 1097-1110.

PETERSON, R.L., EGER, J.L. & MITCHELL, L. 1995. Chiroptères. *Faune de Madagascar.* Vol 84. Paris: Natural History Museum.

PINE, R.H. 1993. A new species of *Thyroptera* Spix (Mammalia: Chiroptera: Thyropteridae) from the Amazon Basin of northwestern Perú. *Mammalia* 57: 213-225.

PRADHAN, M.S., MONDAL, A.K., BHAGWAT, A.M. & AGRAWAL, V.C. 1993. Taxonomic studies of Indian Bandicoot rats (Rodentia: Muridae: Murinae) with description of a new species. *Rec. Zool. Survey India* 93(1-2).

QUEIROZ, H.L. 1992. A new species of capuchin monkey, genus *Cebus* Erxleben 1977 (Cebidae, Primates), from eastern Brazilian Amazonia. *Goeldiana Zoologia* (15): 1-3.

RASOLOARISON, R.M., GOODMAN, S.M. & GANZHORN, J.U. 2000. Taxonomic revision of mouse lemurs (*Microcebus*) in the western portions of Madagascar. *Int. J. Primatology* 21(6): 963-1019.

RAUSCH, R.L. & RAUSCH, V.R. 1995. The taxonomic status of the shrew of St. Lawrence Island, Bering Sea (Mammalia: Soricidae). *Proc. Biol. Soc. Washington*, 108: 717-728.

RAUSCH, R.L. & RAUSCH, V.R. 1997. Evidence for specific independence of the shrew (Mammalia: Soricidae) of St. Paul Island (Pribilof Islands, Bering Sea). *Mamm. Biol. (Z. Säugert.)* 62: 193-202.

RAY, J.C. & HUTTERER, R. 1996. Structure of a shrew community in the Central African Republic based on the analysis of carnivore scats, with the description of a new *Sylvisorex* (Mammalia: Soricidae). *Ecotropica* 1(2)[1995]: 85-97.

REDFORD, K.H. & EISENBERG, J.F. 1992. *Mammals of the Neotropics. Volume 2. The Southern Cone: Chile, Argentina, Uruguay, Paraguay.* Chicago and London: University of Chicago Press.

REEVES, R.R., STEWART, B.S., CLAPHAM, P.J., POWELL, J.A. & FOLKENS, P.A. 2002. *Sea Mammals of the World.* London: A & C Black.

REID, F.A. 1997. *A Field Guide to the Mammals of Central America and Southeast Mexico.* Oxford: Oxford University Press.

REYES, J.C., VAN WAEREBEEK, K., CARDENAS, J.-C. & YANEZ, J.L. 1995. *Mesoplodon bahamondi* sp. n. (Cetacea: Ziphiidae), a new beaked whale from the Juan Fernandez Archipelago. *Boletin del Museo Nacional de Historia Natural de Chile* 45(1): 31-44.

RICE, D.W. 1998. *Marine Mammals of the World: systematics and distribution.* Society of Marine Mammalogy: Special Publication Number 4.

RICHARDS, G.C. & HALL, L.S. 2002. A new flying fox of the genus *Pteropus* (Chiroptera: Pteropodidae) from Torres Strait, Australia. *Australian Zoologist* 32(1): 69-75.

RICKART, E.A., HEANEY, L.R., BALETE, D.S. & TABARANZA, B.R., Jr. 1998. A review of the genera *Crunomys* and *Archboldomys* (Rodentia, Muridae, Murinae), with descriptions of two new species for the Philippines. *Fieldiana: Zoology* (n.s.) 89: 1-24.

RICKART, E.A., HEANEY, L.R. & TABARANZA, B.R., Jr. 2002. Review of *Bullimus* (Muridae: Murinae) and description of a new species from Camiguin Island, Philippines. *J. Mammalogy* 83(2): 421-436.

RUEDAS, L.A. 1995. Description of a new large-bodied species of *Apomys* Mearns, 1905 (Mammalia: Rodentia: Muridae) from Mindoro Island, Philippones. *Proc. Biol. Soc. Washington* 108(2): 302-318.

RUEDI, M. 1995. Taxonomic revision of shrews of the genus *Crocidura* from the Sunda Shelf and Sulawesi with description of two new species (Mammalia: Soricidae). *Zool. J. Linn. Soc.* 115: 211-265.

SARMIENTO, E. 1997-1998. The validity of "*Pseudopotto martini*". *African Primates* 3(1-2): 44-45.

SCHWARTZ, J.H. 1996. *Pseudopotto martini*: A new genus and species of extant lorisiforme primate. *Anthropological Papers of the American Museum of Natural History* 78: 1-14. [See also Groves, 1997-1998; Bearder, 1997-1998; Sarmiento, 1997-1998].

SIMMONS, N.B. 1996. A new species of *Micronycteris* (Chiroptera: Phyllostomidae) from northeastern Brazil, with comments on phylogenetic relationships. *Amer. Mus. Novitates* 3158.

SIMMONS, N.B. & VOSS, R.S. 1998. The mammals of Paracou, French Guiana: a neotropical lowland rainforest fauna, Part 1. Bats. *Bull. Amer. Mus. Nat. Hist.* 237: 1-219.

SIMMONS, N.B., VOSS, R.S. & FLECK, D.W. 2002. A new Amazonian species of *Micronycteris* (Chiroptera: Phyllostomidae) with notes on the roosting behavior of sympatric congeners. *Amer. Mus. Novitates* 3158.

SOKOLOV, V.E., ROZHNOV, V.V. & ANH, P.T. 1997. New species of viverrids of the genus *Viverra* (Mammalia, Carnivora) from Vietnam. *Zoologischeskii Zhurnal* 76: 585-589. [See also Walston & Veron, 2001]

SPITZENBERGER, F., HARING, E & TVRTKOVIC, N. 2002. *Plecotus microdontus* (Mammalia, Vespertilionidae), a new bat species from Austria. *Natura Croatica* 11(1): 1-18.

SPITZENBERGER, F., PIALEK, J. & HARING, E. 2001. Systematics of the genus *Plecotus* (Mammalia, Vespetilionidae) in Austria based on morphometric and molecular investigations. *Folia Zool.* 50(3): 161-172.

STANLEY, W.T. & HUTTERER, R. 2000. A new species of *Myosorex* Gray, 1832 (Mammalia: Soricidae) from the Eastern Arc mountains, Tanzania. *Bonn. Zool. Beitr.* 49(1-4): 19-29.

STEPPAN, S.J. 1995a. Phylogenetic relationships of the phyllotine rodents (Sigmodontinae) and the evolution of phenotypic patterns of covariation in *Phyllotis*. Ph.D. dissertation, University of Chicago.

STEPPAN, S.J. 1995b. Revision of the leaf-eared mice Phyllotini (Rodentia: Sigmodontinae) with a phylogenetic hypothesis for the Sigmodontinae. *Fieldiana: Zoology* (n.s.) 80: 1-112.

STEWART, D.T., PERRY, N.D. & FUMAGALLI, L.. 2002. The maritime shrew, *Sorex maritimensis* (Insectivora: Soricidae): a newly recognized, Canadian endemic. *Canadian J. Zool.* 80: 94-99.

STUART, C. & STUART, T. 2001. *Field Guide to Mammals of Southern Africa*. 3rd edn. Cape Town: Struik.

TAYLOR, P.J., MEESTER, J. & KEARNEY, T. 1993. The taxonomic status of Saunders' vlei rat, *Otomys saundersiae* Roberts (Rodentia: Muridae: Otomyinae). *J. African Zool.* 107(6): 571-596.

THALMANN, U. & GEISSMANN, T. 2000. Distributions and geographic variation in the western woolly lemur (*Avahi occidentalis*) with description of a new species (*A. unicolor*). *Int. J. Primatology* 21(6): 915-941.

TOPÁL, G. 1997. A new mouse-eared bat species from Nepal, with statistical analyses of some other species of subgenus *Leuconoe* (Chiroptera, Vespertilionidae). *Acta Zoologica Academiae Scientiarum Hungaricae* 43: 375-402.

TUOC, D., DUNG, V.V., DAWSON, S., ARCTANDER, P. & MACKINNON, J. 1994. Introduction of a new large mammal species in Vietnam. *Science and Technology News*. Forest Inventory and Planning Institute (Hanoi), March 1994: 4-13 [in Vietnamese].

VAN DER STRAETEN, E. & KERBIS PETERHANS, J.-C. 1999. *Praomys degraafi*, a new species of Muridae (Mammalia) from central Africa. *South African J. Zool.* 34(2): 80-90.

VAN DYCK, S. 1982. The relationships of *Antechinus stuartii* and *A. flavipes* (Dasyuridae, Marsupialia) with special reference to Queensland. pp. 723-766. *In:* M. Archer (ed.) *Carnivorous Marsupials*. Royal Zoological Society of New South Wales, pp. 804.

VAN DYCK, S. & CROWTHER, M.S. 2000. Reassessment of northern representatives of the *Antechinus stuartii* complex (Marsupialia: Dasyuridae): *A. subtropicus* sp. nov. and *A. adustus* new status. *Memoirs of the Queensland Museum* 45(2): 611-635.

VAN DYCK, S, WOINARSKI, J.C.Z. & PRESS, A.J. 1994. The Kakadu dunnart, *Sminthopsis bindi* (Marsupialia: Dasyuridae), a new species from the stony woodlands of the Northern Territory. *Memoirs of the Queensland Museum* 37(1): 311-323.

VAN HELDEN, A.L., BAKER, A.N., DALEBOUT, M.L., REYES, J.C., VAN WAEREBEEK, K. & BAKER, C.S. 2002. Resurrection of *Mesoplodon traversii* (Gray, 1874), senior synonym for *M. bahamondi* Reyes, van Waerebeek, Cárdenas and Yáñez, 1995 (Cetacea: Ziphiidae). *Marine Mammal Science* 18(3): 609-621.

VAN ROOSMALEN, M.G.M., VAN ROOSMALEN, T. & MITTERMEIER, R.A. 2002. A taxonomic review of the titi monkeys, genus *Callicebus* Thomas, 1903, with the description of two new species, *Callicebus bernhardi* and *Callicebus stephennashi*, from Brazilian Amazonia. *Neotropical Primates* 10(suppl.): 1-52.

VAN ROOSMALEN, M.G.M., VAN ROOSMALEN, T., MITTERMEIER, R.A. & DA FONSECA, G.A.B. 1998. A new and distinctive species of marmoset (Callitrichidae, Primates) from the lower Rio Aripuana, state of Amazonas, central Brazilian Amazonia. *Goeldiana Zoologia* (22): 1-27.

VAN ROOSMALEN, M.G.M., VAN ROOSMALEN, T., MITTERMEIER, R.A. & RYLANDS, A.B. 2000. Two new species of marmoset, genus *Callithrix* Erxleben, 1777 (Callitrichidae, Primates), from the Tapajos/Madeira interfluvium, south central Amazonia, Brazil. *Neotropical Primates* 8(1): 1-18.

VENTURA, J., PEREZ-HERNANDEZ, R. & LOPEZ-FUSTER, M.J. 1998. Morphometric assessment of the *Monodelphis brevicaudata* group in Venezuela. *J. Mammalogy* 79: 104-117.

VERHEYEN, W.N., COLYN, M. & HULSELMANS, J. 1996. Re-evaluation of the *Lophuromys nudicaudus* (Heller, 1911) species-complex with a description of a new species from Zaire (Muridae-Rodentia). *Bulletin de l'Institut Royal des Sciences Naturelles de Belgique: Biologie* 66: 241-273.

VERHEYEN, W.N., DIERCKX, T. & HULSELMANS, J. 2000. The brush-furred rats of Angola and southern Rep. Congo: description of a new taxon of the *Lophuromys sikapusi* species complex. *Bulletin de l'Institut Royal des Sciences Naturelles de Belgique: Biologie* 70: 253-267.

VERHEYEN, W.N., HULSELMANS, J., COLYN, M. & HUTTERER, R. 1997. Systematics and zoogeography of the small mammal fauna of Cameroun: description of two new *Lophuromys* (Rodentia: Muridae) endemic to Mount Cameroun and Mount Oku. *Bulletin de l'Institut Royal des Sciences Naturelles de Belgique: Biologie* 67: 163-186.

VERHEYEN, W.N., HULSELMANS, J.L.J., DIERCKX, T. & VERHEYEN, E. 2002. The *Lophuromys flavopunctatus* Thomas 1888 s.l. species complex: a craniometric study, with the description and genetic characterization of two new species (Rodentia – Muridae – Africa). *Bulletin de l'Institut Royal des Sciences Naturelles de Belgique: Biologie* 72: 141-182.

VIÉ, J.C., VOLOBOUEV, V., PATTON, J.L. & GRANJON, L. 1996. A new species of *Isothrix* (Rodentia: Echimyidae) from French Guiana. *Mammalia* 60(3): 393-406.

VIVAR, E., PACHECO, V. & VALQUI, M. 1997. A new species of *Cryptotis* (Insectivora: Soricidae) from northern Peru. *Amer. Mus. Novitates* 3202.

VIZCAÍNO, S.F. 1995. Identificación especifica de las "mulitas", Género *Dasypus* L. (Mammalia, Dasypodidae), del noroeste argentino. Descripción de una nueva especie. *Mastozoologia Neotropical* 2(1): 5-13.

VON HELVERSEN, O., HELLER, K.-G., MAYER, F., NEMETH, A., VOLLETH, M. & GOMBKÖTÖ, P. 2001. Cryptic mammalian species: a new species of whiskered bat (*Myotis alcathoe* n. sp.) in Europe. *Die Naturwissenschaften* 88(5): 217-223.

VOSS, R.S. & ANGERMANN, R. 1997. Revisionary notes on neotropical porcupines (Rodentia: Erethizontidae). 1. Type material described by Olfers (1818) and Kuhl (1820) in the Berlin Zoological Museum. *Amer. Mus. Novitates* 3214.

VOSS, R.S. & CARLETON, M.D. 1993. A new genus for *Hesperomys molitor* Winge and *Holochilus magnus* Hershkovitz, with an analysis of its phylogenetic relationships. *Amer. Mus. Novitates* 3085.

VOSS, R.S. & DA SILVA, M.N.F. 2001. Revisionary notes on neotropical porcupines (Rodentia, Erethizontidae). 2, A review of the *Coendou vestitus* group with a description of two new species from Amazonia. *Amer. Mus. Novitates* 3351.

VOSS, R.S., GÓMEZ-LAVERDE, M. & PACHECO, V. 2002. A new genus for *Aepeomys fuscatus* Allen, 1912, and *Oryzomys intectus* Thomas, 1921: enigmatic murid rodents from Andean cloud forests. *Amer. Mus. Novitates* 3373.

VOSS, R.S., LUNDE, D.P. & SIMMONS, N.B. 2001. The mammals of Paracou, French Guiana: a Neotropical lowland rainforest fauna. Part 2. Nonvolant species. *Bull. Amer. Mus. Nat. Hist.* 263: 1-236.

WALSTON, J. & VERON, G., 2001. Questionable status of the "Taynguyen civet", *Viverra tainguensis* Sokolov, Rozhnov and Pham Trong Anh, 1997 (Mammalia: Carnivora: Viverridae). *Mamm. Biol. (Z. Säugert.)* 66: 181-184.

WEKSLER, M., BONVICINO, C.R., OTAZU, I.B. & SILVA, J.S., Jr. 2001. Status of *Proechimys roberti* and *P. oris* (Rodentia: Echimyidae) from eastern Amazonia and central Brazil. *J. Mammalogy* 82(1): 109-122.

WEKSLER, M., GEISE, L. & CERQUEIRA, R. 1999. A new species of *Oryzomys* (Rodentia, Sigmondontinae) from southeast Brazil, with comments on the classification of the *O. capito* species group. *Zool. J. Linnaean Soc.* 125: 445-462.

WILLIAMS, D.F., GENOWAYS, H.H. & BRAUN, J.K. 1993. Taxonomy. Pp. 38-196, *In:* H.H. Genoways & J.H. Brown (eds.) *Biology of the Heteromyidae.* American Society of Mammalogists, Spec. Publ. 10.

WILLIAMS, S.L., WILLIG, M.R. & REID, F.A. 1995. A review of the *Tonatia bidens*-complex (Mammalia: Chiroptera), with descriptions of two new subspecies. *J. Mammalogy* 76: 612-626.

WILSON, D.E. & COLE, F.R. 2000. *Common Names of Mammals of the World.* Washington and London: Smithsonian Institution Press.

WILSON, D.E. & REEDER, D.M. (eds.) 1993. *Mammal Species of the World: a taxonomic and geographic reference.* 2nd edition. Washington and London: Smithsonian Institution Press.

WOODMAN, N. 1996. Taxonomic status of the enigmatic *Cryptotis avia* (Mammalia: Insectivora: Soricidae), with comments on the distribution of the Colombian small-eared shrew, *Cryptotis colombiana. Proc. Biol. Soc. Washington* 109(3): 409-418.

WOODMAN, N. 2002. A new species of small-eared shrew from Colombia and Venezuela (Mammalia: Soricomorpha: Soricidae: Genus *Cryptotis*). *Proc. Biol. Soc. Washington* 115(2): 249-272.

WOODMAN, N. & TIMM, R.M. 1993. Intraspecific and interspecific variation in the *Cryptotis nigrescens* species complex of small-eared shrews (Insectivora: Soricidae), with the description of a new species from Colombia. *Fieldiana: Zoology* (n.s.) 74: 1-30.

ZIMMERMAN, E., CEPOK, S., RAKOTOARISON, N., ZIETEMANN, V. & RADESPIEL, U. 1998. Sympatric mouse lemurs in north-west Madagascar: a new rufous mouse lemur species. *Folia Primatol.* 69: 106-114.

ZUNINO, G.E., VACCARO, O.B., CANEVARI, M. & GARDNER, A.L. 1995. Taxonomy of the genus *Lycalopex* (Carnivora, Canidae) in Argentina. *Proc. Biol. Soc. Washington* 108(4): 729-747.

INDEX OF SCIENTIFIC NAMES

INDEX OF ENGLISH NAMES